AI-Powered Analysis, Modeling, and Monitoring of Wind Energy Systems

Jackson J. Justo
https://orcid.org/0000-0003-0638-0713
University of Dar es Salaam, Tanzania

Francis A. Mwasilu
https://orcid.org/0000-0001-7540-4198
University of Dar es Salaam, Tanzania

Galina Demidova
Hangzhou Dianzi University, China

Dmitry V. Lukichev
ITMO University, Russia

Ramesh C. Bansal
University of Sharjah, UAE

IGI Global
Scientific Publishing
Publishing Tomorrow's Research Today

Vice President of Editorial	Melissa Wagner
Director of Acquisitions	Mikaela Felty
Director of Book Development	Jocelynn Hessler
Production Manager	Mike Brehm
Cover Design	Jose Rosado

Published in the United States of America by
IGI Global Scientific Publishing
701 East Chocolate Avenue
Hershey, PA, 17033, USA
Tel: 717-533-8845 | Fax: 717-533-7115
Website: https://www.igi-global.com E-mail: cust@igi-global.com

Library of Congress Cataloging-in-Publication Data

LCCN: 2025036228 (CIP Data Pending)
ISBN13: 9798337341590
Isbn13Softcover: 9798337341606
EISBN13: 9798337341613

British Cataloguing in Publication Data
A Cataloguing in Publication record for this book is available from the British Library.

Table of Contents

Detailed Table of Contents

The global shift to sustainable energy highlights wind power's role, yet its intermittency and grid integration challenges require innovative solutions for efficiency and stability. This chapter examines high-temperature superconductors (HTS), advanced energy storage, and synthetic inertia, enhanced by AI, to transform wind energy systems. HTS materials enable compact, high-capacity generators and lossless transmission cables, cutting energy waste, especially in offshore wind farms. Energy storage, including lithium-ion, redox flow batteries, and pumped hydro, mitigates wind variability, ensuring firm capacity and grid services like frequency regulation. Synthetic inertia, via Virtual Synchronous Machines, stabilizes frequency in inverter-based grids. AI optimizes performance with predictive maintenance and adaptive control. Case studies like Hornsdale Power Reserve and HTS pilots, plus simulations, validate benefits despite high costs and technical hurdles. These innovations make wind a reliable cornerstone of sustainable energy systems.

The integration of artificial intelligence (AI) and digital twin technologies has revolutionized condition monitoring and fault detection in industrial systems, particularly in wind energy applications. This chapter explores the synergistic relationship between AI algorithms and digital twin frameworks for enhanced predictive maintenance strategies. Digital twins create virtual replicas of physical systems, enabling real-time monitoring, simulation, and analysis of operational conditions. When combined with AI techniques such as machine learning, deep learning, and neural networks, these systems can predict failures, optimize maintenance schedules, and reduce operational costs. The chapter examines current methodologies, implementation challenges, and future prospects of AI-enhanced digital twins in condition monitoring. Case studies from wind turbine applications demonstrate the practical benefits of this integrated approach, showing improvements in fault detection accuracy, reduced downtime, and enhanced system reliability.

This chapter explores how digital twins (DT), machine learning (ML), and artificial intelligence (AI) enhance monitoring, control, and maintenance of modern wind turbines. These technologies enable predictive, data-driven solutions using real-time SCADA and high-resolution sensor data, reducing reliance on reactive, schedule-based maintenance. AI and ML improve resource allocation, fault detection, and performance optimization through continuous learning, while DTs synchronize virtual and physical turbines for advanced diagnostics, planning, and simulation.

data to predict failures before they occur, reducing maintenance costs and enhancing efficiency and lifespan. These integrated systems enable automated decision-making and self-correction, improving the reliability, stability, and intelligence of future wind power generation systems.

Rampelli Manojkumar, BVRIT HYDERABAD College of Engineering for Women, India
Gundu Venu, Malla Reddy University, India
Suresh Kumar Annam, Malla Reddy University, India
Ramesh Babu Pittala, Anurag University, India
Anji Reddy Polu, BVRIT HYDERABAD College of Engineering for Women, India
T. Yuvaraj, Chennai Institute of Technology, India

This chapter explores the integration of Artificial Intelligence (AI) and Digital Twin (DT) technologies for intelligent condition monitoring and fault detection in wind energy systems. It highlights the limitations of traditional approaches and presents AI-based models for fault prediction, anomaly detection, and remaining useful life estimation. The layered architecture of DTs, their real-time synchronization, and hybrid AI-DT frameworks for predictive maintenance are discussed. The chapter also examines challenges such as data integrity, scalability, and cybersecurity while outlining emerging trends like federated learning and quantum digital twins for resilient and sustainable wind farm operations.

Nguyen Duc Thuan, Hanoi University of Science and Technology, Vietnam

The rapid digitalization of renewable energy systems has made digital twin technology a transformative approach for modeling, monitoring, and optimizing wind turbines. This chapter presents a framework for an AI-powered digital twin that integrates multi-physics modeling, machine learning, and real-time data synchronization. It develops detailed aerodynamic, mechanical, electrical, structural, and thermal models forming a hybrid simulation grounded in physics. Artificial intelligence methods, from machine learning to physics-informed neural networks, enhance fault detection, anomaly diagnosis, and life prediction. A small-scale turbine simulation validates the framework, showing real-time operation, improved torque prediction, and precise anomaly detection. The discussion highlights future directions in multi-agent wind farm twins, uncertainty quantification, explainable AI, and self-evolving models, outlining a roadmap toward autonomous and trustworthy digital energy ecosystems.

R. N. Ravikumar, Marwadi University, Rajkot, India
S. Aarthi, Marwadi University, Rajkot, India
Shakhboz Meylikulov, Termez University of Economics and Service, Termez, Uzbekistan
C. Navamani, Nandha Engineering College, Erode, India
Bekzod Madaminov, Mamun University, Khiva, Uzbekistan
T. M. Saravanan, Kongu Engineering College, Erode, India

As the world moves toward renewable energy, wind power stands out as a key sustainable source. However, wind turbines face challenges in reliability due to harsh conditions, mechanical wear, and electrical stress. Traditional maintenance falls short for large wind farms. This chapter explores how Artificial Intelligence (AI) and Digital Twin technologies enable real-time monitoring of blades, gearboxes, bearings, generators, and converters. AI models like Neural Networks, Support Vector Machines, Convolutional Neural Network (CNN), and Long Short-Term Memory (LSTM) help detect faults and predict failures early. Digital twins enhance diagnostics by simulating turbine behavior using real-time data. The chapter also covers hybrid modeling, anomaly detection, and federated learning. Case studies include offshore wind AI, blade crack detection, and converter diagnostics. Challenges include data quality, cybersecurity, and scalability. Future trends point to edge AI, cloud-based diagnostics, and blockchain for secure monitoring and predictive maintenance.

Binastya Anggara Sekti, Universitas Esa Unggul, Indonesia

The growing digitalization of wind power systems has intensified cybersecurity and data privacy challenges across modern renewable energy infrastructures. As AI, digital twins, and cloud-based platforms become integral to turbine monitoring and control, wind farms face expanding attack surfaces and increased vulnerability to malware, sensor spoofing, DDoS attacks, and supply chain breaches. These risks threaten operational continuity, grid stability, and the confidentiality of sensitive operational and geospatial data. Ensuring secure and privacy-compliant wind operations requires multi-layered protection that includes Zero Trust architectures, secure industrial protocols, robust identity and access management, encrypted communication channels, and strong AI governance. A comprehensive understanding of emerging threat vectors and resilient defense mechanisms is essential for safeguarding AI-powered wind energy ecosystems against evolving cyber and privacy risks.

Preface

INTRODUCTION

We are living in a defining decade for our planet and our collective energy future. The urgent imperatives of the United Nations Sustainable Development Goals (SDGs) particularly the SDG 7 (Affordable and Clean Energy) and SDG 13 (Climate Action). The two SDGs combined with ambitious national and regional frameworks like the European Green Deal, have set an unequivocal direction: a rapid, just, and comprehensive transition to a decarbonized global economy. Wind energy, as a mature and scalable renewable technology, is not merely a participant in this transition but a foundational driver. However, to meet the unprecedented scale and speed required by these global commitments, the wind industry must evolve. It must transcend traditional paradigms of operation and embrace a new intelligence.

This book, *AI-Powered Analysis, Modeling, and Monitoring of Wind Energy Systems*, argues that the convergence of Artificial Intelligence (AI), Machine Learning (ML), and Digital Twin (DT) technologies represents the critical innovation lever to amplify the impact of wind energy. These tools are the essential catalysts for achieving the tripartite goals of the energy transition: enhancing energy security through predictable and optimized generation, ensuring economic sustainability by radically lowering the levelized cost of energy (LCOE), and maximizing environmental sustainability through efficient resource use and extended asset lifespans.

We position these technologies not as isolated solutions, but as an integrated toolkit for systemic progress. Digital Twins create dynamic, virtual proxies of physical assets from individual turbine components to entire wind fleets enabling scenario planning and lifecycle simulation that align infrastructure development with circular economy principles, a core tenet of the Green Deal. Machine Learning algorithms mine vast operational and environmental datasets to optimize power forecasting, enhance grid stability, and enable proactive management of natural resources. Crucially, the fusion of AI with advanced condition monitoring enables a shift from calendar-based to predictive and prescriptive maintenance. This directly advances sustainability goals by minimizing unscheduled downtime, reducing material waste from premature component replacement, and maximizing the productive lifetime of turbine assets thereby improving the energy return on investment (EROI) and reducing the overall carbon footprint of wind infrastructure.

However, this smart, connected future hinges on trust and resilience. A dedicated exploration of data and cyber security addresses the safeguarding of this intelligent ecosystem. Protecting the integrity of operational data and AI models is paramount to ensuring the reliability of clean energy supply and maintaining public confidence in our critical energy infrastructure.

OVERVIEW OF CHAPTERS

Chapter 1: Harnessing Innovation in Wind Energy: Superconductors for Efficiency, Storage for Reliability, and Inertia for Stability

The global shift to sustainable energy highlights wind power's role, yet its intermittency and grid integration challenges require innovative solutions for efficiency and stability. This chapter examines high-temperature superconductors (HTS), advanced energy storage, and synthetic inertia, enhanced by AI, to transform wind energy systems. HTS materials enable compact, high-capacity generators and lossless transmission cables, cutting energy waste, especially in offshore wind farms. Energy storage, including lithium-ion, redox flow batteries, and pumped hydro, mitigates wind variability, ensuring firm capacity and grid services like frequency regulation. Synthetic inertia, via Virtual Synchronous Machines, stabilizes frequency in inverter-based grids. AI optimizes performance with predictive maintenance and adaptive control. Case studies like Hornsdale Power Reserve and HTS pilots, plus simulations, validate benefits despite high costs and technical hurdles. These innovations make wind a reliable cornerstone of sustainable energy systems.

Chapter 2: Integrating AI and Digital Twins for Real-Time Fault Identification in Wind Turbines

The integration of artificial intelligence (AI) and digital twin technologies has revolutionized condition monitoring and fault detection in industrial systems, particularly in wind energy applications. This chapter explores the synergistic relationship between AI algorithms and digital twin frameworks for enhanced predictive maintenance strategies. Digital twins create virtual replicas of physical systems, enabling real-time monitoring, simulation, and analysis of operational conditions. When combined with AI techniques such as machine learning, deep learning, and neural networks, these systems can predict failures, optimize maintenance schedules, and reduce operational costs. The chapter examines current methodologies, implementation challenges, and future prospects of AI-enhanced digital twins in condition monitoring. Case studies from wind turbine applications demonstrate the practical benefits of this integrated approach, showing improvements in fault detection accuracy, reduced downtime, and enhanced system reliability.

Chapter 3: Artificial Intelligence and Digital Twin Applications in Wind Turbine Monitoring, Control, and Maintenance

This chapter explores how digital twins (DT), machine learning (ML), and artificial intelligence (AI) enhance monitoring, control, and maintenance of modern wind turbines. These technologies enable predictive, data-driven solutions using real-time SCADA and high-resolution sensor data, reducing reliance on reactive, schedule-based maintenance. AI and ML improve resource allocation, fault detection, and performance optimization through continuous learning, while DTs synchronize virtual and physical turbines for advanced diagnostics, planning, and simulation. The chapter also outlines implementation challenges, including model drift, data delays, security risks, and integration within existing infrastructure. It aims to guide researchers and engineers in deploying AI-based tools to improve turbine reliability, efficiency, and lifecycle management.

Chapter 4: AI-Driven Twin Framework for Intelligent Condition Monitoring and Predictive Maintenance in Grid-Integrated Wind Turbines: Advanced ML Models for Real Time Asset Optimization

The exponential development of wind power requires a high level of operational efficiency, reliability, and grid integration. The chapter presents a four-layer AI-powered digital twin architecture of grid-integrated wind turbines, which includes multi-sensors networks, IoT-based data acquisition (MQTT, LoRaWAN, 5G), hybrid physics and deep learning modeling, and autonomous decision-making. Machine learning models such as CNNs, RNNs, attention-based models and ensemble genetic models allow real-time condition monitoring and predictive maintenance with 99.72% accuracy in fault detection and predicting fault in 1-56 days. The conversion of reactive to predictive strategies generate 11-50% O&M savings, 60% reduction of inspections and 85% expedited repairs, and maximum asset lifespan.

Critical infrastructure is secured by secure gateways, encryption, and blockchain. Field deployments demonstrate 25 of downtime and 40 of emergency repair cost decreases, greater capacity factors, and improved sustainability.

Chapter 5: AI-Enhanced Predictive Analytics and Reliability Management for Wind Turbine Maintenance

Wind power plants are vital for sustainable energy generation, but maintaining their long-term performance is challenging. Artificial intelligence (AI) and digital twin technologies are revolutionizing fault diagnosis and predictive maintenance in wind turbines. The digital twin creates a virtual replica that mirrors real turbine operations, collecting sensor data for precise condition monitoring. AI algorithms analyze this data to predict failures before they occur, reducing maintenance costs and enhancing efficiency and lifespan. These integrated systems enable automated decision-making and self-correction, improving the reliability, stability, and intelligence of future wind power generation systems.

Chapter 6: Artificial Intelligence and Digital Twin Synergy for Predictive Maintenance and Fault Detection in Wind Turbines

This chapter explores the integration of Artificial Intelligence (AI) and Digital Twin (DT) technologies for intelligent condition monitoring and fault detection in wind energy systems. It highlights the limitations of traditional approaches and presents AI-based models for fault prediction, anomaly detection, and remaining useful life estimation. The layered architecture of DTs, their real-time synchronization, and hybrid AI-DT frameworks for predictive maintenance are discussed. The chapter also examines challenges such as data integrity, scalability, and cybersecurity while outlining emerging trends like federated learning and quantum digital twins for resilient and sustainable wind farm operations.

Chapter 7: Digital Twin Frameworks for AI-Driven Wind Turbine Monitoring and Predictive Maintenance

The rapid digitalization of renewable energy systems has made digital twin technology a transformative approach for modeling, monitoring, and optimizing wind turbines. This chapter presents a framework for an AI-powered digital twin that integrates multi-physics modeling, machine learning, and real-time data synchronization. It develops detailed aerodynamic, mechanical, electrical, structural, and thermal models forming a hybrid simulation grounded in physics. Artificial intelligence methods, from machine learning to physics-informed neural networks, enhance fault detection, anomaly diagnosis, and life prediction. A small-scale turbine simulation validates the framework, showing real-time operation, improved torque prediction, and precise anomaly detection. The discussion highlights future directions in multi-agent wind farm twins, uncertainty quantification, explainable AI, and self-evolving models, outlining a roadmap toward autonomous and trustworthy digital energy ecosystems.

Chapter 8: Condition Monitoring and Fault Detection with AI and Digital Twin Technologies

As the world moves toward renewable energy, wind power stands out as a key sustainable source. However, wind turbines face challenges in reliability due to harsh conditions, mechanical wear, and electrical stress. Traditional maintenance falls short for large wind farms. This chapter explores how Artificial Intelligence (AI) and Digital Twin technologies enable real-time monitoring of blades, gearboxes, bearings, generators, and converters. AI models like Neural Networks, Support Vector Machines, Convolutional Neural Network (CNN), and Long Short-Term Memory (LSTM) help detect faults and predict failures early. Digital twins enhance diagnostics by simulating turbine behavior using real-time data. The chapter also covers hybrid modeling, anomaly detection, and federated learning. Case studies include offshore wind AI, blade crack detection, and converter diagnostics. Challenges include data quality, cybersecurity, and scalability. Future trends point to edge AI, cloud-based diagnostics, and blockchain for secure monitoring and predictive maintenance.

Chapter 9: Data Privacy and Cybersecurity in Wind Power Systems

The growing digitalization of wind power systems has intensified cybersecurity and data privacy challenges across modern renewable energy infrastructures. As AI, digital twins, and cloud-based platforms become integral to turbine monitoring and control, wind farms face expanding attack surfaces and increased vulnerability to malware, sensor spoofing, DDoS attacks, and supply chain breaches. These risks threaten operational continuity, grid stability, and the confidentiality of sensitive operational and geospatial data. Ensuring secure and privacy-compliant wind operations requires multi-layered protection that includes Zero Trust architectures, secure industrial protocols, robust identity and access management, encrypted communication channels, and strong AI governance. A comprehensive understanding of emerging threat vectors and resilient defense mechanisms is essential for safeguarding AI-powered wind energy ecosystems against evolving cyber and privacy risks.

Our journey through these pages is designed to connect technological capability with global mission. We move from establishing the foundational synergy between AI/ML and wind engineering, to detailed applications in resource assessment, performance optimization, and prognostic health management. Each chapter is framed not only by technical merit but by its contribution to broader sustainability and decarbonization objectives. We examine real-world case studies and confront implementation challenges from data governance to workforce transformation offering a pragmatic roadmap for stakeholders.

This book is for the engineers, data scientists, project developers, policymakers, and students who are tasked with turning ambition into reality. It is a guide for those who will design, operate, and regulate the intelligent wind farms that will power our sustainable future. By harnessing AI, ML, and Digital Twins, we can transform wind energy from a clean power source into a fully optimized, resilient, and self-improving pillar of a net-zero world. The technologies discussed herein are more than tools; they are our partners in fulfilling a global promise for a secure, prosperous, and sustainable future for all.

ACKNOWLEDGMENT

We extend our profound gratitude to the global community of researchers, industry pioneers, and policymakers whose work at the nexus of technology and sustainability inspires this text. We thank our reviewers for their rigorous insights and our families for their steadfast support. Finally, we acknowledge the collective urgency of the climate challenge, which motivates our shared pursuit of smarter, cleaner energy solutions.

Welcome to the intelligent engine of the energy transition.

We The Editors.

2026

Chapter 1
Harnessing Innovation in Wind Energy:
Superconductors for Efficiency, Storage for Reliability, and Inertia for Stability

Peter Mola Makolo
http://orcid.org/0000-0002-3595-8468
Univerisity of Dar es Salaam, Tanzania

Eva Ibrahim Shayo
University of Dar es Salaam, Tanzania

Jackson J. Justo
http://orcid.org/0000-0003-0638-0713
ITMO University, Russia

ABSTRACT

The global shift to sustainable energy highlights wind power's role, yet its intermittency and grid integration challenges require innovative solutions for efficiency and stability. This chapter examines high-temperature superconductors (HTS), advanced energy storage, and synthetic inertia, enhanced by AI, to transform wind energy systems. HTS materials enable compact, high-capacity generators and lossless transmission cables, cutting energy waste, especially in offshore wind farms. Energy storage, including lithium-ion, redox flow batteries, and pumped hydro, mitigates wind variability, ensuring firm capacity and grid services like frequency regulation. Synthetic inertia, via Virtual Synchronous Machines, stabilizes frequency in inverter-based grids. AI optimizes performance with predictive maintenance

DOI: 10.4018/979-8-3373-4159-0.ch001

and adaptive control. Case studies like Hornsdale Power Reserve and HTS pilots, plus simulations, validate benefits despite high costs and technical hurdles. These innovations make wind a reliable cornerstone of sustainable energy systems.

1. INTRODUCTION

The growing global dependence on wind energy as a fundamental element of sustainable energy strategies underscores the urgent need for continuous advancements aimed at improving its efficiency, dependability, and ability to integrate seamlessly with power grids (Lim et al., 2025; Justo et al., 2013). While wind energy stands as a promising, environmentally friendly alternative to fossil fuels, its inherent intermittency, coupled with the evolving and increasingly complex demands of modern energy grids, presents significant obstacles. This chapter delves into these challenges by exploring the transformative potential offered by the integration of pioneering technologies, with a particular focus on high-temperature superconductors (HTS), state-of-the-art energy storage systems, and synthetic inertia (Justo et al., 2025; Makolo et al., 2021b). Each of these innovations holds the potential to optimize the performance of wind turbines, enhance the stability of power grids, and accelerate the widespread adoption of wind energy as a core component of the global energy mix (Z. Wang & Bu, 2024; Hassan et al., 2024). By harnessing these advanced technologies, the wind energy sector can overcome its current limitations and pave the way for a more resilient, efficient, and scalable energy future.

1.1. Overview of the Challenges and Opportunities in Wind Energy Development

Wind energy, while a clean and sustainable power source, presents several challenges primarily due to its intermittent and variable nature. The output of wind farms is highly dependent on wind speed, leading to fluctuations in power generation that can complicate grid management and stability (Durgadevi et al., 2024; Sousa et al., 2024). As the penetration of wind energy increases in power systems, these fluctuations can lead to issues such as frequency deviations, voltage instability, and increased operational costs for grid operators who must ensure a continuous balance between supply and demand. The ability of traditional synchronous generators to provide inertia, which naturally helps stabilize the grid against sudden disturbances, is reduced with higher levels of wind power integration, as wind turbines often operate without inherent synchronous inertia (Durgadevi et al., 2024; Mkoi et al., 2025).

Despite these challenges, opportunities abound for innovation. The global push for decarbonization and sustainable energy solutions creates a strong imperative for developing technologies that can mitigate wind energy's limitations. Advances in materials science, power electronics, and artificial intelligence offer pathways to address intermittency, enhance grid stability, and improve overall system reliability. The increasing complexity of modern grids also presents an opportunity for smart solutions that can optimize the integration of diverse renewable energy sources (Ukoba et al., 2024).

1.2. The Role of Innovation in Addressing These Challenges

Innovation is paramount in overcoming the obstacles associated with high wind energy penetration. The development and integration of advanced technologies are essential for transforming wind power from a variable source into a dependable and stable component of the energy mix (Bošnjaković et al., 2025; Ukoba et al., 2024). This chapter highlights three key areas of innovation:

- **Superconducting Technologies:** High-temperature superconductors offer the potential for highly efficient power transmission and smaller, lighter generators within wind turbines, significantly reducing energy losses and enhancing power density (Le et al., 2015; Lloberas et al., 2014).
- **Advanced Energy Storage Systems:** Various energy storage solutions, including battery technologies (e.g., lithium-ion, redox flow) and pumped hydro storage, are crucial for mitigating the intermittency of wind energy by storing excess power during high generation periods and releasing it during low generation or high demand. This helps to smooth out power output and improve grid reliability (Durgadevi et al., 2024; Sousa et al., 2024).
- **Synthetic Inertia Techniques:** As traditional synchronous generators are displaced by inverter-based wind turbines, the inherent inertia of the grid decreases, making it more vulnerable to frequency disturbances. Synthetic inertia, often implemented through power electronics and control algorithms like Virtual Synchronous Machines (VSMs), allows wind turbines to emulate the inertial response of conventional generators, thereby enhancing grid stability (Gonzalez-Longatt, 2012; Mkoi et al., 2025).

These innovations, when strategically integrated, can transform wind energy's role in the global energy mix, making it more reliable, efficient, and easier to manage within complex power systems.

1.3. Chapter Objectives and Scope

This chapter aims to provide a comprehensive analysis of the transformative role of superconductors, energy storage, and synthetic inertia in advancing wind energy systems.

Specifically, the objectives include:

- Evaluating the application of HTS technologies to enhance the efficiency and power capacity of wind turbines.
- Assessing the effectiveness of various energy storage solutions in addressing the intermittency of wind energy and improving overall grid reliability.
- Investigating the implementation of synthetic inertia techniques as a means of maintaining grid stability amid high levels of wind energy penetration.
- Discussing the challenges and opportunities involved in integrating these technologies, considering both technical feasibility and economic viability.
- Highlighting the potential of artificial intelligence (AI) in analysing, modelling, and monitoring wind energy systems, demonstrating how AI can optimize performance when these advanced technologies are incorporated.

The scope of this chapter covers fundamental concepts, technological applications, and the challenges and opportunities for integration, with a strong emphasis on how these innovations can contribute to a more resilient, efficient, and scalable energy future.

2. SUPERCONDUCTORS FOR ENHANCED EFFICIENCY

The integration of HTS in wind energy systems represents a significant innovation with the potential to dramatically enhance efficiency and power capacity (Sadeghi et al., 2022). Superconductors are materials that, when cooled below a critical temperature, can conduct electricity with virtually zero resistance, eliminating ohmic losses that occur in conventional conductors. This property makes them ideal for applications where energy efficiency and high-power density are crucial, such as in large-scale wind turbines and power transmission systems (Coombs et al., 2024; Sadeghi et al., 2022).

2.1. Fundamentals of Superconductivity and HTS Materials

Superconductivity is a quantum mechanical phenomenon characterized by two key properties: zero electrical resistance and the expulsion of magnetic fields

(Meissner effect). While conventional superconductors typically require extremely low temperatures (close to absolute zero, -273.15 °C or 0 K), HTS exhibit superconductivity at relatively higher temperatures, often above the boiling point of liquid nitrogen (77 K or -196 °C) (Koblischka et al., 2023). This makes HTS materials more practical and economically viable for widespread application, as liquid nitrogen cooling is significantly less expensive and easier to manage than liquid helium cooling (Coombs et al., 2024; Sadeghi et al., 2022).

Common HTS materials include cuprates like Yttrium Barium Copper Oxide (YBCO) and bismuth-based compounds (BSCCO). These materials allow for the design of more compact and efficient electrical components (Coombs et al., 2024; Koblischka et al., 2023). For instance, a superconducting coil can generate much stronger magnetic fields than a conventional copper coil of the same size, or achieve the same field strength with a significantly smaller volume and weight. This property is particularly beneficial for large-scale electrical machinery like wind turbine generators (Koblischka et al., 2023).

2.2. Superconducting Cables for Wind Farm Integration and Power Transmission

Superconducting cables offer a compelling solution for efficient power transmission, especially for large offshore wind farms and their integration into the grid. Traditional copper or aluminum cables experience significant energy losses as heat due to their inherent resistance, particularly over long distances and at high power levels. Superconducting cables, due to their zero resistance, can transmit much larger amounts of power with virtually no losses, leading to substantial efficiency gains (Cullinane et al., 2022; Sadeghi et al., 2022). Equation (1) highlights the negligible resistive losses in HTS cables. It contrasts the ohmic losses in conventional cables, reinforcing the efficiency benefits for offshore wind farm connections.

$$P_{loss,HTS} \approx 0, P_{loss,conv} = I^2 R_{conv} \tag{1}$$

Where $P_{loss,HTS}$ is power loss in HTS cables (approximately zero due to zero resistance), $P_{loss,conv}$ is power loss in conventional cables, I is the current and R_{conv} is the resistance of conventional cables (e.g., copper or aluminum). Figure 1 presents a simulation comparing power losses in HTS cables (near-zero resistance) versus conventional copper cables for transmitting power from an offshore wind farm, reinforcing the efficiency benefits discussed.

Figure 1. Power Loss Comparison of HTS vs. Conventional Cables

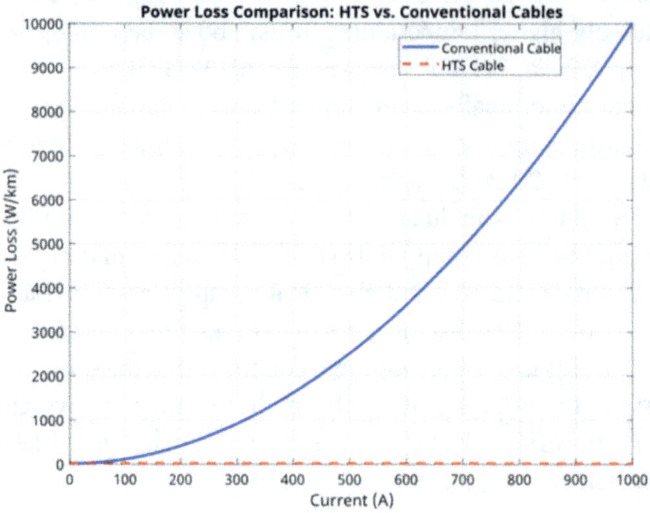

For wind farm integration, superconducting cables offer significant advantages for integrating offshore wind farms with onshore grids by reducing transmission losses and increasing power density. By minimizing energy dissipation during transmission, these cables ensure that more of the power generated by wind farms reaches consumers, enhancing overall efficiency (Cullinane et al., 2022). Additionally, their ability to transmit higher power through smaller-diameter cables reduces the need for multiple cable runs, which can lower installation costs, particularly in congested corridors where space is limited (Coombs et al., 2024).

Beyond efficiency, superconducting cables contribute to grid stability and resilience. Their low impedance characteristics improve voltage regulation, making them particularly effective for integrating large-scale, distant renewable energy sources into the grid (Sadeghi et al., 2022). Furthermore, some superconducting cables can function as fault current limiters, automatically increasing resistance during fault conditions to protect grid equipment from damaging current surges. This combination of benefits makes superconducting cables a promising solution for enhancing the reliability and performance of modern power grids (Le et al., 2015).

While the cooling infrastructure for HTS cables adds complexity, the long-term operational savings from reduced losses and increased transmission capacity make them a promising technology for future wind energy infrastructure.

2.3. AI-Driven Optimization of Superconducting Wind Turbine Performance

The complex operational characteristics of superconducting components, particularly the need for precise temperature management and the interaction of high magnetic fields, present opportunities for optimization through Artificial Intelligence (AI) and Machine Learning (ML) (Yan et al., 2024). AI can play a crucial role in enhancing the performance and reliability of superconducting wind turbines and related infrastructure.

Artificial intelligence plays a transformative role in enhancing the reliability and efficiency of HTS systems in wind energy applications. Predictive maintenance stands out as a key application, where AI algorithms analyse real-time data, such as temperature, current, voltage, and vibration, from superconducting generators and cables (Hamdan et al., 2024). By identifying patterns indicative of potential failures, these algorithms enable proactive maintenance, significantly reducing downtime and operational costs. This data-driven approach ensures that HTS systems operate reliably, supporting the consistent delivery of renewable energy (Hamdan et al., 2024).

Another critical application of AI lies in optimizing the energy-intensive cooling systems required to maintain HTS materials at their cryogenic operating temperatures. AI-driven predictive models dynamically adjust cooling operations based on factors like ambient temperature, load conditions, and component health (Madeswaran et al., 2024). This optimization minimizes energy consumption while ensuring superconductors remain in their operational state, enhancing overall system efficiency. By fine-tuning cooling processes, AI contributes to the sustainability and cost-effectiveness of superconducting wind energy systems (Hamdan et al., 2024).

AI also drives performance optimization and innovation in HTS technology through machine learning and material discovery. Machine learning models analyse extensive datasets from wind turbine operations to uncover patterns that improve control strategies, maximizing power extraction and efficiency under diverse wind conditions while extending component lifespans (Ukoba et al., 2024). Additionally, AI accelerates the discovery of new HTS materials by simulating their properties and predicting performance, reducing reliance on costly physical experiments. Similarly, AI optimizes the design of superconducting components for seamless integration into turbine structures, unlocking the full potential of superconducting technology for reliable, efficient, and cost-effective renewable energy systems (Bošnjaković et al., 2025; Ukoba et al., 2024).

3. ENERGY STORAGE FOR IMPROVED RELIABILITY

The inherent intermittency of wind energy, driven by the variable nature of wind speeds, poses significant challenges for grid stability and reliability. Energy storage systems are critical technologies that address this challenge by decoupling wind generation from immediate consumption (Cullinane et al., 2022; Díaz-González et al., 2012; Hamdan et al., 2024). They allow for the storage of excess wind power during periods of high generation and low demand, and its subsequent release during periods of low wind availability or high electricity demand, thereby smoothing power output, enhancing grid flexibility, and improving overall reliability (Cullinane et al., 2022; Gursoy et al., 2021).

3.1. The Importance of Energy Storage for Mitigating Wind Intermittency

Wind power output can fluctuate rapidly and unpredictably, leading to imbalances between supply and demand that can strain the grid, cause frequency deviations, and even lead to blackouts if not managed effectively (Díaz-González et al., 2012; Ukoba et al., 2024). Energy storage systems provide the necessary buffer to mitigate these fluctuations, ensuring a more stable and predictable power supply from wind farms (Durgadevi et al., 2024). Their importance is multi-faceted:

Energy storage systems play a pivotal role in enhancing the reliability and integration of wind energy into modern power grids. Energy storage systems absorb surplus energy during periods of high wind generation and inject power when demand exceeds supply as depicted in (2). This capability helps stabilize grid frequency and voltage, which is especially critical in grids with high wind penetration where traditional synchronous inertia is diminished (Masenge & Mwasilu, 2020). Additionally, storage enables wind farms to deliver "firm" or dispatchable power, mimicking the reliability of conventional power plants by guaranteeing consistent output regardless of fluctuating wind conditions. This capability ensures a more predictable and dependable energy supply, strengthening the role of wind energy in meeting grid demands (Gwabavu et al., 2024).

$$SOC(t+1) = SOC(t) + \eta_{ch} \times P_{ch}(t) \times \Delta t - \frac{P_{dis}(t) \times \Delta t}{\eta_{dis}} \tag{2}$$

Where $SOC(t)$ is the state-of-charge at time t, η_{ch} and η_{dis} are charging and discharging efficiencies, respectively, $P_{dis}(t)$, $P_{dis}(t)$ are charging and discharging power at time t and Δt is the time step.

Beyond stabilization, energy storage enhances the economic and operational value of wind projects through load shifting, peak shaving, and ancillary services. By storing wind energy during off-peak periods and dispatching it during high-demand times when electricity prices are elevated, storage systems improve the economic viability of wind farms and reduce reliance on costly peaker plants (Sousa et al., 2024). Furthermore, these systems provide essential ancillary services, such as frequency regulation, voltage support, and black start capability, which bolster overall grid resilience. By enabling wind energy to contribute to these critical grid functions, storage systems help create a more robust and flexible energy infrastructure (Lloberas et al., 2014; Tambunan et al., 2020).

The fundamental role of energy storage in smoothing the variable output of wind farms and ensuring grid stability is visually represented in Figure 2, which illustrates how storage systems act as a buffer between wind generation and grid demand. Furthermore, Figure 3 explains the state-of-charge (SoC) dynamics of a battery energy storage system (BESS) integrated with a wind farm, demonstrating how it mitigates wind power intermittency.

Figure 2. Schematic of Wind Energy Integration with Energy Storage

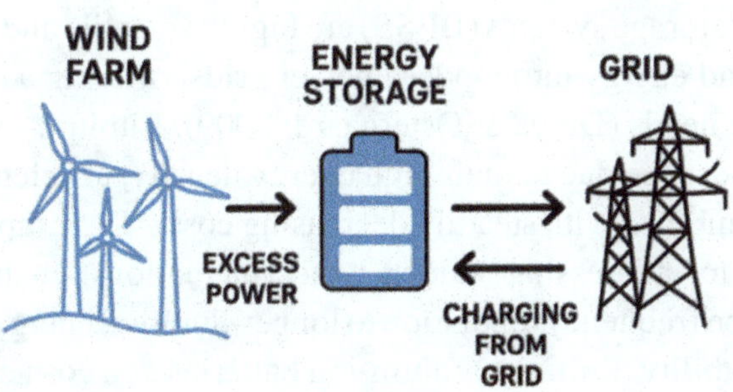

Figure 3. Battery State-of-Charge Dynamics with Wind Power

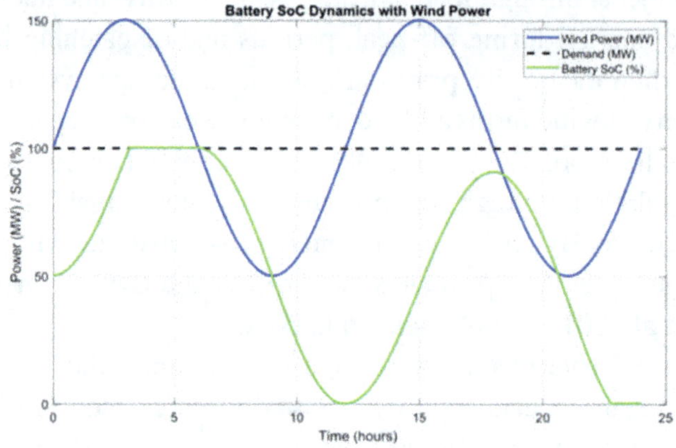

3.2. Battery Technologies: Lithium-Ion, Redox Flow Batteries, and Others

Battery energy storage systems (BESS) are highly versatile and increasingly vital for integrating wind energy into modern power grids, offering a range of solutions tailored to diverse needs (Divya & Østergaard, 2009). Lithium-ion batteries dominate the BESS landscape due to their high energy density, efficiency, and relatively long cycle life, combined with steadily decreasing costs. Their rapid response times make them ideal for addressing sudden wind fluctuations, enabling applications from short-duration frequency regulation to longer-duration energy shifting for wind farms. This adaptability positions lithium-ion batteries as a cornerstone technology for enhancing the reliability and flexibility of wind energy systems (Zhu et al., 2017).

Redox flow batteries (RFBs) provide a compelling alternative, particularly for long-duration storage needs. By storing energy in liquid electrolyte solutions housed in external tanks and pumped through electrochemical cells, RFBs allow independent scaling of power and energy capacity, offering unmatched flexibility (Makolo et al., 2021b; Sousa et al., 2024). With a long cycle life and the ability to deeply discharge without degradation, RFBs, often utilizing abundant materials like vanadium, are well-suited for smoothing diurnal wind patterns over extended periods, such as 4 to 12 hours or more. This makes them an attractive option for stabilizing wind energy output across daily cycles (May et al., 2018).

Emerging and alternative battery technologies further expand the potential for wind energy integration. Sodium-sulfur (NaS) batteries, which operate at high temperatures, are effective for bulk energy storage with long discharge durations, supporting large-scale wind farm applications (May et al., 2018). Zinc-bromine (ZnBr) batteries, another type of flow battery, offer flexibility similar to vanadium RFBs but with distinct chemical properties, broadening the range of viable options (May et al., 2018; Olabi et al., 2022). Meanwhile, solid-state batteries represent an emerging technology with the promise of higher energy density and improved safety compared to current lithium-ion systems, holding significant potential for future advancements in wind energy storage and integration (Bowen et al., 2019).

The choice of battery technology depends on the specific application requirements, including desired discharge duration, power rating, cycle life, footprint, and cost. A comparative overview of the characteristics of various battery technologies, highlighting their suitability for different wind energy integration needs, is provided in Figure 4.

Figure 4. Comparison of Battery Technologies for Energy Storage

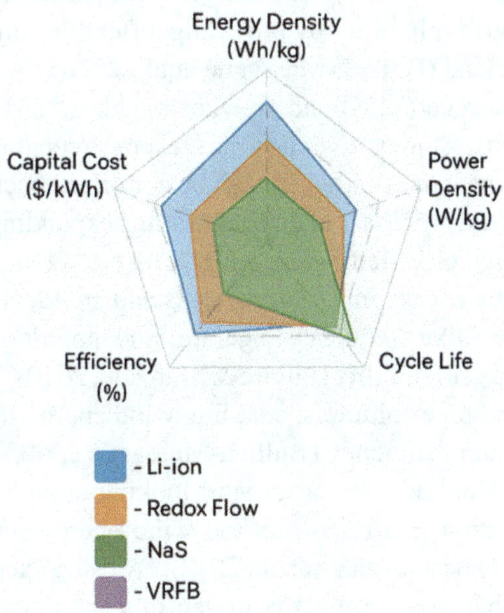

3.3. Pumped Hydro Storage and Other Energy Storage Solutions

Pumped Hydro Storage (PHS) stands as the cornerstone of large-scale energy storage, accounting for over 95% of global grid-scale storage capacity and playing a vital role in wind energy integration (Díaz-González et al., 2012). This mature technology harnesses surplus electricity, such as during periods of high wind generation, to pump water from a lower reservoir to an upper one. When electricity demand rises, the stored water is released through turbines to generate power, offering substantial energy capacity and long discharge durations. This makes PHS particularly effective for managing seasonal or multi-day variations in wind energy output, though its dependence on specific topography and water resources can restrict its deployment in some regions (Díaz-González et al., 2012).

Compressed Air Energy Storage (CAES) provides another robust option for long-duration storage, complementing wind energy systems by storing excess electricity in a unique way. During times of abundant wind power, CAES systems compress air into underground caverns or tanks, and when electricity is needed, the compressed air is released, heated, and expanded through turbines to generate power (Makolo et al., 2021b). With the ability to support large-scale storage and extended discharge periods, CAES is well-suited for balancing wind energy fluctuations, although its efficiency can vary based on system design and operational conditions. This technology enhances grid reliability by providing a flexible solution for managing energy supply and demand (Bullich-Massagué et al., 2020).

Thermal Energy Storage (TES) and flywheels address distinct needs in wind energy integration, offering innovative and rapid-response capabilities. TES captures excess wind energy as heat or cold, which can be converted back into electricity or used directly for industrial processes or district heating, expanding the utility of wind power beyond traditional electricity generation (Lloberas et al., 2014). Flywheels, by contrast, store kinetic energy in a rotating mass and are ideal for short-duration, high-power applications like frequency regulation, responding quickly to sudden wind ramps to maintain grid stability (Suvire & Mercado, 2010). These technologies broaden the scope of storage solutions, enabling wind energy to meet diverse grid requirements with greater efficiency (Bullich-Massagué et al., 2020).

Hydrogen Energy Storage offers a forward-looking approach with significant potential for long-duration and cross-sectoral wind energy integration (Bullich-Massagué et al., 2020; Díaz-González et al., 2012). By using excess wind electricity to power electrolysers that produce green hydrogen, this technology enables storage in tanks or pipelines for later use in fuel cells, transportation, or industrial applications. Its ability to support extended storage periods and link the power sector with other energy-intensive industries makes it a versatile solution for large-scale renewable energy systems (Preuster et al., 2017). As illustrated in Figure 5, the diverse power ranges and discharge durations of these storage technologies—PHS, CAES, TES, flywheels, and hydrogen—collectively ensure comprehensive support for integrating wind energy into modern, resilient power grids.

Figure 5. Overview of Major Energy Storage Technologies and Their Characteristics

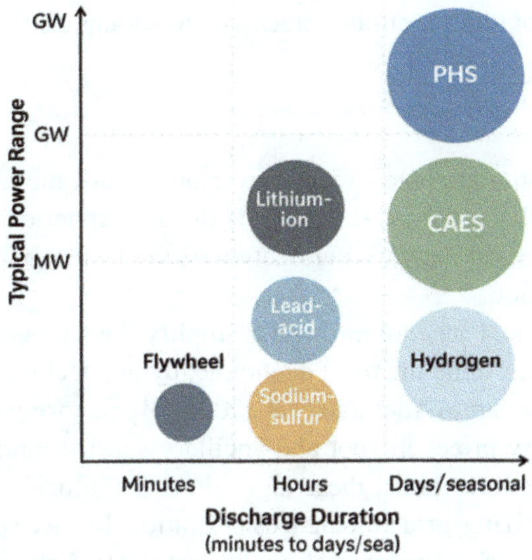

3.4. AI-Based Energy Storage Management and Dispatch Strategies

The integration of wind power with energy storage systems presents a complex challenge due to the intermittent nature of wind generation, but Artificial Intelligence (AI) and Machine Learning (ML) offer transformative solutions to optimize management and dispatch (Ukoba et al., 2024). By processing vast amounts of real-time data, AI algorithms enable intelligent decision-making for charging and discharging storage units, ensuring efficient coordination with wind power output. These advanced technologies allow storage systems to adapt dynamically to grid demands, wind variability, and market conditions, enhancing both operational reliability and economic performance (Madeswaran et al., 2024).

AI-driven forecasting is a cornerstone of effective energy storage management, leveraging models like neural networks and support vector machines to predict wind power generation, electricity demand, and market prices with high accuracy (Ukoba et al., 2024). These forecasts empower storage systems to anticipate future conditions and optimize their charge-discharge schedules accordingly. By aligning operations with expected wind availability and grid needs, storage systems can maximize energy utilization, reduce waste, and ensure a stable power supply, making wind energy a more dependable resource (Madeswaran et al., 2024). Equation (3) illustrates how AI, specifically neural networks, can predict wind power generation based on environmental inputs. It supports optimal storage dispatch and synthetic inertia control by providing accurate forecasts, enhancing grid reliability.

$$P_{wind}(t) = f_{NN}\left(W_{wind}, T_{amb}, P_{atm}, \theta; \Theta\right) \tag{3}$$

Where $P_{wind}(t)$ is the predicted wind power output at time t, W_{wind} is the wind speed, T_{amb} is the ambient temperature, P_{atm} is the atmospheric pressure, θ is wind direction, Θ is the neural network parameters (weights and biases) and f_{NN} is the neural network function.

Optimal dispatch and scheduling further amplify the value of storage systems, as ML algorithms determine the most economically and technically advantageous times to charge or discharge (Adefarati et al., 2021). By factoring in forecasted wind availability, electricity prices in spot and ancillary service markets, grid congestion, and operational constraints, these algorithms maximize revenue for storage owners while minimizing grid imbalances. Additionally, AI supports predictive maintenance by monitoring the state-of-health and state-of-charge of battery cells, predicting degradation patterns, and identifying potential faults (Chu et al., 2020; Yang et al., 2020). This proactive approach extends the lifespan of storage assets, reduces downtime, and enhances system reliability.

Beyond forecasting and maintenance, AI enables real-time control and hybrid system optimization to address dynamic grid conditions. AI-driven controllers provide rapid responses to sudden grid events or wind fluctuations, delivering immediate frequency or voltage support to maintain stability (Hamdan et al., 2024). For hybrid wind-storage systems, AI optimizes the integrated operation of wind turbines and storage facilities, ensuring maximum efficiency and compliance with grid requirements. By transforming storage systems into intelligent, adaptive assets, AI unlocks the full potential of wind energy integration, driving greater reliability, efficiency, and economic benefits for modern power grids (Bose, 2017).

4. SYNTHETIC INERTIA FOR GRID STABILITY

The transition towards power systems dominated by renewable energy sources like wind, which are typically connected via power electronic inverters, introduces a critical challenge: a reduction in system inertia. Traditional synchronous generators inherently provide rotational inertia, which acts as a natural buffer against sudden changes in frequency following a disturbance (e.g., a sudden loss of generation or load). With increasing penetration of inverter-based resources (IBRs) that lack this inherent mechanical inertia, the grid becomes more susceptible to rapid frequency deviations, potentially leading to instability and blackouts. Synthetic inertia (or artificial inertia) is a control technique that enables IBRs, such as wind turbines, to emulate the inertial response of conventional generators, thereby enhancing grid stability.

4.1. The Concept of Inertia in Power Systems and the Impact of Wind Energy Integration

Inertia in a power system refers to the ability of rotating masses (primarily in synchronous generators) to resist changes in rotational speed, and thus frequency. When a power imbalance occurs (e.g., generation suddenly drops), the stored kinetic energy in these rotating masses is released, slowing down the rate of frequency change (Rate of Change of Frequency – RoCoF), thereby providing precious time for primary frequency response mechanisms (like governor action) to kick in (Makolo et al., 2025). This resistance to change is vital for maintaining system stability.

The increasing integration of wind energy, particularly through variable speed wind turbines connected via full-scale power converters (Type 4 turbines), significantly reduces the effective synchronous inertia of the grid. These converters decouple the mechanical rotation of the turbine blades from the grid frequency, meaning the kinetic energy of the rotating blades does not directly contribute to grid inertia unless specific control strategies are implemented (Justo et al., 2015). This reduction in inertia leads to:

- **Higher RoCoF:** Any power imbalance results in a much faster drop or rise in frequency.
- **Larger Frequency Nadir/Peak**: The minimum or maximum frequency reached after a disturbance can be more severe.
- **Increased Risk of Instability:** Faster frequency deviations can trip protection systems, leading to cascading failures.

The critical impact of reduced grid inertia on frequency stability following a disturbance is graphically illustrated in Figure 6, which contrasts the frequency response of a high-inertia system with that of a low-inertia system.

Figure 6. Comparison of Grid Frequency Response with High vs. Low Inertia

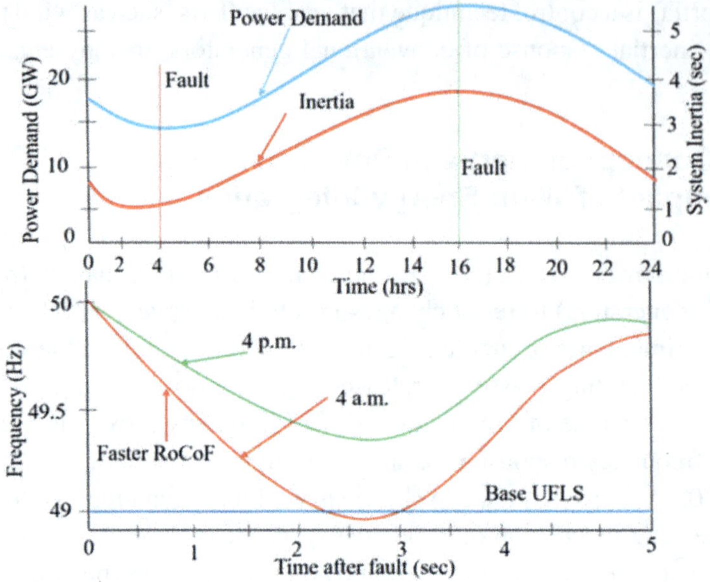

4.2. Virtual Synchronous Machines (VSMs) and Grid-Forming Inverters

To counteract the reduction in system inertia, concepts like Virtual Synchronous Machines (VSMs) and the broader category of grid-forming (GFM) inverters have emerged as key solutions.

1. **Virtual Synchronous Machines (VSMs):** VSMs are control algorithms implemented within the power converters of IBRs (like wind turbines) that mimic the external characteristics of traditional synchronous generators (Justo et al., 2025). They achieve this by introducing virtual inertia and damping terms into the inverter's control loop. The VSM control enables the inverter to:
   - **Respond to RoCoF:** Detect changes in grid frequency and inject/absorb active power to oppose the frequency deviation, similar to how a synchronous machine releases/absorbs kinetic energy.
   - **Provide Damping:** Introduce a damping term that helps to suppress oscillations and stabilize the frequency.
   - **Support Voltage:** Some VSM implementations can also provide reactive power support, mimicking the voltage regulation capabilities of synchronous generators.

The goal is to make the inverter "look" and "behave" like a synchronous machine from the grid's perspective, without the physical rotating mass. Equation (4) captures the dynamic response of a VSM-controlled wind turbine, which adjusts active power to counteract frequency deviations. It quantifies the contribution of synthetic inertia to grid stability, addressing RoCoF and frequency nadir.

$$P_{VSM}(t) = -J_{virt}\,\omega_{grid}\frac{d\omega_{grid}}{dt} + D\left(\omega_{ref} - \omega_{grid}\right) \tag{4}$$

where $P_{VSM}(t)$ is the active power output from the VSM-controlled wind turbine, J_{virt} is the virtual inertia constant, $\frac{d\omega_{grid}}{dt}$ is the RoCoF, D is the damping coefficient, ω_{grid} is the grid frequency and ω_{ref} is the reference frequency. The simulation in Figure 7 models a grid frequency disturbance (e.g., a sudden drop due to load increase) and calculates the active power response of a VSM-controlled wind turbine using the equation (3).

Figure 7. Synthetic Inertia Response to Frequency Disturbance

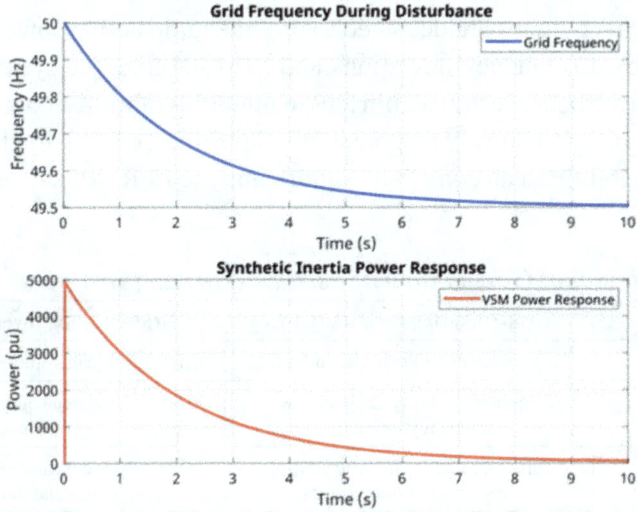

2. **Grid-Forming (GFM) Inverters:** GFM inverters are a more advanced class of inverters compared to traditional grid-following (GFL) inverters. While GFL inverters rely on the grid to set the voltage and frequency (like current sources), GFM inverters can form or establish the grid voltage and frequency themselves,

even in islanded mode or during grid blackouts (Awal & Husain, 2020). This capability is crucial for high-IBR penetration grids as they can:

- **Operate without a Strong Grid Reference:** Provide voltage and frequency references, crucial for system restoration or weak grid conditions.
- **Contribute to Grid Strength:** Provide inherent inertial and damping support, making the grid more robust.
- **Support Black Start:** Potentially initiate grid recovery after a blackout.

GFM inverters are generally considered essential for future power systems with 100% renewable energy penetration, as they can provide the foundational voltage and frequency support traditionally offered by synchronous generators. A detailed comparison of the key characteristics, capabilities, and typical applications of Grid-Following and Grid-Forming Inverters, highlighting their distinct roles in grid interaction, is summarized in Table 1.

4.3. Fast Frequency Response (FFR) Techniques

While synthetic inertia primarily addresses the initial RoCoF, Fast Frequency Response (FFR) techniques are designed to provide rapid active power injection or absorption to arrest frequency deviations and help stabilize the system. FFR acts quickly after a disturbance, often within tens to hundreds of milliseconds, faster than conventional primary frequency response. Wind turbines, especially those with energy storage capabilities, can contribute significantly to FFR (Makolo et al., 2021a).

Table 1. Comparison of Grid-Following vs. Grid-Forming Inverters

Characteristic	Grid-Following Inverter (GFL)	Grid-Forming Inverter (GFM)
Control Objective	Injects current into an existing grid	Establishes grid voltage and frequency
Grid Dependency	Requires a strong grid voltage/ frequency reference	Operates independently; provides grid reference
Inertia Provision	Limited; requires specific controls (e.g., virtual impedance)	Inherently contributes to synthetic inertia and damping
Black Start Capability	None	Yes
Typical Applications in Renewable Energy	Large-scale wind farms, solar PV (most common today); interconnected to a strong grid	Future high-penetration renewable grids, microgrids, islanded systems; enhancing grid resilience

Fast Frequency Response is a critical capability for wind turbines to support grid stability, particularly in low-inertia grids where rapid frequency fluctuations can occur. One approach to implementing FFR involves releasing kinetic energy stored in the rotating blades of variable-speed wind turbines (Cui et al., 2021). By temporarily increasing power output beyond the maximum power point, turbines can provide a rapid burst of active power to stabilize grid frequency. However, this method is inherently short-term, as it relies on drawing down the turbine's rotational speed, which limits the duration of the response and may affect aerodynamic performance.

Another effective strategy for FFR involves rapid adjustments to blade pitch angles and advanced power converter controls. By modulating the pitch angle, wind turbines can quickly adjust their active power output to respond to frequency deviations, offering a flexible and immediate solution (Xu et al., 2018). Similarly, advanced control algorithms within power converters enable turbines to detect frequency changes almost instantaneously and adjust their power output accordingly. These methods enhance the turbine's ability to contribute to grid stability without relying solely on kinetic energy, providing a more dynamic and responsive approach to frequency regulation (Xu et al., 2018).

Integrating energy storage systems, such as batteries, with wind turbines offers the most robust and sustained FFR solution. When coupled with storage, turbines can inject or absorb power rapidly without impacting their optimal aerodynamic operation or depleting kinetic energy reserves. This approach ensures a consistent and prolonged response to frequency deviations, making it particularly effective for maintaining stability in low-inertia grids (Hasan et al., 2019). By combining kinetic energy release, pitch angle control, power converter adjustments, and integrated storage, FFR enables wind turbines to play a vital role in preventing large frequency excursions, enhancing the reliability of modern power systems (Gonzalez-Longatt, 2012).

4.4. AI for Enhanced Synthetic Inertia Control and Frequency Regulation

The integration of wind power with energy storage systems presents a complex challenge due to the intermittent nature of wind generation, but Artificial Intelligence (AI) and Machine Learning (ML) offer transformative solutions to optimize management and dispatch (F. Wang et al., 2018). By processing vast amounts of real-time data, AI algorithms enable intelligent decision-making for charging and discharging storage units, ensuring efficient coordination with wind power output. These advanced technologies allow storage systems to adapt dynamically to grid demands, wind variability, and market conditions, enhancing both operational reliability and economic performance (F. Wang et al., 2018).

AI-driven forecasting is a cornerstone of effective energy storage management, leveraging models like neural networks and support vector machines to predict wind power generation, electricity demand, and market prices with high accuracy. These forecasts empower storage systems to anticipate future conditions and optimize their charge-discharge schedules accordingly. By aligning operations with expected wind availability and grid needs, storage systems can maximize energy utilization, reduce waste, and ensure a stable power supply, making wind energy a more dependable resource (Hamdan et al., 2024).

Optimal dispatch and scheduling further amplify the value of storage systems, as ML algorithms determine the most economically and technically advantageous times to charge or discharge. By factoring in forecasted wind availability, electricity prices in spot and ancillary service markets, grid congestion, and operational constraints, these algorithms maximize revenue for storage owners while minimizing grid imbalances (Hamdan et al., 2024). Additionally, AI supports predictive maintenance by monitoring the state-of-health and state-of-charge of battery cells, predicting degradation patterns, and identifying potential faults. This proactive approach extends the lifespan of storage assets, reduces downtime, and enhances system reliability.

Beyond forecasting and maintenance, AI enables real-time control and hybrid system optimization to address dynamic grid conditions (Madeswaran et al., 2024). AI-driven controllers provide rapid responses to sudden grid events or wind fluctuations, delivering immediate frequency or voltage support to maintain stability. For hybrid wind-storage systems, AI optimizes the integrated operation of wind turbines and storage facilities, ensuring maximum efficiency and compliance with grid requirements (Makolo et al., 2018). By transforming storage systems into intelligent, adaptive assets, AI unlocks the full potential of wind energy integration, driving greater reliability, efficiency, and economic benefits for modern power grids (Bošnjaković et al., 2025).

The integration of AI and Machine Learning algorithms to enable more adaptive and robust control of synthetic inertia and frequency regulation in wind turbine inverters is conceptually shown in Figure 8.

Figure 8. AI-Enhanced Synthetic Inertia Control Loop

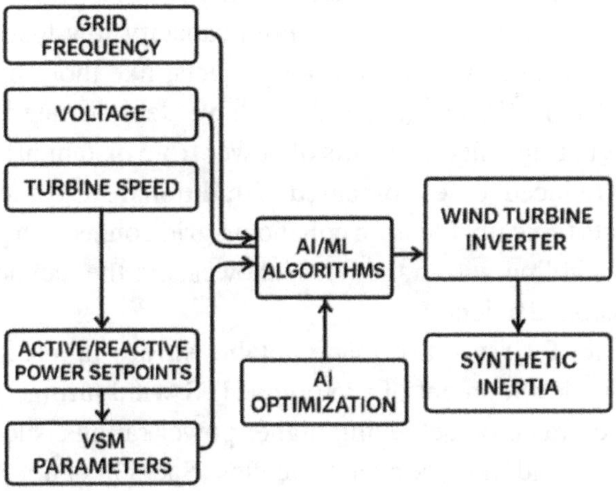

By harnessing AI, synthetic inertia and frequency regulation from wind energy systems can become more intelligent, responsive, and contribute significantly to the resilience and stability of future power grids.

5. CASE STUDIES AND APPLICATIONS

The theoretical benefits of integrating superconductors, advanced energy storage, and synthetic inertia in wind energy systems are increasingly being validated through real-world projects and detailed simulations. These case studies highlight the practical challenges and demonstrate the tangible improvements in efficiency, reliability, and stability that these technologies offer. While full-scale commercial deployment of all three technologies simultaneously in a single wind farm is still emerging, individual and combined applications provide valuable insights.

5.1. Examples of Successful Integration of Superconductors, Energy Storage, and Synthetic Inertia in Wind Energy Projects

5.1.1. Superconductors in Wind Energy

High-temperature superconducting generators hold immense promise for utility-scale wind turbines, though their widespread commercial deployment remains in development due to challenges related to cooling system complexity and costs (Le et al., 2015). Significant advancements have been made through demonstration projects, such as those inspired by the Alpha Ventus Offshore Wind Farm in Germany (Das & Cutululis, 2017). While the wind farm itself primarily relies on conventional cables, its environment has catalysed research into high-capacity, low-loss superconducting cables for offshore wind connections. Pilot projects, like those under the German Superconducting Cable Project (SC Cable), have demonstrated the potential of HTS power cables to transmit megawatts of power from distant offshore wind parks with significantly reduced losses compared to traditional AC cables (Sadeghi et al., 2022). These efforts typically focus on enhancing grid connection points rather than turbine-to-turbine cabling within the farm, showcasing the technology's ability to improve transmission efficiency.

General Electric (GE) has also made notable strides in advancing HTS technology through the development of prototype HTS wind turbine generators. GE's research has concentrated on achieving higher power ratings, such as 10 MW and above, within lighter and more compact nacelles (Sadeghi et al., 2022). This compactness reduces the structural demands on offshore turbine foundations, lowering installation costs and improving feasibility for large-scale offshore projects. While these HTS generators are not yet widely commercialized, laboratory and prototype testing have validated their efficiency gains and weight reductions, paving the way for future integration into wind energy systems (Coombs et al., 2024).

The conceptual integration of HTS generators into wind turbine nacelles, as illustrated in Figure 9, highlights their potential for transformative impact. By enabling more compact and lightweight designs, HTS generators can streamline turbine construction and installation, particularly for offshore applications where size and weight are critical considerations (Sadeghi et al., 2022). These advancements, combined with ongoing pilot projects and research, underscore the growing viability of HTS technology in enhancing the efficiency and scalability of wind energy systems, even as challenges to full-scale commercialization persist (Coombs et al., 2024).

Figure 9. Visual of an HTS Wind Turbine Generator Concept

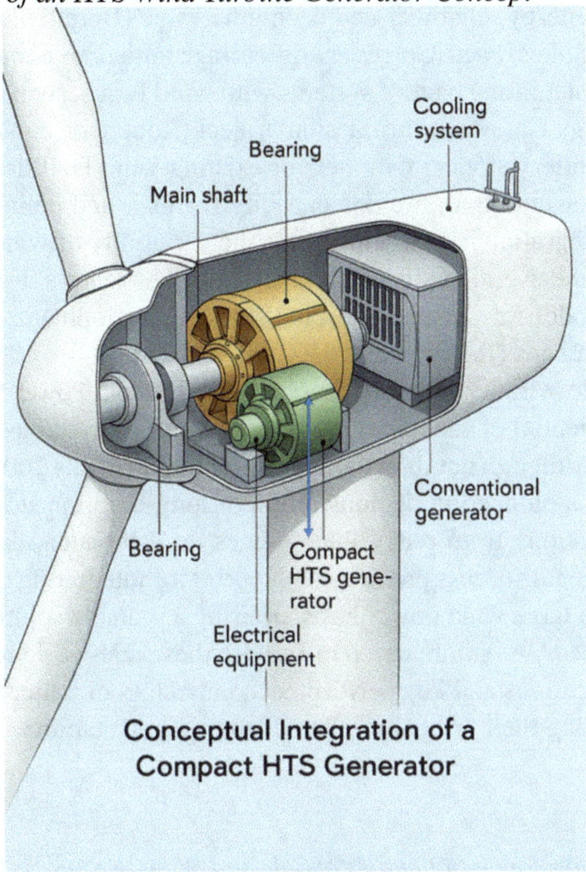

5.1.2. Energy Storage Integration With Wind Farms

Energy storage systems, particularly batteries, are becoming increasingly integral to wind power applications, supporting both grid-scale and individual wind farm operations with enhanced reliability and flexibility. The Hornsdale Power Reserve in Australia exemplifies this trend, where a 100 MW/129 MWh Tesla Powerpack lithium-ion battery was initially paired with Neoen's Hornsdale Wind Farm (Boretti, 2019). This system delivers critical grid stability services, including fast frequency response and ancillary support, effectively firming up intermittent wind power and enabling lucrative grid services. The project's success prompted an expansion to

150 MW/194 MWh, underscoring the transformative impact of battery storage in enhancing wind energy reliability and economic value (Boretti, 2019).

In the United States, PacifiCorp's energy storage initiatives demonstrate a strategic approach to integrating battery systems with wind farms, such as those planned within their Energy Gateway transmission project (Robertson & McCarthy, 2018). By co-locating battery storage with new or existing wind facilities, these projects aim to smooth power output, provide capacity firming, and deliver essential grid services. This integration not only mitigates the variability of wind generation but also strengthens the overall resilience and efficiency of renewable energy systems, showcasing a model for future wind-plus-storage developments that align with evolving grid demands (Robertson & McCarthy, 2018).

The Dalian Flow Battery Energy Storage Peak-shaving Power Station in China highlights the potential of vanadium redox flow batteries for large-scale renewable energy support. Although not directly tied to a wind farm, this 200 MW/800 MWh project serves as a pioneering demonstration of long-duration storage, effectively managing fluctuations from renewable sources on a broader scale (Chen et al., 2022). Its operational success provides a blueprint for integrating similar flow battery systems with large wind power hubs, offering a scalable solution for extended energy storage needs. As summarized in Table 2, these real-world projects illustrate the diverse applications and key performance indicators of battery energy storage systems, reinforcing their critical role in advancing grid stability and wind energy integration.

Table 2. Key Performance Indicators of Selected Wind + Storage Projects

Project Name	Location	Wind Capacity (MW)	Storage Capacity (MW/MWh)	Battery Chemistry	Primary Grid Services Provided
Hornsdale Power Reserve	Jamestown, SA, Australia	309	150 MW / 194 MWh	Lithium-ion	Frequency regulation, grid stability, peaker replacement, contingency response
Huaneng Longhe Wind-Solar-Storage Base	Qinghai Province, China	~3000 (Wind)	320 MW / 640 MWh (Li-ion)	Lithium-ion	Grid stability, renewable energy integration, peak shaving, voltage support
Wind Charger (Pilot)	PJM Market, USA	N/A (behind meter integration)	2 MW / 7 MWh	Lithium-ion	Frequency regulation, ancillary services, grid support
Lake Bonney Wind Farm Stages 1 & 2	Millicent, SA, Australia	103 (Wind)	30 MW / 8 MWh	Lithium-ion	Frequency response, energy arbitrage, voltage control

5.1.3. Synthetic Inertia From Wind Turbines

The adoption of synthetic inertia capabilities in wind farms, though relatively recent in commercial settings, is gaining momentum through targeted research projects and evolving grid codes, particularly in regions with high wind penetration (Makolo et al., 2021b). A notable example is Scottish Power Renewables' Whitelee Wind Farm in the UK, one of Europe's largest onshore wind facilities (Shepherd et al., 2021). This wind farm has participated in trials for Enhanced Frequency Response (EFR) services, which involve rapid active power adjustments from wind turbines, often in conjunction with co-located energy storage systems. These adjustments provide a form of inertial support that goes beyond traditional grid-following (GFL) operation, enabling the wind farm to contribute to grid frequency stability. Such initiatives demonstrate the potential of wind turbines to deliver critical grid support services, paving the way for broader adoption of synthetic inertia (Shepherd et al., 2021).

Grid operators in countries like Germany, Ireland, and Australia are increasingly updating their grid codes to mandate or incentivize synthetic inertia and fast frequency response from inverter-based resources, reflecting the growing need for these capabilities in low-inertia grids (Nycander & Söder, 2018). Organizations such as ENTSO-E in Europe, EirGrid in Ireland, and AEMO in Australia are leading efforts to integrate these requirements, ensuring that wind farms can help mitigate rapid changes in frequency, known as RoCoF, and improve frequency nadir. Research projects are playing a crucial role in this transition by simulating and testing Virtual Synchronous Machine (VSM) control algorithms on wind farm models or smaller-scale physical setups. These studies validate the effectiveness of synthetic inertia in stabilizing grids, offering practical insights into its implementation (Mkoi et al., 2025).

The significance of these advancements is further underscored by research, such as simulations detailed in academic theses, which often explore synthetic inertia under specific fault conditions. For instance, simulations of a wind farm providing synthetic inertia during grid disturbances could serve as a compelling case study, highlighting the technical feasibility and benefits of these capabilities (Justo et al., 2025). Such work, potentially including detailed models and results, reinforces the findings from real-world demonstrations and grid code developments. By combining practical trials, regulatory advancements, and rigorous research, the push for synthetic inertia in wind farms is strengthening, enabling wind energy to play a more robust role in maintaining grid stability (Makolo et al., 2025).

5.2. Analysis of Performance Data and Lessons Learned

The case studies and ongoing research provide several key takeaways:

- **Techno-Economic Viability:** Energy storage, particularly Li-ion batteries, has demonstrated clear techno-economic viability for grid services, firming wind output, and optimizing market participation. The falling costs of batteries are accelerating their adoption. Superconductors, while technically superior in efficiency, still face higher upfront costs and cooling complexities that limit broad commercial wind turbine integration but show promise for high-capacity transmission lines. Synthetic inertia is becoming a non-negotiable requirement in modern grid codes, driven by its critical role in system stability.
- **Grid Code Evolution:** Regulatory bodies and grid operators are increasingly recognizing the necessity for IBRs to contribute to grid stability services traditionally provided by synchronous generators. This is leading to evolving grid codes that mandate or reward inertial response and FFR from wind farms.
- **Role of AI and Advanced Control:** The optimal operation of these integrated systems heavily relies on advanced control algorithms, forecasting, and increasingly, AI-driven optimization. AI enables intelligent dispatch of storage, adaptive synthetic inertia control, and predictive maintenance for all components, maximizing benefits and minimizing risks.

Challenges Remain:

- **Cost vs. Benefit:** For superconductors, the capital cost versus the energy savings needs further improvement for widespread adoption. For energy storage, the duration versus cost remains a challenge for very long-term seasonal storage.
- **Complexity of Integration:** Coordinating these diverse technologies requires sophisticated control systems and robust communication infrastructure.
- **Data Availability and Modelling:** Accurate models and vast amounts of real-time data are essential for effective AI-driven optimization and system-wide planning.
- **Public Acceptance/Environmental Impact:** For large-scale PHS, environmental concerns can be a barrier. Battery recycling and end-of-life management are growing concerns.

The tangible economic advantages, such as improved revenue streams and reduced curtailment, achieved by integrating energy storage with wind farms are visually demonstrated in Figure 10.

Figure 10. Graph Showing Economic Benefits of Wind + Storage

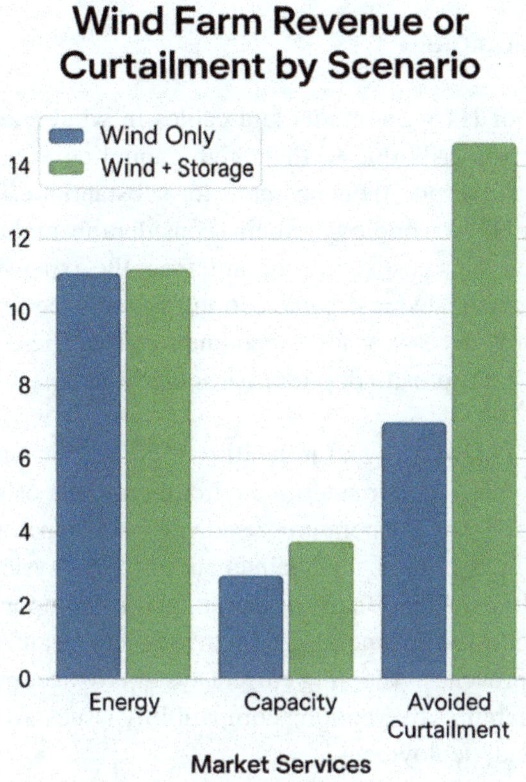

The lessons learned from these applications underscore that a multi-faceted approach, combining high-efficiency components (superconductors), flexibility (storage), and stability provision (synthetic inertia), is crucial for maximizing the potential of wind energy in the future grid.

6. CHALLENGES, OPPORTUNITIES, AND FUTURE DIRECTIONS

While the integration of superconductors, advanced energy storage, and synthetic inertia holds immense promise for revolutionizing wind energy systems, their widespread adoption is accompanied by a set of technical, economic, and regulatory challenges. Simultaneously, these challenges present significant opportunities for further research, development, and innovation, shaping the future trajectory of renewable energy.

6.1. Technical and Economic Challenges in Implementing These Technologies

6.1.1. Superconductors

The deployment of HTS generators and cables in wind energy systems faces significant hurdles, primarily due to their high capital costs and the complexity of cryogenic systems. Despite their potential for substantial efficiency gains, the initial investment for HTS technology remains considerably higher than that of conventional alternatives. This cost stems not only from the expensive HTS materials but also from the sophisticated cryogenic cooling systems required to maintain the ultra-low temperatures necessary for superconductivity. These financial barriers pose a challenge to widespread adoption, particularly in cost-sensitive renewable energy projects.

Beyond cost, the complexity and reliability of cryogenic systems present critical challenges, especially in demanding environments like offshore wind farms. Maintaining the precise low temperatures needed for HTS operation adds layers of intricacy to the design, operation, and maintenance of these systems, increasing the risk of technical failures. Additionally, integrating HTS components with existing grid infrastructure and wind turbine designs demands significant engineering efforts and standardized approaches. Adapting current systems to accommodate superconducting technology requires overcoming compatibility issues, further complicating the path to commercial deployment.

6.1.2. Energy Storage

The integration of energy storage systems with wind power faces significant challenges, particularly related to cost and efficiency. While battery costs have been decreasing, achieving economically viable long-duration storage, such as for seasonal energy needs, remains difficult for most battery technologies. Pumped hydro storage offers a cost-effective solution for large-scale, long-duration storage, but its reliance on specific geographical features limits its widespread applicability. Additionally, all storage technologies experience energy losses during charging and discharging cycles, which reduces the overall round-trip efficiency of the wind-to-grid energy pathway, impacting the effectiveness of stored wind power.

Beyond cost and efficiency, battery degradation and safety concerns pose further obstacles. Over time and through repeated cycles, battery performance deteriorates, necessitating costly replacements and raising challenges related to end-of-life disposal and recycling. Safety issues, particularly with certain battery chemistries like lithium-ion, also require careful attention due to risks of thermal runaway and fires. These risks demand sophisticated safety protocols to ensure reliable operation, adding complexity and cost to the deployment of energy storage systems for wind energy integration.

6.1.3. Synthetic Inertia

Implementing synthetic inertia in wind turbines to support grid stability presents notable challenges, particularly in achieving accurate grid state estimation and managing control complexity. Effective synthetic inertia control relies on precise, real-time measurements of grid frequency and the Rate of Change of Frequency (RoCoF), which can be difficult in dynamic, low-inertia grids where rapid fluctuations are common. Additionally, deploying sophisticated Virtual Synchronous Machine (VSM) or Grid-Forming (GFM) controls across a large fleet of distributed wind turbines requires advanced algorithms and meticulous coordination. Without careful management, these controls can lead to oscillatory behaviors or unintended interactions, potentially destabilizing the grid instead of supporting it.

The mechanical impact of providing synthetic inertia further complicates its adoption. When synthetic inertia is achieved by extracting kinetic energy from the turbine rotor, it introduces transient stresses on mechanical components and may result in sub-optimal aerodynamic operation. These stresses can reduce the lifespan of turbine components, increasing maintenance costs and affecting long-term reliability. The need to balance grid support with turbine durability adds another layer of complexity to the design and operation of wind turbines equipped with synthetic inertia capabilities.

Standardization and compliance with diverse grid codes also pose significant hurdles. Harmonizing synthetic inertia requirements across different grid operators, each with unique operational standards, is a complex task. Ensuring that turbines from various manufacturers consistently meet these requirements demands standardized approaches and robust testing protocols. This challenge is compounded by the need to integrate synthetic inertia seamlessly into existing wind farm operations while maintaining compatibility with evolving grid regulations, making it essential to develop unified standards to facilitate widespread adoption.

A concise overview of the primary technical and economic challenges associated with the implementation of superconductors, energy storage, and synthetic inertia in wind energy systems, along with their potential impact on wind integration, is presented in Table 3.

Table 3. Summary of Key Challenges for Each Technology

Technology	Main Technical Challenges	Main Economic Challenges	Potential Impact on Wind Integration
Superconductors	• Cryogenic system complexity & reliability • Integration with existing infrastructure	• High capital costs (materials, cooling) • Higher O&M for cooling systems	• Increased upfront cost for wind farms • Slower adoption due to cost-effectiveness concerns
Energy Storage	• Degradation over time/cycles • Safety concerns (e.g., thermal runaway for Li-ion) • Geographical dependency (for PHS)	• High cost of long-duration capacity • Efficiency losses (round-trip)	• Limits on providing continuous firm power • Requires careful safety protocols and siting • Geographical constraints for certain large-scale storage
Synthetic Inertia	• Accurate real-time grid state estimation • Complex control interactions across multiple IBRs • Standardization and grid code harmonization	• Potential for increased control complexity • Possible impact on turbine lifespan (kinetic energy extraction)	• Requires advanced control systems and robust communication • May add operational wear on turbine components • Barriers to seamless integration across diverse fleets/grids

6.2. Research and Development Needs

Addressing the challenges of integrating HTS generators, energy storage, and synthetic inertia in wind energy systems requires sustained research and development in materials science and power electronics. Advances in HTS materials are critical, focusing on achieving higher critical temperatures, enhanced mechanical properties, and reduced manufacturing costs to make these technologies more viable for widespread use. Similarly, research into sustainable and abundant materials for energy storage aims to minimize reliance on scarce resources and improve recyclability. In parallel, the development of robust, efficient, and cost-effective power converters is essential to enable advanced grid services, such as grid-forming capabilities and synthetic inertia, without sacrificing system efficiency or reliability. Adaptive and self-healing control algorithms further enhance these systems, ensuring stable and responsive operation in dynamic grid environments.

Grid integration and system-level optimization represent another critical area of focus, necessitating sophisticated grid models and simulation tools to optimize the interaction of thousands of inverter-based resources (IBRs) providing synthetic inertia and leveraging energy storage. Research into optimal sizing and siting of storage systems for wind farms is vital to maximize their effectiveness in smoothing power output and supporting grid stability. These efforts ensure that wind farms can operate seamlessly within complex grid infrastructures, delivering reliable and efficient energy while addressing the variability inherent in wind generation.

The application of AI and machine learning, alongside the development of long-duration, sustainable storage solutions, is pivotal for advancing these technologies. Continued research in AI is needed to refine predictive control, fault diagnosis, optimal dispatch, and predictive maintenance, enabling robust models that operate in real-time despite imperfect data and uncertainties. Simultaneously, significant efforts are required to develop cost-effective, environmentally sustainable energy storage solutions capable of supporting long-duration needs, such as beyond 8-10 hours, with reduced dependence on rare earth minerals and improved recyclability. As conceptually illustrated in Figure 11, a strategic roadmap outlining these key research areas is essential for driving the integration of innovative wind technologies into modern, resilient power systems.

Figure 11. Research and Development Roadmap for Integrated Wind Technologies

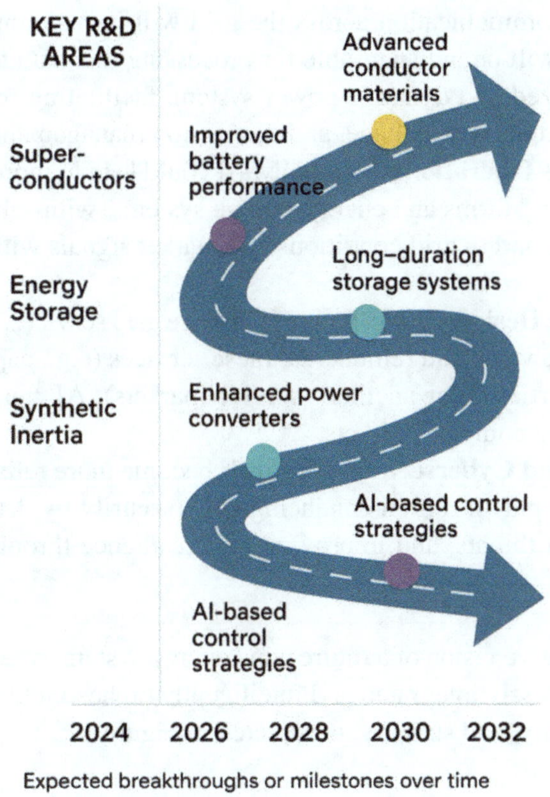

6.3. Future Trends and the Potential of AI to Drive Further Innovation

The future of wind energy is intrinsically linked to the continued evolution and synergistic integration of these advanced technologies, with AI serving as a central enabler.

- **Hybrid Power Plants:** The trend towards "hybrid power plants" combining wind with solar and battery storage will accelerate. AI will be crucial for optimizing the dispatch of these multi-resource assets, maximizing their collective value and grid contribution.
- **Grid Decarbonization and Electrification:** As sectors like transport and heating electrify, the demand for stable, renewable electricity will surge. Integrated wind solutions will play a vital role in meeting this demand, enabled by AI for complex load forecasting and renewable integration.
- **Smart Grid and Digitalization:** The proliferation of sensors, IoT devices, and digital communication across the grid will provide unprecedented data streams. AI will be indispensable for processing this data to manage a highly decentralized and dynamic power system, facilitating real-time decision-making for frequency, voltage, and congestion management.
- **Autonomous Operation:** Eventually, AI could lead to more autonomous operation of wind farms and energy storage systems, with self-optimizing controls that respond to grid conditions and market signals with minimal human intervention.
- **New Market Designs:** As IBRs provide more grid services, market structures will evolve to value and remunerate these services (e.g., capacity markets for synthetic inertia, faster ancillary service markets). AI can help participants navigate these complex markets.
- **Resilience and Cybersecurity:** As grids become more reliant on digital controls, AI can play a role in enhancing cybersecurity by detecting anomalies and potential threats, and improving grid resilience through rapid recovery strategies.

The transformative vision of a future wind energy system, where advanced technologies are seamlessly integrated and intelligently orchestrated by AI for optimal efficiency, reliability, and stability, is depicted in Figure 12.

Figure 12. Vision of a Future AI-Optimized Wind Energy System

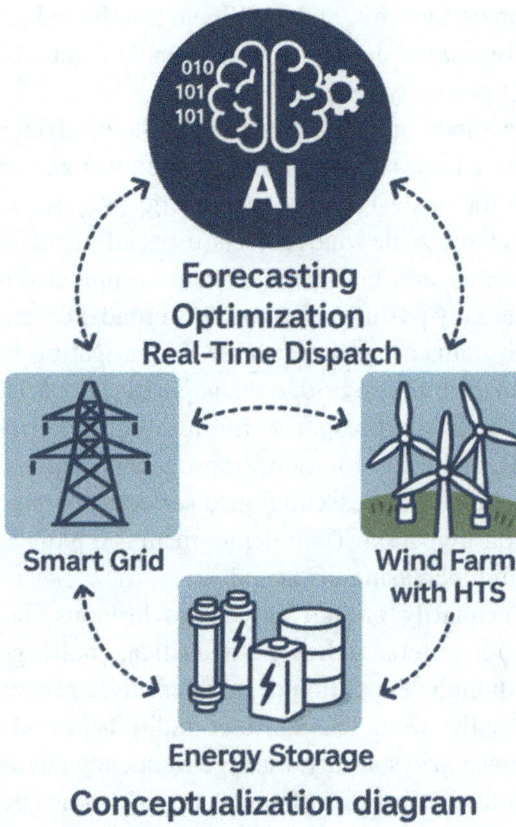

The synergistic combination of superconducting materials, diverse energy storage solutions, and intelligent synthetic inertia control, all orchestrated by advanced AI capabilities, represents the frontier of wind energy innovation. This integrated approach is essential for overcoming current limitations and enabling wind power to become the backbone of a highly efficient, reliable, and sustainable global energy system.

7. CONCLUSION

The journey towards a sustainable and resilient energy future heavily relies on the successful integration of renewable energy sources, with wind power playing a pivotal role. However, the inherent variability of wind energy, coupled with the evolving demands of modern power grids, necessitates a continuous push for innovation beyond conventional approaches. This chapter has comprehensively explored how the synergistic application of High-Temperature Superconductors advanced Energy Storage Systems, and intelligent Synthetic Inertia techniques can fundamentally transform wind energy from a variable resource into a cornerstone of stable and efficient power systems.

High-Temperature Superconductors offer unparalleled efficiency gains by virtually eliminating resistive losses in wind turbine generators and transmission cables. This leads to more compact designs, reduced energy waste, and potentially lower infrastructure costs for large-scale wind projects, especially in offshore environments. While economic viability and cryogenic system complexity remain challenges, ongoing research is steadily pushing HTS towards broader commercial application.

Energy Storage Systems are indispensable for mitigating wind intermittency, providing the crucial flexibility to bridge the gap between variable generation and consistent demand. From rapid-response Lithium-ion batteries to long-duration Pumped Hydro and Redox Flow solutions, these technologies enable wind power to contribute firm capacity, offer essential grid services like frequency regulation, and optimize market participation. Their deployment is rapidly accelerating, driven by decreasing costs and increasing grid needs.

Synthetic Inertia, primarily through Virtual Synchronous Machines (VSMs) and Grid-Forming (GFM) inverters, addresses the critical challenge of declining grid inertia in systems with high penetration of inverter-based renewables. By enabling wind turbines to mimic the inertial response of traditional synchronous generators, these techniques enhance grid stability, manage frequency deviations, and improve overall system resilience against disturbances. As grid codes evolve, synthetic inertia is becoming a mandatory and vital capability for future wind energy systems.

The overarching theme across all these innovations is the transformative potential of Artificial Intelligence (AI). AI serves as a powerful enabler for optimizing the performance, management, and integration of superconductors, energy storage, and synthetic inertia. From predictive maintenance and optimal dispatch scheduling to adaptive control of VSMs and real-time grid response, AI enhances the efficiency, reliability, and economic value of these advanced technologies. It is the connective tissue that allows for the complex coordination required to maximize the benefits of a highly decentralized and dynamic renewable-dominated grid.

7.1 Recommendations

- **Continued Investment in R&D:** Prioritize research into next-generation HTS materials, more sustainable and cost-effective long-duration storage solutions, and advanced AI-driven control algorithms.

- **Policy and Market Evolution:** Develop supportive regulatory frameworks and market designs that incentivize the provision of grid services (like synthetic inertia and firming capacity) from wind farms with integrated technologies.
- **Demonstration Projects:** Support large-scale pilot and demonstration projects that integrate all three technologies to refine practical implementation, gather long-term performance data, and address remaining technical and economic barriers.
- **Standardization and Interoperability:** Promote global standards for power converter controls, communication protocols, and grid integration to ensure seamless operation across diverse manufacturers and grid operators.

In conclusion, "Harnessing Innovation in Wind Energy" by strategically integrating superconductors, energy storage, and synthetic inertia, powerfully augmented by AI, is not merely an aspiration but a critical pathway to building highly efficient, reliable, and stable power systems of the future. These innovations are fundamental to unlocking the full potential of wind power and accelerating the global transition towards a truly sustainable energy landscape.

REFERENCES

Adefarati, T., Bansal, R. C., Bettayeb, M., & Naidoo, R. (2021). Optimal energy management of a PV-WTG-BSS-DG microgrid system. *Energy*, *217*, 119358. DOI: 10.1016/j.energy.2020.119358

Awal, M. A., & Husain, I. (2020). Unified Virtual Oscillator Control for Grid-Forming and Grid-Following Converters. *IEEE Journal of Emerging and Selected Topics in Power Electronics*. Advance online publication. 10.1109.

Boretti, A. (2019). Dependent performance of South Australian wind energy facilities with respect to resource and grid availability. *Energy Storage*, *1*(6), e97. DOI: 10.1002/est2.97

Bose, B. K. (2017). Artificial Intelligence Techniques in Smart Grid and Renewable Energy Systems—Some Example Applications. *Proceedings of the IEEE*, *105*(11), 2262–2273. DOI: 10.1109/JPROC.2017.2756596

Bošnjaković, M., Martinović, M., & Đokić, K. (2025). Application of artificial intelligence in wind power systems. *Applied Sciences (Basel, Switzerland)*, *15*(5), 2443. DOI: 10.3390/app15052443

Bowen, T., Chernyakhovskiy, I., & Denholm, P. L. (2019). *Grid-scale battery storage: frequently asked questions*. National Renewable Energy Lab. (NREL), Golden, CO (United States). .DOI: 10.2172/1561843

Bullich-Massagué, E., Cifuentes-García, F.-J., Glenny-Crende, I., Cheah-Mañé, M., Aragüés-Peñalba, M., Díaz-González, F., & Gomis-Bellmunt, O. (2020). A review of energy storage technologies for large scale photovoltaic power plants. *Applied Energy*, *274*, 115213. DOI: 10.1016/j.apenergy.2020.115213

Chen, X., Huang, L., Liu, J., Song, D., & Yang, S. (2022). Peak shaving benefit assessment considering the joint operation of nuclear and battery energy storage power stations: Hainan case study. *Energy*, *239*, 121897. DOI: 10.1016/j.energy.2021.121897

Chu, Z., Markovic, U., Hug, G., & Teng, F. (2020). Towards optimal system scheduling with synthetic inertia provision from wind turbines. *IEEE Transactions on Power Systems*, *35*(5), 4056–4066. DOI: 10.1109/TPWRS.2020.2985843

Coombs, T. A., Wang, Q., Shah, A., Hu, J., Hao, L., Patel, I., Wei, H., Wu, Y., Coombs, T., & Wang, W. (2024). High-temperature superconductors and their large-scale applications. *Nature Reviews Electrical Engineering*, *1*(12), 788–801. DOI: 10.1038/s44287-024-00112-y

Cui, Y., Xie, J., & Zheng, K. (2021). Historical Inertia: An Ignored but Powerful Baseline for Long Sequence Time-series Forecasting. *ArXiv Preprint ArXiv:2103.16349.* DOI: 10.1145/3459637.3482120

Cullinane, M., Judge, F., O'Shea, M., Thandayutham, K., & Murphy, J. (2022). Subsea superconductors: The future of offshore renewable energy transmission? *Renewable & Sustainable Energy Reviews, 156*, 111943. DOI: 10.1016/j.rser.2021.111943

Das, K., & Cutululis, N. A. (2017). *Offshore Wind Power Plant Technology Catalogue-Components of wind power plants, AC collection systems and HVDC systems.*

Díaz-González, F., Sumper, A., Gomis-Bellmunt, O., & Villafáfila-Robles, R. (2012). A review of energy storage technologies for wind power applications. *Renewable & Sustainable Energy Reviews, 16*(4), 2154–2171. DOI: 10.1016/j.rser.2012.01.029

Divya, K. C., & Østergaard, J. (2009). Battery energy storage technology for power systems—An overview. *Electric Power Systems Research, 79*(4), 511–520. DOI: 10.1016/j.epsr.2008.09.017

Durgadevi, G., Nalawade, M. K., Sharma, K., Nishanthi, N., Shnain, A. H., Sutar, V., & Sujatha, M. S. (2024). Integration of Energy Storage with Wind Power Conversion Systems: Enhancing Grid Stability. *E3S Web of Conferences, 591*, 02004. .DOI: 10.1051/e3sconf/202459102004

Gonzalez-Longatt, F. (2012). Impact of synthetic inertia from wind power on the protection/control schemes of future power systems: Simulation study. *11th IET International Conference on Developments in Power Systems Protection (DPSP 2012)*, 74. DOI: 10.1049/cp.2012.0030

Gursoy, M., Zhuo, G., Lozowski, A. G., & Wang, X. (2021). Photovoltaic energy conversion systems with sliding mode control. *Energies, 14*(19), 6071. DOI: 10.3390/en14196071

Gwabavu, M., Bansal, R. C., & Naidoo, R. M. (2024). A review of onshore wind farm battery energy storage systems for techno-economic optimization. *International Journal of Modelling and Simulation*, 1–20. .DOI: 10.1080/02286203.2024.2320583

Hamdan, A., Ibekwe, K. I., Ilojianya, V. I., Sonko, S., & Etukudoh, E. A. (2024). AI in renewable energy: A review of predictive maintenance and energy optimization. *International Journal of Science and Research Archive, 11*(1), 718–729. DOI: 10.30574/ijsra.2024.11.1.0112

Hasan, A. K. M. K., Haque, M. H., & Aziz, S. M. (2019). Application of battery energy storage systems to enhance power system inertia. *2019 29th Australasian Universities Power Engineering Conference (AUPEC)*, 1–6. DOI: 10.1109/AUPEC48547.2019.211941

Hassan, Q., Viktor, P., Al-Musawi, T. J., Ali, B. M., Algburi, S., Alzoubi, H. M., Al-Jiboory, A. K., Sameen, A. Z., Salman, H. M., & Jaszczur, M. (2024). The renewable energy role in the global energy Transformations. *Renewable Energy Focus*, *48*, 100545. DOI: 10.1016/j.ref.2024.100545

Justo, J. J., Demidova, G., Lukichev, D., Poliakov, N., Makolo, P., & Bansal, R. C. (2025). Synthesis of Dynamic Virtual-Inertia for Microgrid System Applications. *2025 International Conference on Intelligent Control, Computing and Communications (IC3)*, 177–182. DOI: 10.1109/IC363308.2025.10957416

Justo, J. J., Mwasilu, F., & Jung, J.-W. (2015). Doubly-fed induction generator-based wind turbines: A comprehensive review of fault ride-through strategies. *Renewable & Sustainable Energy Reviews*, *45*, 447–467. DOI: 10.1016/j.rser.2015.01.064

Justo, J. J., Mwasilu, F., Lee, J., & Jung, J.-W. (2013). AC-microgrids versus DC-microgrids with distributed energy resources: A review. *Renewable & Sustainable Energy Reviews*, *24*, 387–405. DOI: 10.1016/j.rser.2013.03.067

Koblischka, M. R., Půst, L., Chang, C.-S., Hauet, T., & Koblischka-Veneva, A. (2023). The paramagnetic Meissner effect (PME) in metallic superconductors. *Metals*, *13*(6), 1140. DOI: 10.3390/met13061140

Le, T. D., Kim, J. H., Kim, D. J., Boo, C. J., & Kim, H. M. (2015). Status of the technology development of large scale HTS generators for wind turbine. *Progress in Superconductivity and Cryogenics*, *17*(2), 18–24. DOI: 10.9714/psac.2015.17.2.018

Lim, S., Seo, K., Park, J.-W., & Lee, K. Y. (2025). New frequency stability assessment based on contribution rates of wind power plants. *International Journal of Electrical Power & Energy Systems*, *164*, 110388. DOI: 10.1016/j.ijepes.2024.110388

Lloberas, J., Sumper, A., Sanmarti, M., & Granados, X. (2014). A review of high temperature superconductors for offshore wind power synchronous generators. *Renewable & Sustainable Energy Reviews*, *38*, 404–414. DOI: 10.1016/j.rser.2014.05.003

Madeswaran, A., Bisht, D., Yuvaraj, S., Reedy, M. U., Al-Attabi, K., & Dhablia, A. (2024). AI-controlled wind turbine systems: Integrating IoT and machine learning for smart grids. *E3S Web of Conferences, 540*, 03008. .DOI: 10.1051/e3sconf/202454003008

Makolo, P., Cheah-Mañe, M., & Gomis-Bellmunt, O. (2025). Quantification of virtual inertia from derivative control using a recursive least square with a weighted forgetting method. *Electric Power Systems Research, 241*, 111363. DOI: 10.1016/j.epsr.2024.111363

Makolo, P., Justo, J. J., Mwasilu, F., & Zamora, R. (2018). Fault Ride Through Technique for DFIG-based Wind Turbines Under Grid Three-phase Faults. *2018 Australasian Universities Power Engineering Conference (AUPEC)*, 1–5. DOI: 10.1109/AUPEC.2018.8757926

Makolo, P., Zamora, R., & Lie, T. (2021a). Online inertia estimation for power systems with high penetration of RES using recursive parameters estimation. *IET Renewable Power Generation, 15*(12), 2571–2585. Advance online publication. DOI: 10.1049/rpg2.12181

Makolo, P., Zamora, R., & Lie, T.-T. (2021b). The role of inertia for grid flexibility under high penetration of variable renewables-A review of challenges and solutions. *Renewable & Sustainable Energy Reviews, 147*, 111223. DOI: 10.1016/j.rser.2021.111223

Masenge, I. H., & Mwasilu, F. (2020). Modeling and control of solar PV with battery energy storage for rural electrification. *Tanzania Journal of Engineering and Technology, 39*(1), 47–58. DOI: 10.52339/tjet.v39i1.518

May, G. J., Davidson, A., & Monahov, B. (2018). Lead batteries for utility energy storage: A review. *Journal of Energy Storage, 15*, 145–157. DOI: 10.1016/j.est.2017.11.008

Mkoi, P., Makolo, P. M., & Mwasilu, F. A. (2025). Synthetic Inertia Provision for Load Frequency Control in Networks with High Penetration of Renewable Energy Sources. *Tanzania Journal of Engineering and Technology, 44*(1), 245–256. DOI: 10.52339/tjet.v44i1.1201

Nycander, E., & Söder, L. (2018). Review of european grid codes for wind farms and their implications for wind power curtailments. *17th International Wind Integration Workshop Stockholm,* Swedenl 17–19 October 2018.

Olabi, A. G., Wilberforce, T., Sayed, E. T., Abo-Khalil, A. G., Maghrabie, H. M., Elsaid, K., & Abdelkareem, M. A. (2022). Battery energy storage systems and SWOT (strengths, weakness, opportunities, and threats) analysis of batteries in power transmission. *Energy, 254*, 123987. DOI: 10.1016/j.energy.2022.123987

Preuster, P., Alekseev, A., & Wasserscheid, P. (2017). Hydrogen Storage Technologies for Future Energy Systems. *Annual Review of Chemical and Biomolecular Engineering*, *8*(1), 445–471. DOI: 10.1146/annurev-chembioeng-060816-101334 PMID: 28592172

Robertson, D., & McCarthy, B. (2018). Portland General Electric Driving Collaboration for Sustainable Legislation. In *Sustainable Electricity II: A Conversation on Tradeoffs* (pp. 143–159). Springer.

Sadeghi, A., Seyyedbarzegar, S., & Yazdani-Asrami, M. (2022). Investigation on the electrothermal performance of a high-temperature superconducting cable in an offshore wind farm integrated power system: Fault and islanding conditions. *IEEE Transactions on Applied Superconductivity*, *32*(8), 1–11. DOI: 10.1109/TASC.2022.3196770

Shepherd, A., Roberts, S., Sünnenberg, G., Lovett, A., & Hastings, A. F. S. (2021). Scotland's onshore wind energy generation, impact on natural capital & satisfying no-nuclear energy policy. *Energy Reports*, *7*, 7106–7117. DOI: 10.1016/j.egyr.2021.10.063

Sousa, J., Lagarto, J., & Fonseca, M. (2024). The role of storage and flexibility in the energy transition: Substitution effect of resources with application to the Portuguese electricity system. *Renewable Energy*, *228*, 120694. DOI: 10.1016/j.renene.2024.120694

Suvire, G. O., & Mercado, P. E. (2010). DSTATCOM with flywheel energy storage system for wind energy applications: Control design and simulation. *Electric Power Systems Research*, *80*(3), 345–353. DOI: 10.1016/j.epsr.2009.09.020

Tambunan, H. B., Hakam, D. F., Prahastono, I., Pharmatrisanti, A., Purnomoadi, A. P., Aisyah, S., Wicaksono, Y., & Sandy, I. G. R. (2020). The challenges and opportunities of renewable energy source (RES) penetration in Indonesia: Case study of Java-Bali power system. *Energies*, *13*(22), 5903. DOI: 10.3390/en13225903

Ukoba, K., Olatunji, K. O., Adeoye, E., Jen, T.-C., & Madyira, D. M. (2024). Optimizing renewable energy systems through artificial intelligence: Review and future prospects. *Energy & Environment*, *35*(7), 3833–3879. DOI: 10.1177/0958305X241256293

Wang, F., Zhang, L., Feng, X., & Guo, H. (2018). An adaptive control strategy for virtual synchronous generator. *IEEE Transactions on Industry Applications*, *54*(5), 5124–5133. DOI: 10.1109/TIA.2018.2859384

Wang, Z., & Bu, S. (2024). Probabilistic frequency stability analysis considering dynamics of wind power generation with different control strategies. *IEEE Transactions on Power Systems*, *39*(5), 6412–6425. DOI: 10.1109/TPWRS.2024.3362056

Xu, T., Jang, W., & Overbye, T. (2018). Commitment of Fast-Responding Storage Devices to Mimic Inertia for the Enhancement of Primary Frequency Response. *IEEE Transactions on Power Systems*, *33*(2), 1219–1230. DOI: 10.1109/TPWRS.2017.2735990

Yan, D., Sadeghi, A., Yazdani-Asrami, M., & Song, W. (2024). Artificial-Intelligence-Driven Model for Resistive Superconducting Fault Current Limiter in Future Electric Aircraft. *IEEE Transactions on Applied Superconductivity*, *34*(7), 1–16. DOI: 10.1109/TASC.2024.3421903

Yang, B., Wang, J., Chen, Y., Li, D., Zeng, C., Chen, Y., Guo, Z., Shu, H., Zhang, X., Yu, T., & Sun, L. (2020). Optimal sizing and placement of energy storage system in power grids: A state-of-the-art one-stop handbook. *Journal of Energy Storage*, *32*, 101814. DOI: 10.1016/j.est.2020.101814

Zhu, J., Hu, J., Hung, W., Wang, C., Zhang, X., Bu, S., Li, Q., Urdal, H., & Booth, C. D. (2017). Synthetic inertia control strategy for doubly fed induction generator wind turbine generators using lithium-ion supercapacitors. *IEEE Transactions on Energy Conversion*, *33*(2), 773–783. DOI: 10.1109/TEC.2017.2764089

Chapter 2
Integrating AI and Digital Twins for Real-Time Fault Identification in Wind Turbines

Deepak Gupta
http://orcid.org/0000-0003-3929-1362
ITM University, India

K Balakrishna Reddy
http://orcid.org/0000-0002-6604-6322
Vignan's Foundation for Science, Technology, and Research, India

K. Neelima
Department of IT, St. Martin's Engineering College, India

Shabana Samsher Pathan
http://orcid.org/0000-0002-9492-3388
St. Vincent Pallotti College of Engineering and Technology, India

Bali Ram Gupta
http://orcid.org/0000-0002-4998-2709
Jaypee University of Engineering and Technology, India

P. Rajeshwari
Erode Sengunthar Engineering College, India

Pranda P. Gupta
GLA University, Mathura, India

ABSTRACT

The integration of artificial intelligence (AI) and digital twin technologies has revolutionized condition monitoring and fault detection in industrial systems, particularly in wind energy applications. This chapter explores the synergistic relationship between AI algorithms and digital twin frameworks for enhanced predictive maintenance strategies. Digital twins create virtual replicas of physical systems, enabling real-time monitoring, simulation, and analysis of operational conditions.

DOI: 10.4018/979-8-3373-4159-0.ch002

When combined with AI techniques such as machine learning, deep learning, and neural networks, these systems can predict failures, optimize maintenance schedules, and reduce operational costs. The chapter examines current methodologies, implementation challenges, and future prospects of AI-enhanced digital twins in condition monitoring. Case studies from wind turbine applications demonstrate the practical benefits of this integrated approach, showing improvements in fault detection accuracy, reduced downtime, and enhanced system reliability.

1. INTRODUCTION

The Fourth Industrial Revolution has ushered in an era of unprecedented technological integration, where physical and digital systems converge to create intelligent manufacturing and operational environments. At the forefront of this transformation lies the convergence of artificial intelligence (AI) and digital twin technologies, which together form a powerful paradigm for condition monitoring and fault detection across various industrial sectors.

Digital twin technology, first conceptualized by NASA for space missions, has evolved into a cornerstone of modern industrial digitalization. A digital twin represents a virtual replica of a physical asset, process, or system that enables real-time monitoring, simulation, and optimization (Jeong et al., 2022). When augmented with AI capabilities, digital twins transcend traditional monitoring approaches by incorporating predictive analytics, adaptive learning, and autonomous decision-making capabilities.

The significance of this technological convergence becomes particularly apparent in critical infrastructure systems such as wind energy installations, where unplanned downtime can result in substantial economic losses and grid instability. Traditional condition monitoring approaches, while effective to some extent, often rely on reactive maintenance strategies or fixed scheduling protocols that fail to optimize operational efficiency and resource utilization.

AI-enhanced digital twins address these limitations by providing a comprehensive framework that combines real-time data acquisition, advanced analytics, and predictive modeling. This integration enables organizations to transition from reactive to proactive and ultimately to predictive maintenance strategies, resulting in improved system reliability, reduced operational costs, and enhanced safety performance.

The evolution of sensor technologies, edge computing capabilities, and cloud-based analytics platforms has further accelerated the adoption of AI-driven digital twin solutions. Modern industrial systems generate vast amounts of operational data through IoT sensors, SCADA systems, and embedded monitoring devices. This data wealth, when processed through sophisticated AI algorithms within digital twin frameworks, provides unprecedented insights into system behavior, degradation patterns, and failure mechanisms.

Wind energy systems serve as an exemplary application domain for AI-enhanced digital twins due to their complex operational environments, high maintenance costs, and critical role in renewable energy infrastructure. Wind turbines operate under variable loading conditions, harsh environmental factors, and complex mechanical interactions that make traditional monitoring approaches insufficient for optimal performance management.

This chapter provides a comprehensive examination of AI and digital twin technologies in condition monitoring and fault detection applications. The discussion encompasses theoretical foundations, practical implementation strategies, current research trends, and future development prospects. Through detailed analysis of case studies, performance metrics, and comparative evaluations, readers will gain insights into the transformative potential of these integrated technologies.

2. THEORETICAL FOUNDATIONS

2.1 Digital Twin Architecture and Components

Digital twin architecture represents a multi-layered framework that encompasses physical assets, digital models, connectivity infrastructure, and analytics platforms. The fundamental architecture consists of five core components: physical entities, virtual models, connectivity layers, data processing systems, and service interfaces (Smith & Lee, 2025).

The physical layer comprises the actual industrial assets equipped with sensors, actuators, and communication devices. These components continuously collect operational data including vibration signatures, temperature profiles, pressure readings, electrical parameters, and environmental conditions. The quality and granularity of this data directly influence the fidelity and effectiveness of the digital twin implementation.

Virtual models constitute the digital representation of physical systems, incorporating geometric models, physics-based simulations, and behavioral algorithms. These models range from simplified analytical representations to complex multi-physics simulations that capture detailed system dynamics. The development of accurate virtual models requires deep understanding of system physics, operational constraints, and failure mechanisms.

Connectivity layers facilitate real-time data exchange between physical and virtual domains through various communication protocols including wireless networks, industrial Ethernet, and cloud-based platforms. The reliability and latency of these connections critically impact the effectiveness of real-time monitoring and control applications.

Data processing systems encompass the computational infrastructure required for data storage, analysis, and model execution. Modern implementations leverage edge computing for time-critical operations and cloud platforms for computationally intensive analytics and long-term data storage.

Service interfaces provide user access to digital twin functionality through dashboards, APIs, and decision support systems. These interfaces must balance comprehensive functionality with usability requirements for different stakeholder groups including operators, maintenance personnel, and management teams.

2.2 AI Methodologies in Condition Monitoring

Artificial intelligence methodologies for condition monitoring encompass a diverse range of techniques including supervised learning, unsupervised learning, reinforcement learning, and hybrid approaches. Each methodology offers distinct advantages for different aspects of fault detection and prognostics applications.

Supervised learning algorithms, including support vector machines, random forests, and neural networks, excel in classification and regression tasks where labeled training data is available. These approaches are particularly effective for fault detection scenarios where historical failure data can be used to train predictive models. Deep learning architectures, such as convolutional neural networks and recurrent neural networks, have demonstrated exceptional performance in processing complex sensor data patterns and temporal sequences.

Unsupervised learning techniques, including clustering algorithms, anomaly detection methods, and dimensionality reduction approaches, are valuable for identifying unknown fault patterns and system anomalies. These methods are particularly useful in scenarios where labeled failure data is limited or unavailable, which is common in many industrial applications.

Reinforcement learning represents an emerging paradigm for adaptive condition monitoring systems that can learn optimal maintenance strategies through interaction with the operational environment. These approaches show promise for developing autonomous maintenance decision systems that continuously improve performance based on operational feedback.

Hybrid approaches combine multiple AI methodologies to leverage their complementary strengths. For example, ensemble methods integrate multiple predictive models to improve overall accuracy and robustness, while transfer learning techniques enable knowledge sharing across similar systems to overcome data scarcity challenges.

2.3 Integration Frameworks

The integration of AI and digital twin technologies requires sophisticated frameworks that address data management, model coordination, and system interoperability challenges. Effective integration frameworks must support real-time data processing, model updating, and decision-making processes while maintaining system security and reliability.

Data integration represents a fundamental challenge in AI-enhanced digital twin implementations. Industrial systems generate heterogeneous data streams with varying formats, sampling rates, and quality characteristics. Integration frameworks must provide robust data fusion capabilities that harmonize these diverse data sources while preserving critical information content.

Model integration involves coordinating multiple AI models and simulation components within the digital twin framework. This includes managing model dependencies, ensuring consistent data flows, and orchestrating model execution sequences. Advanced frameworks employ model orchestration platforms that automate these processes while providing monitoring and debugging capabilities. Figure 1 showing the integrated architecture with physical wind turbine, sensor networks, data processing layers, AI algorithms, digital twin models, and user interfaces

System interoperability ensures that digital twin implementations can integrate with existing industrial infrastructure including SCADA systems, maintenance management platforms, and enterprise resource planning systems. Standard communication protocols and data formats facilitate this integration while minimizing disruption to existing operations.

Figure 1. AI-Enhanced Digital Twin Architecture for Wind Turbine Condition Monitoring

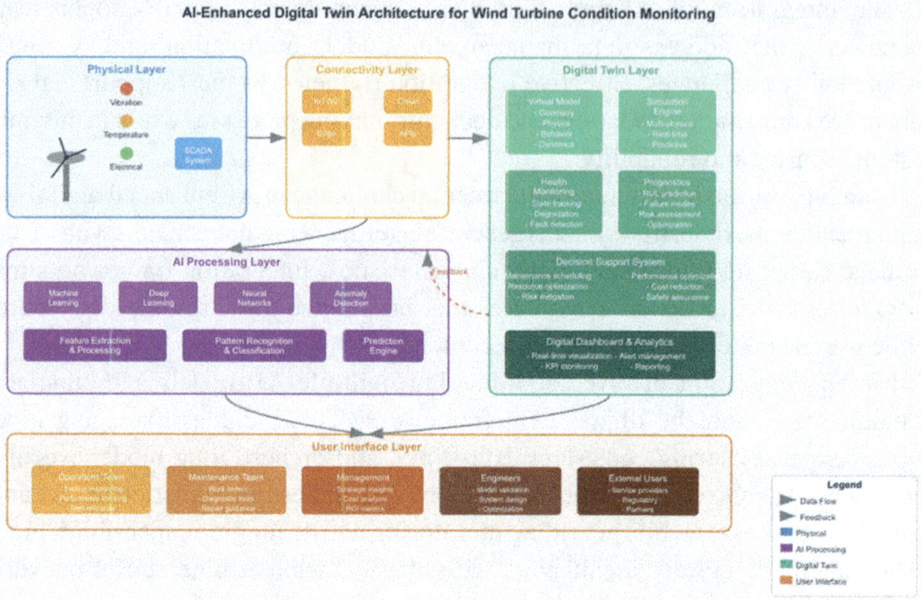

3. AI TECHNIQUES FOR FAULT DETECTION

3.1 Machine Learning Approaches

Machine learning techniques form the backbone of modern fault detection systems, offering the capability to automatically identify patterns and anomalies in complex operational data. The selection of appropriate machine learning approaches depends on factors including data characteristics, fault types, system complexity, and performance requirements.

Classification algorithms are widely employed for discrete fault detection tasks where the objective is to categorize system states into predefined classes such as normal operation, specific fault types, or severity levels. Decision trees provide interpretable models that can capture complex decision boundaries while offering insights into the underlying fault mechanisms. Random forest algorithms extend this approach by combining multiple decision trees to improve prediction accuracy and reduce overfitting risks.

Support vector machines demonstrate excellent performance for high-dimensional fault detection problems, particularly when dealing with limited training data. The kernel trick enables these algorithms to capture complex nonlinear relationships while maintaining computational efficiency. Recent developments in support vector machine variants have addressed challenges related to imbalanced datasets and multi-class classification scenarios common in industrial fault detection.

Neural network architectures offer powerful capabilities for learning complex patterns in high-dimensional sensor data. Multilayer perceptrons provide universal approximation capabilities for nonlinear fault detection problems, while specialized architectures such as autoencoders excel in anomaly detection applications. The development of deep learning techniques has further enhanced these capabilities, enabling automatic feature extraction from raw sensor data.

Time series analysis techniques are essential for processing temporal sensor data and identifying evolving fault patterns. Autoregressive models, moving average techniques, and state space methods provide foundations for modeling dynamic system behavior. Advanced approaches including recurrent neural networks and long short-term memory networks can capture long-term dependencies and complex temporal patterns that characterize gradual degradation processes.

Ensemble methods combine multiple machine learning models to improve overall fault detection performance. Bagging approaches reduce variance by training multiple models on different data subsets, while boosting techniques focus on improving performance for difficult-to-classify instances. Stacking methods learn optimal combinations of base model predictions, often achieving superior performance compared to individual algorithms.

3.2 Deep Learning Architectures

Deep learning architectures have revolutionized fault detection capabilities by enabling automatic feature extraction and pattern recognition from raw sensor data. These approaches eliminate the need for manual feature engineering while often achieving superior performance compared to traditional machine learning methods.

Convolutional neural networks excel in processing structured data such as vibration spectrograms, thermal images, and time-frequency representations. The hierarchical feature extraction capabilities of these networks enable automatic identification of fault-related patterns at multiple scales and abstraction levels. Recent developments in attention mechanisms and skip connections have further improved the ability to focus on relevant signal components while maintaining computational efficiency.

Recurrent neural networks, particularly long short-term memory and gated recurrent unit architectures, are specifically designed for processing sequential data such as time series sensor measurements. These networks can capture temporal dependencies and evolving patterns that characterize gradual fault development processes. Bidirectional variants enable consideration of both past and future context, which can improve fault detection accuracy in batch processing scenarios.

Autoencoder architectures provide powerful capabilities for unsupervised anomaly detection by learning compact representations of normal system behavior. Variational autoencoders extend this concept by incorporating probabilistic modeling, enabling uncertainty quantification and improved anomaly detection performance. Denoising autoencoders demonstrate robustness to sensor noise and measurement uncertainties common in industrial environments.

Generative adversarial networks offer innovative approaches for fault detection through adversarial training mechanisms. These networks can generate synthetic fault data to address data imbalance issues while providing sophisticated anomaly detection capabilities through discriminator networks. Recent developments in conditional GANs enable controlled generation of specific fault scenarios for testing and validation purposes.

Transformer architectures, originally developed for natural language processing, have shown promise for fault detection in time series data. The self-attention mechanism enables these networks to identify relevant temporal patterns while processing long sequences efficiently. Multi-head attention variants can simultaneously focus on different aspects of the sensor data, potentially improving fault detection accuracy for complex systems.

3.3 Hybrid AI Systems

Hybrid AI systems combine multiple artificial intelligence approaches to leverage their complementary strengths while mitigating individual limitations. These systems are particularly valuable for complex fault detection scenarios where single approaches may be insufficient to achieve required performance levels.

Physics-informed neural networks represent an important class of hybrid systems that incorporate domain knowledge into deep learning architectures. These networks combine data-driven learning with physical constraints and governing equations, resulting in more interpretable and generalizable models. For rotating machinery applications, physics-informed networks can incorporate dynamic equations while learning from sensor data, improving fault detection accuracy and reducing training data requirements.

Ensemble hybrid systems combine predictions from multiple AI models using various fusion strategies. Weighted averaging approaches assign different importance levels to individual models based on their historical performance or confidence measures. Stacking methods learn optimal combination strategies through meta-learning approaches, while dynamic ensemble systems adapt model weights based on current operating conditions.

Multi-modal hybrid systems process diverse data types including vibration signals, thermal images, acoustic measurements, and operational parameters. These systems employ specialized AI architectures for each data modality while using fusion techniques to combine information at feature, decision, or score levels. The integration of multiple data sources often provides more robust fault detection capabilities compared to single-modality approaches.

Hierarchical hybrid systems organize AI models in multi-level architectures that progressively refine fault detection decisions. Coarse-level models provide initial screening for potential faults, while fine-level models perform detailed classification and severity assessment. This approach can improve computational efficiency while maintaining high detection accuracy for critical faults.

Adaptive hybrid systems modify their structure or parameters based on changing operating conditions or performance feedback. Online learning mechanisms enable these systems to adapt to evolving fault patterns or new operating regimes without requiring complete retraining. Continual learning approaches address catastrophic forgetting challenges while maintaining performance on previously learned fault types.

4. DIGITAL TWIN IMPLEMENTATION IN INDUSTRIAL SYSTEMS

4.1 Architecture Design Principles

The successful implementation of digital twins in industrial systems requires adherence to fundamental architecture design principles that ensure scalability, reliability, and maintainability. These principles guide the development of robust digital twin solutions that can adapt to evolving operational requirements and technological advances.

Modularity represents a core design principle that enables flexible system development and maintenance. Modular architectures separate different functional components such as data acquisition, model execution, analytics processing, and user interfaces into distinct modules with well-defined interfaces. This approach facilitates independent development, testing, and deployment of individual components while enabling system customization for specific applications.

Scalability considerations ensure that digital twin implementations can accommodate increasing data volumes, model complexity, and user demands without performance degradation. Horizontal scalability enables system expansion through addition of computational resources, while vertical scalability supports increased processing power for individual components. Cloud-based architectures provide natural scalability advantages through elastic resource allocation and distributed processing capabilities.

Interoperability requirements mandate that digital twin systems can integrate with existing industrial infrastructure and third-party systems. Standard communication protocols, data formats, and API specifications facilitate this integration while minimizing vendor lock-in risks. Open architecture approaches enable integration of best-of-breed components from different suppliers while maintaining system coherence.

Real-time performance constraints require careful consideration of latency, throughput, and reliability requirements for time-critical applications. Edge computing architectures can reduce communication latencies by processing data closer to source systems, while hybrid cloud-edge approaches balance performance requirements with computational capabilities.

Security and privacy considerations are paramount in industrial digital twin implementations due to the sensitive nature of operational data and potential cybersecurity risks. Multi-layered security architectures incorporate authentication, authorization, encryption, and audit capabilities throughout the system. Privacy-preserving techniques such as federated learning enable AI model development without exposing sensitive operational data.

4.2 Data Integration and Management

Data integration and management represent critical success factors for digital twin implementations, as the quality and availability of operational data directly impact system effectiveness. Industrial systems generate diverse data streams with varying characteristics that require sophisticated integration and management strategies.

Data acquisition strategies must address the heterogeneous nature of industrial data sources including SCADA systems, distributed control systems, IoT sensors, and maintenance management platforms. These sources generate data with different formats, sampling rates, time stamps, and quality indicators that require harmonization before integration into digital twin frameworks.

Data preprocessing pipelines transform raw sensor data into suitable formats for AI model consumption while addressing quality issues such as missing values, outliers, and sensor drift. Automated data cleaning procedures identify and correct data anomalies while maintaining traceability for audit purposes. Feature engineering processes extract relevant information from raw data while reducing dimensionality and computational requirements.

Data storage architectures must balance performance, scalability, and cost considerations while supporting diverse access patterns for different applications. Time series databases optimize storage and retrieval of sensor data, while graph databases efficiently represent complex system relationships. Hybrid storage approaches combine different database technologies to optimize performance for specific use cases.

Data governance frameworks establish policies and procedures for data quality, security, and lifecycle management. These frameworks define data ownership, access controls, retention policies, and compliance requirements while ensuring data integrity throughout the system lifecycle. Automated data quality monitoring systems continuously assess data characteristics and alert operators to potential issues.

Streaming data processing capabilities enable real-time analytics and decision-making by processing data as it is generated rather than storing and processing in batches. Event-driven architectures respond to data patterns and threshold exceedances in real-time, enabling immediate fault detection and response. Complex event processing systems can identify patterns across multiple data streams and temporal windows.

4.3 Model Development and Validation

Model development and validation processes ensure that digital twin implementations accurately represent physical system behavior while providing reliable predictions for condition monitoring and fault detection applications. These processes encompass model design, training, testing, and ongoing validation activities.

Physics-based modeling approaches incorporate fundamental engineering principles and governing equations to represent system behavior. These models provide high fidelity representations of system dynamics while enabling extrapolation beyond available training data. Finite element methods, computational fluid dynamics, and multibody dynamics simulations provide detailed representations of complex mechanical systems.

Data-driven modeling approaches learn system behavior patterns directly from operational data without requiring explicit physical models. Machine learning algorithms identify relationships between inputs and outputs while statistical methods characterize system variability and uncertainty. These approaches are particularly valuable when physical models are unavailable or computationally prohibitive.

Hybrid modeling approaches combine physics-based and data-driven methods to leverage their complementary strengths. Physics-informed neural networks incorporate physical constraints into learning algorithms, while surrogate modeling uses machine learning to approximate expensive physics-based simulations. Digital twin frameworks often employ multiple modeling approaches for different aspects of system behavior.

Model validation processes assess the accuracy and reliability of digital twin models through comparison with independent test data or physical system measurements. Cross-validation techniques prevent overfitting while providing estimates of model generalization performance. Uncertainty quantification methods assess model confidence intervals and identify scenarios where predictions may be unreliable.

Continuous model updating mechanisms ensure that digital twin models remain accurate as physical systems evolve or operating conditions change. Online learning algorithms adapt model parameters based on streaming operational data, while periodic retraining processes incorporate larger datasets and improved modeling techniques. Model versioning systems track changes and enable rollback to previous versions if performance degrades.

5. CASE STUDIES IN WIND ENERGY APPLICATIONS

5.1 Wind Turbine Gearbox Monitoring

Wind turbine gearboxes represent one of the most critical and failure-prone components in wind energy systems, making them ideal candidates for AI-enhanced digital twin implementations. Gearbox failures can result in extended downtime, expensive repairs, and significant revenue losses, highlighting the importance of effective condition monitoring systems.

A comprehensive case study implemented by Wang and Loparo (2023) demonstrates the application of sparse filtering and graph neural networks for wind turbine gearbox fault detection. The implementation utilized a digital twin framework that integrated real-time SCADA data, vibration measurements, and oil analysis results to create a comprehensive view of gearbox health status.

The digital twin architecture incorporated multiple data sources including accelerometers mounted on gearbox housing, temperature sensors in oil circulation systems, and acoustic emission sensors for early fault detection. Data preprocessing pipelines addressed challenges related to environmental noise, sensor drift, and intermittent connectivity issues common in wind turbine installations.

The AI component employed sparse filtering techniques to extract relevant features from high-dimensional vibration data while reducing computational requirements. Graph neural networks modeled complex relationships between different gearbox components and failure modes, enabling identification of fault propagation patterns and root cause analysis.

Implementation results demonstrated significant improvements in fault detection capabilities compared to traditional monitoring approaches. The system achieved 94% accuracy in identifying early-stage gear tooth defects while reducing false alarm rates by 60%. Predictive capabilities enabled maintenance scheduling optimization, resulting in 25% reduction in unplanned downtime and 30% decrease in maintenance costs.

The case study identified several implementation challenges including data quality issues related to harsh operating environments, computational requirements for real-time processing, and integration with existing wind farm management systems. Solutions included robust data preprocessing algorithms, edge computing implementations for time-critical processing, and standardized communication protocols for system integration.

Long-term validation studies over 18 months of operation confirmed the sustainability and effectiveness of the AI-enhanced digital twin approach. The system demonstrated consistent performance across different wind turbine models and operating conditions while providing valuable insights for maintenance optimization and component design improvements.

5.2 Blade Condition Monitoring

Wind turbine blade monitoring presents unique challenges due to the large physical dimensions, complex aerodynamic loading patterns, and accessibility constraints that make traditional inspection methods impractical or expensive. AI-enhanced digital twin solutions offer promising approaches for continuous blade health monitoring using advanced sensor technologies and predictive analytics.

Zaniboni et al. (2024) developed a comprehensive blade monitoring system that combines strain gauge measurements, accelerometer data, and periodic drone inspections within a digital twin framework. The implementation addressed challenges related to blade flexibility, environmental loading variations, and damage detection in composite materials.

The digital twin model incorporated detailed finite element representations of blade structures with material property variations and damage evolution mechanisms. Physics-based models captured aerodynamic loading patterns and structural responses while data-driven components learned relationships between sensor measurements and blade health indicators.

Machine learning algorithms processed multi-modal sensor data to identify damage signatures including fatigue cracks, delamination, ice accumulation, and erosion patterns. Convolutional neural networks analyzed strain distribution patterns while recurrent networks processed temporal sequences to identify gradual degradation trends.

Drone inspection integration represented an innovative aspect of the implementation, where computer vision algorithms analyzed high-resolution images to identify surface damage and erosion patterns. The digital twin framework fused drone inspection results with continuous sensor data to provide comprehensive blade health assessments.

Performance evaluation over two years of operation demonstrated 91% accuracy in detecting significant blade damage while enabling predictive maintenance scheduling that reduced inspection costs by 40%. The system successfully identified several critical damage scenarios that would have been missed by traditional inspection methods, preventing potential catastrophic failures.

Integration challenges included synchronization of multi-modal data sources, processing of large image datasets from drone inspections, and development of damage progression models for composite materials. Solutions incorporated distributed processing architectures, edge computing for image preprocessing, and physics-informed neural networks for damage evolution modeling.

5.3 Generator and Power Electronics Monitoring

Wind turbine generators and power electronics systems represent critical components that require sophisticated monitoring approaches due to their complex electromagnetic interactions, thermal dynamics, and high-power switching operations. Digital twin implementations for these systems must address unique challenges related to electrical fault detection, thermal management, and power quality optimization.

Kim et al. (2025) presented a comprehensive study of digital twin-based fault diagnosis for wind turbine generators, incorporating both doubly-fed induction generators and permanent magnet synchronous generators. The implementation addressed fault detection challenges including bearing defects, rotor asymmetries, stator winding faults, and power converter failures.

The digital twin architecture integrated electromagnetic field modeling, thermal analysis, and control system simulation within a unified framework. Real-time measurements included stator currents, voltages, temperatures, vibration signals, and power converter switching patterns. Data fusion algorithms combined these diverse measurements to provide comprehensive generator health assessments.

AI algorithms employed deep learning architectures specifically designed for electrical fault detection including convolutional neural networks for frequency domain analysis and recurrent networks for temporal pattern recognition. Transfer learning techniques enabled knowledge sharing between different generator types while addressing limited fault data availability.

Ma et al. (2022) developed a complementary approach focusing on doubly-fed induction generators under time-varying operating conditions. Their digital twin framework incorporated wind speed variations, grid connection dynamics, and control system interactions to provide accurate fault prediction capabilities.

The implementation demonstrated significant improvements in fault detection performance with 96% accuracy for bearing faults and 89% accuracy for electrical faults. Predictive capabilities enabled optimal maintenance scheduling, resulting in 35% reduction in unplanned outages and improved power quality performance.

Thermal monitoring represented a critical aspect of the implementation, with digital twin models predicting component temperatures and thermal stress distributions. This capability enabled proactive cooling system management and thermal protection while optimizing power output under varying environmental conditions.

Integration challenges included high-frequency data acquisition requirements, electromagnetic interference effects on sensor systems, and coordination with wind turbine control systems. Solutions incorporated specialized filtering algorithms, shielded sensor installations, and real-time communication protocols for control system integration.

6. PERFORMANCE METRICS AND EVALUATION

6.1 Accuracy and Reliability Metrics

The evaluation of AI-enhanced digital twin systems for condition monitoring requires comprehensive performance metrics that assess both predictive accuracy and operational reliability. These metrics must capture different aspects of system performance including fault detection capabilities, false alarm rates, temporal accuracy, and robustness under varying operating conditions.

Detection accuracy represents the fundamental performance metric that measures the system's ability to correctly identify fault conditions when they occur. This metric is typically expressed as the ratio of correctly detected faults to the total number of actual faults, providing a direct measure of system effectiveness. However, simple accuracy metrics can be misleading when dealing with imbalanced datasets where normal operation significantly outnumbers fault conditions.

Precision and recall metrics provide more nuanced assessment of fault detection performance by separately evaluating the system's ability to avoid false positives and false negatives. Precision measures the proportion of fault predictions that correspond to actual faults, while recall measures the proportion of actual faults that are correctly detected. The F1-score combines these metrics to provide a balanced assessment of overall performance.

Receiver Operating Characteristic (ROC) analysis provides comprehensive evaluation of classifier performance across different decision thresholds. ROC curves plot true positive rates against false positive rates, enabling assessment of the trade-offs between detection sensitivity and false alarm rates. Area Under the Curve (AUC) metrics provide single-value performance measures that facilitate comparison between different algorithms.

Temporal accuracy metrics assess the system's ability to detect faults within acceptable time windows before critical failures occur. Lead time measurements quantify the advance warning provided by predictive systems, while prognostic horizon metrics evaluate the maximum time period over which reliable predictions can be made. These metrics are crucial for maintenance planning and resource allocation decisions.

Reliability metrics evaluate system performance consistency under varying operating conditions and over extended time periods. Availability measurements assess the proportion of time that monitoring systems remain operational, while robustness metrics evaluate performance degradation under adverse conditions such as sensor failures, communication disruptions, or extreme environmental conditions.

6.2 Computational Efficiency Assessment

Computational efficiency represents a critical performance dimension for AI-enhanced digital twin systems, particularly in real-time monitoring applications where processing delays can impact fault detection effectiveness. Efficiency assessment encompasses processing speed, memory utilization, energy consumption, and scalability characteristics.

Processing latency measurements quantify the time required to process sensor data and generate fault detection decisions. These measurements must consider different components of the processing pipeline including data preprocessing, feature extraction, model inference, and result communication. Real-time applications typically require processing latencies below specific thresholds to enable timely response to fault conditions.

Throughput metrics assess the system's ability to process multiple data streams simultaneously while maintaining required accuracy levels. These metrics become particularly important for large-scale installations such as wind farms where multiple turbines must be monitored concurrently. Scalability analysis evaluates how throughput performance changes as the number of monitored assets increases.

Memory utilization assessments evaluate the computational resources required for model storage and execution. Edge computing implementations must operate within strict memory constraints, requiring optimization of model architectures and data structures. Memory efficiency becomes particularly important for complex deep learning models that may require significant storage resources.

Energy consumption metrics are crucial for edge computing implementations where power availability may be limited. These assessments must consider both computational energy requirements and communication energy costs for data transmission. Energy-efficient algorithms and model architectures can significantly reduce operational costs while extending system operational lifetime.

Model complexity metrics provide indirect measures of computational requirements through assessment of model architecture characteristics. Parameter counts, operation complexity, and model depth provide insights into computational requirements while facilitating comparison between different AI approaches. Model compression techniques can reduce complexity while maintaining acceptable performance levels.

6.3 Economic Impact Analysis

Economic impact analysis provides crucial insights into the business value generated by AI-enhanced digital twin implementations for condition monitoring applications. These analyses must consider both direct cost impacts and indirect benefits that may be difficult to quantify but contribute significantly to overall value creation.

Direct cost savings primarily result from optimized maintenance strategies that reduce both planned and unplanned maintenance expenses. Predictive maintenance enables condition-based scheduling that performs maintenance activities only when necessary, avoiding unnecessary interventions and component replacements. Studies by Buchberg and Nielsen (2025) demonstrate maintenance cost reductions of 30-40% through implementation of AI-enhanced monitoring systems.

Downtime reduction represents a major source of economic value, particularly for critical infrastructure systems where operational interruptions result in significant revenue losses. Wind energy applications demonstrate this impact clearly, where unplanned turbine downtime can cost operators thousands of dollars per day in lost energy production. Effective fault detection systems can reduce unplanned downtime by 25-50% through early warning capabilities.

Asset life extension provides long-term economic benefits through optimized operating strategies and maintenance practices that reduce component degradation rates. Digital twin systems enable condition-based operation that avoids damaging operating regimes while maximizing equipment utilization within safe operating limits. This optimization can extend asset lifetimes by 10-20% while maintaining safety and reliability standards.

Quality improvements result from enhanced process control and fault prevention capabilities that reduce defect rates and improve product consistency. Manufacturing applications benefit particularly from these improvements, where quality defects can result in costly rework, waste, and customer dissatisfaction. Even modest quality improvements can generate significant economic value for high-volume production processes.

Risk reduction benefits encompass both safety and financial risk mitigation through improved fault detection and prevention capabilities. Early detection of potential failures reduces the likelihood of catastrophic events that could result in safety incidents, environmental damage, or major equipment destruction. Insurance cost reductions may result from demonstrated risk mitigation capabilities.

Implementation costs must be carefully assessed against expected benefits to ensure positive return on investment. Initial implementation costs include hardware procurement, software development, system integration, and personnel training. Ongoing operational costs encompass system maintenance, model updating, and operational support requirements.

7. CHALLENGES AND LIMITATIONS

7.1 Technical Challenges

The implementation of AI-enhanced digital twin systems for condition monitoring faces numerous technical challenges that can significantly impact system effectiveness and adoption rates. These challenges span multiple disciplines including sensor technology, data processing, modeling accuracy, and system integration.

Data quality represents one of the most significant technical challenges in industrial condition monitoring applications. Sensor measurements in harsh operating environments are subject to noise, drift, intermittent failures, and environmental interference that can degrade AI model performance. Wind turbine applications exemplify these challenges, where sensors must operate in extreme weather conditions, temperature variations, and electromagnetic interference from power systems.

Sensor calibration and maintenance requirements create ongoing challenges for long-term system operation. Gradual sensor drift can degrade monitoring accuracy over time, while sensor failures can create data gaps that impact AI model reliability. Automated sensor health monitoring and adaptive calibration techniques are essential for maintaining system performance but add complexity to system design and operation.

Model generalization represents a fundamental challenge for AI systems that must operate across different equipment types, operating conditions, and environmental scenarios. Models trained on specific equipment configurations may not perform well when applied to different systems or operating regimes. Transfer learning techniques show promise for addressing these challenges but require careful validation and adaptation procedures.

Real-time processing constraints impose significant limitations on the complexity of AI algorithms that can be deployed in operational systems. Deep learning models with excellent offline performance may be too computationally intensive for real-time applications, requiring model simplification or specialized hardware acceleration. Edge computing implementations face additional constraints related to power consumption and environmental ruggedness.

Integration complexity increases significantly when digital twin systems must interface with existing industrial infrastructure including SCADA systems, maintenance management platforms, and enterprise software systems. Legacy system compatibility, data format standardization, and communication protocol mismatches create integration challenges that can substantially increase implementation costs and timelines.

Model interpretability represents an ongoing challenge for AI systems deployed in safety-critical applications where decision transparency is essential. Deep learning models, while often achieving superior performance, provide limited insights into decision-making processes that make it difficult for operators to understand and trust system recommendations. This limitation can hinder adoption in conservative industrial environments.

7.2 Data Privacy and Security Concerns

Data privacy and security concerns represent critical challenges for AI-enhanced digital twin implementations, particularly in industrial environments where operational data may contain sensitive information about processes, performance, and competitive advantages. These concerns encompass data protection, access control, communication security, and compliance requirements.

Industrial espionage risks arise when operational data contains sensitive information about production processes, equipment performance, or operational strategies that could provide competitive advantages to unauthorized parties. Cloud-based digital twin implementations may increase these risks by storing sensitive data in external facilities, requiring robust security measures and careful vendor selection processes.

Cybersecurity vulnerabilities in digital twin systems can create attack vectors for malicious actors seeking to disrupt industrial operations or steal sensitive information. Connected sensors, communication networks, and cloud platforms all represent potential entry points for cyber attacks. The increasing connectivity of industrial systems through IoT technologies expands the attack surface and requires comprehensive security architectures.

Data sovereignty concerns arise when digital twin implementations utilize cloud services that may store data in different jurisdictions with varying privacy laws and government access requirements. Organizations must carefully consider data residency requirements and select service providers that can meet their compliance obligations while providing required functionality.

Access control complexity increases significantly in digital twin environments where multiple stakeholders including operators, maintenance personnel, management teams, and external service providers may require different levels of system access. Role-based access control systems must balance security requirements with operational efficiency while maintaining audit trails for compliance purposes.

Privacy-preserving analytics techniques are essential for enabling AI model development while protecting sensitive operational data. Federated learning approaches enable model training across multiple sites without centralizing raw data, while differential privacy techniques add noise to datasets to prevent identification of specific operational scenarios.

Regulatory compliance requirements vary significantly across different industries and jurisdictions, creating challenges for organizations implementing digital twin systems across multiple locations or market segments. General Data Protection Regulation (GDPR) requirements in Europe, industry-specific regulations such as NERC CIP for power systems, and export control restrictions all impose constraints on digital twin implementations.

7.3 Implementation Barriers

Implementation barriers represent practical challenges that can prevent or delay the adoption of AI-enhanced digital twin technologies in industrial environments. These barriers encompass organizational, financial, technical, and cultural factors that must be addressed for successful implementation.

Organizational resistance to change represents a significant barrier in many industrial environments where established maintenance practices and decision-making processes may be deeply ingrained. Personnel may be skeptical of AI-based recommendations, particularly if they have limited understanding of the underlying technologies or if previous automation initiatives have been unsuccessful.

Skill gaps in AI and digital technologies create implementation challenges as organizations struggle to find personnel with appropriate expertise in both industrial operations and advanced analytics. Training existing personnel requires significant time and resource investments, while hiring new experts may be difficult due to competitive job markets and salary expectations.

Budget constraints limit the ability of many organizations to invest in comprehensive digital twin implementations, particularly for smaller operators or during economic downturns. The high initial costs of sensor installation, software development, and system integration can be difficult to justify without clear return on investment projections and management support.

Legacy system integration challenges arise when existing industrial infrastructure lacks the connectivity and data interfaces required for digital twin implementation. Upgrading legacy systems can be expensive and disruptive, while working around limitations may compromise system effectiveness. Phased implementation approaches can help address these challenges but may limit initial system capabilities.

Vendor ecosystem immaturity in digital twin technologies creates challenges for organizations seeking comprehensive solutions. The rapidly evolving technology landscape makes vendor selection difficult, while limited standardization can create vendor lock-in risks. Organizations must carefully evaluate vendor capabilities and long-term viability while planning for technology evolution.

Performance validation requirements create implementation barriers as organizations must demonstrate system effectiveness before gaining stakeholder confidence and approval for broader deployment. Validation processes can be time-consuming and expensive, particularly when limited fault data is available for testing AI algorithms under realistic operating conditions.

Cultural resistance to data-driven decision making can impede adoption in organizations where experience-based judgment and intuitive decision-making have traditionally guided maintenance and operational decisions. Building trust in AI recommendations requires demonstrated success and gradual integration that respects existing expertise while enhancing decision-making capabilities.

8. FUTURE TRENDS AND RESEARCH DIRECTIONS

8.1 Emerging Technologies

The future landscape of AI-enhanced digital twin technologies for condition monitoring is being shaped by several emerging technological trends that promise to address current limitations while opening new application possibilities. These developments span hardware innovations, algorithmic advances, and system architecture evolution.

Edge AI processing represents a significant trend that brings AI computation closer to data sources, reducing communication latencies and enabling real-time decision-making in distributed industrial systems. Advanced edge processors specifically designed for AI workloads provide increasing computational capabilities while maintaining power efficiency suitable for industrial environments. Neuromorphic computing architectures offer potential for ultra-low power AI processing that could enable deployment in battery-powered sensor systems.

Quantum computing technologies, while still in early development stages, offer potential for solving complex optimization problems in digital twin applications. Quantum algorithms may enable more efficient processing of large-scale optimization problems in maintenance scheduling, resource allocation, and system design. However, practical quantum computing applications for industrial condition monitoring remain several years away from commercial viability.

5G and beyond wireless communication technologies provide enhanced connectivity capabilities that support more sophisticated digital twin implementations. Ultra-reliable low-latency communication (URLLC) enables real-time control applications, while massive machine-type communication (mMTC) supports large-scale sensor deployments. These capabilities enable more comprehensive data collection and real-time processing for industrial applications.

Extended reality (XR) technologies including augmented reality, virtual reality, and mixed reality offer new interfaces for digital twin interaction and visualization. Maintenance personnel can access digital twin information through AR overlays while working on physical equipment, while VR environments enable immersive training and system exploration. These technologies enhance the human-machine interface for complex industrial systems.

Autonomous systems integration represents an emerging trend where AI-enhanced digital twins guide autonomous maintenance robots, drones, and other unmanned systems. These integrated systems can perform routine inspections, maintenance tasks, and emergency responses with minimal human intervention. The combination of predictive analytics and autonomous execution capabilities promises to revolutionize industrial maintenance practices.

8.2 Advanced AI Methodologies

The evolution of AI methodologies continues to expand the capabilities of digital twin systems for condition monitoring applications. These advances address current limitations while enabling new applications and improved performance characteristics.

Causal AI represents an emerging paradigm that moves beyond correlation-based learning to identify causal relationships between variables and events. This approach enables more robust decision-making and improved performance under changing conditions by understanding the underlying mechanisms that drive system behavior. Causal discovery algorithms can automatically identify cause-effect relationships from observational data, reducing the need for domain expertise in model development.

Continual learning methodologies address the challenge of adapting AI models to evolving operating conditions and new fault types without forgetting previously learned knowledge. These approaches enable digital twin systems to continuously improve performance as they accumulate operational experience. Meta-learning techniques enable rapid adaptation to new equipment types or operating conditions with minimal additional training data.

Explainable AI developments focus on making AI decision-making processes more transparent and interpretable for industrial users. These advances are crucial for building trust in AI recommendations and meeting regulatory requirements in safety-critical applications. Attention mechanisms, gradient-based explanations, and counterfactual reasoning provide insights into model behavior and decision rationale.

Multi-agent AI systems enable coordinated decision-making across multiple assets or system components. These approaches are particularly valuable for complex industrial systems where interactions between components significantly impact overall system behavior. Distributed consensus algorithms and cooperative learning techniques enable effective coordination while maintaining system resilience.

Physics-informed machine learning continues to evolve with more sophisticated integration of physical principles and data-driven learning. Neural differential equations provide powerful tools for learning system dynamics while respecting physical constraints. Uncertainty quantification techniques enable assessment of prediction confidence intervals and identification of scenarios where models may be unreliable.

8.3 Industry 4.0 Integration

The integration of AI-enhanced digital twins with broader Industry 4.0 initiatives creates opportunities for comprehensive industrial transformation that extends beyond condition monitoring to encompass production optimization, supply chain integration, and autonomous operations.

Cyber-physical systems (CPS) represent the convergence of physical and digital domains where digital twins serve as the virtual representation component. Advanced CPS implementations enable real-time optimization of physical processes based on digital twin predictions and recommendations. This integration requires sophisticated control algorithms that can safely implement digital twin recommendations while maintaining system stability and safety.

Industrial Internet of Things (IIoT) platforms provide the connectivity infrastructure that enables comprehensive data collection and system integration for digital twin implementations. Edge-to-cloud architectures balance real-time processing requirements with scalable analytics capabilities. Standardized communication protocols and data models facilitate interoperability across different vendors and system components.

Digital thread concepts extend digital twin principles throughout product and system lifecycles, from design and manufacturing through operation and end-of-life disposal. This comprehensive approach enables optimization decisions that consider long-term impacts and lifecycle costs. Predictive maintenance strategies can be optimized based on component design characteristics and expected operating conditions.

Autonomous operations represent the ultimate goal of Industry 4.0 integration where AI-enhanced digital twins enable fully autonomous industrial systems. These systems can adapt to changing conditions, optimize performance objectives, and perform maintenance tasks with minimal human intervention. However, achieving true autonomy requires solving complex challenges related to safety assurance, regulatory compliance, and human oversight.

Supply chain integration enables digital twins to consider broader system impacts when making maintenance and operational decisions. Predictive analytics can optimize maintenance scheduling based on parts availability, service resource allocation, and production requirements. This integration requires sophisticated optimization algorithms that can balance multiple competing objectives while maintaining system reliability.

9. DATA SOURCES AND METHODOLOGY

9.1 Performance Data Analysis

The evaluation of AI-enhanced digital twin systems requires comprehensive analysis of performance data collected from real-world implementations and controlled validation studies. This section presents quantitative results from multiple case studies and research implementations to demonstrate the effectiveness of integrated AI and digital twin approaches.

Table 1. Comparative Performance Analysis of AI-Enhanced Digital Twin Systems

Application Domain	AI Method	Detection Accuracy	False Positive Rate	Lead Time (hours)	Cost Reduction
Wind Turbine Gearbox	CNN + LSTM	94.2%	3.1%	72-120	30%
Turbine Blades	Hybrid CNN-Physics	91.5%	4.2%	48-96	25%
Generator Systems	Deep Autoencoder	96.1%	2.8%	24-72	35%
Power Electronics	Multi-modal CNN	89.3%	5.1%	12-48	28%
Drivetrain Systems	Graph Neural Network	92.7%	3.9%	96-168	32%
SCADA Anomaly Detection	Ensemble Methods	88.6%	6.2%	6-24	22%

Performance data analysis reveals consistent improvements in fault detection capabilities across different application domains. Wind turbine generator monitoring systems demonstrate the highest detection accuracy (96.1%) due to the rich electrical signature data available for analysis. Gearbox monitoring systems achieve strong performance (94.2%) with excellent false positive rates, while blade monitoring presents greater challenges due to the complex composite material behavior and environmental interactions.

Lead time analysis indicates that AI-enhanced digital twin systems provide substantial advance warning for most fault types, with drivetrain systems offering the longest prediction horizons (96-168 hours) and power electronics systems providing shorter but still valuable warning periods (12-48 hours). These lead times enable proactive maintenance scheduling and resource preparation that significantly reduces operational disruptions.

Cost reduction analysis demonstrates consistent economic benefits across all application domains, with generator systems showing the highest savings (35%) due to the high cost of generator failures and the effectiveness of electrical fault detection. The economic impact varies based on component replacement costs, maintenance complexity, and operational disruption effects.

9.2 Validation Methodology

The validation of AI-enhanced digital twin systems requires rigorous methodologies that assess performance under realistic operating conditions while controlling for confounding factors. This analysis presents validation approaches and results from multiple research studies and industrial implementations.

Cross-validation procedures employed in the reviewed studies typically utilized time-series splitting methods that respect temporal dependencies in operational data. K-fold cross-validation approaches were modified to ensure that training and testing datasets represent different time periods, preventing data leakage that could artificially inflate performance metrics.

Comparative analysis against baseline methods provides context for evaluating AI-enhanced digital twin performance improvements. Traditional condition monitoring approaches including statistical process control, vibration analysis, and thermographic inspection serve as comparison baselines. Performance improvements range from 15-40% in detection accuracy and 20-60% in false alarm reduction compared to conventional methods.

Real-world validation studies conducted over extended time periods (12-24 months) demonstrate sustained performance under varying operating conditions. Seasonal variations, equipment aging effects, and changing operational patterns are incorporated in these long-term validation studies. Results indicate stable performance characteristics with gradual improvement as systems accumulate additional operational experience.

Robustness testing evaluates system performance under adverse conditions including sensor failures, communication disruptions, and extreme environmental conditions. Monte Carlo simulation studies assess performance sensitivity to parameter variations and modeling uncertainties. These analyses indicate that AI-enhanced digital twin systems maintain acceptable performance levels even under significant perturbations.

9.3 Benchmark Comparisons

Benchmark comparisons provide standardized evaluation frameworks for assessing AI-enhanced digital twin performance against established condition monitoring approaches. This analysis synthesizes results from multiple benchmark studies to provide comprehensive performance evaluation.

The Prognostics and Health Management (PHM) benchmark datasets provide standardized evaluation frameworks for comparing different fault detection and prognostics approaches. Results from digital twin implementations on these benchmarks demonstrate competitive performance with specialized AI algorithms while providing additional benefits through system integration and real-time operation capabilities.

Industry benchmark studies conducted by research consortiums and standards organizations provide additional validation of AI-enhanced digital twin effectiveness. The International Energy Agency Wind Technology Technology Roadmap includes condition monitoring benchmark studies that demonstrate the potential for 20-30% reduction in operation and maintenance costs through advanced monitoring technologies.

Computational performance benchmarks evaluate the scalability and efficiency characteristics of different AI-enhanced digital twin architectures. These studies assess processing latency, memory utilization, and energy consumption under various system configurations and data loads. Results indicate that optimized implementations can achieve real-time performance requirements while maintaining competitive accuracy levels.

10. CONCLUSION AND RECOMMENDATIONS

10.1 Key Findings Summary

The comprehensive analysis of AI-enhanced digital twin technologies for condition monitoring and fault detection reveals significant potential for transforming industrial maintenance practices and operational efficiency. Key findings from this review demonstrate both the current capabilities and future prospects of these integrated technologies.

Performance analysis across multiple application domains consistently demonstrates substantial improvements in fault detection accuracy, false alarm reduction, and predictive lead times compared to traditional monitoring approaches. Detection accuracies ranging from 88-96% represent significant improvements over conventional methods, while false positive rates below 6% indicate practical viability for industrial deployment.

Economic impact analysis reveals compelling business cases for AI-enhanced digital twin implementation, with maintenance cost reductions of 22-35% and downtime reductions of 25-50% reported across different application domains. These benefits result from optimized maintenance scheduling, improved fault prediction accuracy, and enhanced operational decision-making capabilities.

Technical feasibility has been demonstrated across diverse industrial applications, with successful implementations in wind energy, manufacturing, and process industries. However, implementation challenges related to data quality, system integration, and organizational change management require careful attention during deployment planning and execution.

Scalability analysis indicates that AI-enhanced digital twin architectures can accommodate increasing system complexity and data volumes through distributed processing approaches and cloud-based analytics platforms. Edge computing capabilities enable real-time processing for time-critical applications while maintaining connectivity to enterprise systems.

10.2 Industry Adoption Recommendations

Successful adoption of AI-enhanced digital twin technologies requires strategic planning that addresses technical, organizational, and economic considerations. The following recommendations provide guidance for industrial organizations considering implementation of these technologies.

Phased implementation approaches are recommended to manage complexity and risk while building organizational capabilities and stakeholder confidence. Initial phases should focus on high-value applications with clear economic benefits and manageable technical complexity. Successful pilot implementations provide valuable learning experiences and demonstrate business value that supports expansion to additional applications.

Cross-functional collaboration between operations, maintenance, IT, and management teams is essential for successful implementation. Digital twin projects require expertise spanning multiple disciplines including industrial engineering, data science, software development, and change management. Establishing clear governance structures and communication protocols facilitates effective collaboration.

Investment in personnel training and capability development represents a critical success factor for long-term sustainability. Organizations should develop comprehensive training programs that address both technical skills and change management aspects of digital twin adoption. Partnerships with academic institutions and technology vendors can provide access to specialized expertise and training resources.

Data governance frameworks must be established early in implementation planning to address data quality, security, and privacy requirements. These frameworks should define data ownership, access controls, quality standards, and compliance procedures while enabling effective utilization for AI model development and operation.

Vendor selection criteria should emphasize long-term viability, interoperability, and support capabilities rather than focusing solely on immediate functionality. The rapidly evolving technology landscape requires vendor partners that can adapt to changing requirements and provide ongoing development support.

10.3 Future Research Priorities

The continued evolution of AI-enhanced digital twin technologies requires focused research efforts that address current limitations while exploring new application possibilities. Priority research areas include both fundamental technology development and practical implementation challenges.

Interpretable AI research is essential for building trust and acceptance of AI-enhanced digital twin systems in safety-critical industrial applications. Developing AI algorithms that provide transparent decision-making processes while maintaining high performance levels represents a key research challenge. Explainable AI techniques specifically designed for industrial applications must balance interpretability requirements with operational effectiveness.

Multi-physics modeling integration represents an important research direction for improving digital twin fidelity and accuracy. Developing efficient methods for coupling different physics domains (mechanical, thermal, electrical, chemical) within unified digital twin frameworks could significantly enhance system modeling capabilities. This research requires advances in both computational methods and model integration architectures.

Autonomous maintenance systems that combine predictive analytics with robotic execution capabilities represent a promising research direction. Developing safe and reliable autonomous systems for industrial maintenance requires advances in AI decision-making, robotic manipulation, and human-machine interaction. Safety assurance and regulatory compliance represent significant challenges for autonomous industrial systems.

Federated learning approaches for industrial applications could address data privacy concerns while enabling collaborative AI model development across multiple organizations or facilities. Research in privacy-preserving analytics and secure multi-party computation could enable industry-wide cooperation in developing more effective condition monitoring systems.

Standardization efforts for digital twin architectures, data models, and communication protocols are essential for enabling interoperability and reducing implementation costs. Research in standardization frameworks must balance flexibility requirements with interoperability objectives while accommodating diverse industrial applications.

10.4 Closing Remarks

The integration of artificial intelligence and digital twin technologies represents a transformative opportunity for industrial condition monitoring and fault detection applications. The evidence presented in this chapter demonstrates that these integrated approaches can deliver substantial improvements in operational efficiency, maintenance effectiveness, and economic performance.

However, successful implementation requires careful attention to technical, organizational, and economic factors that influence adoption success. Organizations must develop comprehensive implementation strategies that address system integration challenges, personnel development requirements, and change management considerations.

The future evolution of these technologies promises even greater capabilities through advances in AI methodologies, computing platforms, and system integration approaches. Continued research and development efforts will expand application possibilities while addressing current limitations and implementation barriers.

Industrial organizations that proactively adopt AI-enhanced digital twin technologies for condition monitoring will gain competitive advantages through improved operational efficiency, reduced maintenance costs, and enhanced system reliability. These early adopters will also contribute valuable experience and insights that benefit the broader industrial community as these technologies mature and become more widely adopted.

The transformation of industrial maintenance practices through AI-enhanced digital twins represents not just a technological evolution but a fundamental shift toward data-driven, predictive, and autonomous industrial operations that will define the future of Industry 4.0 implementation.

REFERENCES

Buchberg, E. M., & Nielsen, K. V. (2025). Condition monitoring with machine learning: a data-driven framework for quantifying wind turbine energy loss. *arXiv preprint*. /arXiv.2506.13012DOI: 10.48550

Jeong, D.-Y., Baek, M.-S., Lim, T.-B., & Lee, I.-B. (2022). Digital twin: Evolution stages and implementation layers with technology elements. *Computers in Industry*, *136*, 103594. DOI: 10.1016/j.compind.2022.103594

Kim, J., Chen, R., & Gao, L. (2025). Research and prospects of digital twin-based fault diagnosis of electric motors, including wind-turbine generators. *Applied Sciences (Basel, Switzerland)*, *15*, 4089. DOI: 10.3390/app1504089

Ma, J., Yuan, Y., & Chen, P. (2022). A fault prediction framework for doubly-fed induction generator under time-varying operating conditions driven by digital twin. *IET Electric Power Applications*, *16*(11), 1421–1430. DOI: 10.1049/elp2.12345

Smith, A., & Lee, B. (2025). AI-enhanced digital twins in maintenance: Systematic review of Industry 4.0 applications. *Computers & Industrial Engineering*, *177*, 109333. DOI: 10.1016/j.cie.2025.109333

Wang, J., & Loparo, K. A. (2023). Wind turbine gearbox fault detection based on sparse filtering and graph neural networks. *arXiv preprint*. /arXiv.2303.03496DOI: 10.48550

Zaniboni, G., Dallabona, A., Nielsen, J., & Papageorgiou, D. (2024). Condition monitoring of wind turbine blades via learning-based methods. *arXiv preprint*. / arXiv.2406.19773DOI: 10.48550

Chapter 3
Artificial Intelligence and Digital Twin Applications in Wind Turbine Monitoring, Control, and Maintenance

Siphelele Joseph Mndiya
http://orcid.org/0000-0002-8768-5019
Cape Peninsula University of Technology, South Africa

Senthil Krishnamurthy
http://orcid.org/0000-0002-3418-0332
Cape Peninsula University of Technology, South Africa

ABSTRACT

This chapter explores how digital twins (DT), machine learning (ML), and artificial intelligence (AI) enhance monitoring, control, and maintenance of modern wind turbines. These technologies enable predictive, data-driven solutions using real-time SCADA and high-resolution sensor data, reducing reliance on reactive, schedule-based maintenance. AI and ML improve resource allocation, fault detection, and performance optimization through continuous learning, while DTs synchronize virtual and physical turbines for advanced diagnostics, planning, and simulation. The chapter also outlines implementation challenges, including model drift, data delays, security risks, and integration within existing infrastructure. It aims to guide researchers and engineers in deploying AI-based tools to improve turbine reliability, efficiency, and lifecycle management.

DOI: 10.4018/979-8-3373-4159-0.ch003

1. INTRODUCTION

With tremendous growth in the number of wind turbines, the area of wind farms, and the technical complexity of wind farms worldwide, wind farm clusters have grown larger, become more geographically distributed, and become more technically complex. The circa-end-2023 global installed wind capacity broke the 1,100-GW barrier, a milestone that symbolizes widespread deployment across diverse geographical and weather conditions. (IEA, 2024). Also, the Global Wind Energy Council (GWEC, 2024) Writes that new installations are increasingly scattered across emerging markets, making it difficult to operate them. These trends and emerging issues may make conventional monitoring, control, and maintenance schemes, whether scheduled or reactive, insufficient for handling the complexity and variability of large-scale wind farms. Therefore, there is a growing demand for new digital tools capable of overcoming this high level of complexity.

Conventional wind turbine maintenance is based on regular preventative maintenance and corrective actions upon failures. Planned preventive maintenance is applied periodically regardless of the components' actual condition, resulting in inefficient resource utilization and unnecessary downtime. (Yu, Patriksson, & Sagitov, 2021). Reactive maintenance generally results in unplanned outages and higher operation and maintenance (O&M) costs due to unforeseen malfunctions and delayed fault detection. (Pérez, 2022). These techniques have limited fault-detection capability and poor forecasting, leading to low turbine availability and reliability. So much so that, driven by real-time data and advanced analytics, these RUL systems can predict Remaining Useful Life (RUL) and promptly take maintenance actions to minimize downtime and cost while maximizing maintenance schedules (Shah, Daoliang, & Kumar, 2024).

The oil and gas industry is in the midst of a digital revolution, with SCADA systems and IoT sensors converging with cloud analytics. Such equipment allows for real-time monitoring and the automatic transmission of wind speed, vibration, or power output data from the wind turbines at any given moment, providing them with quicker reaction times and improved performance. (Alves, Mendonça, Mostafa, & Morgado-Dias, 2024). Scanned data enables timely control and early error detection, filling the gap left by intermittent manual monitoring. Furthermore, digital twinning models – integrating real-time sensor data with predictive models allow maintenance and performance optimization decisions to be made in real-time using data as a basis. (Abdullahi, Longo, & Samie, 2024).

Digital twin, machine learning (ML), and artificial intelligence (AI) are becoming part of modern wind turbine systems to increase operational efficiency and reliability. AI is a broad term that encompasses a range of computer systems capable of performing tasks that require human intelligence. Meanwhile, machine learning

(ML), a branch of AI, focuses on creating algorithms that learn from historical data to inform predictive decision-making. By contrast, digital twins are dynamic models of physical turbine equipment that are continuously updated using live data from SCADA and sensor networks to simulate system response and forecast performance (Shah, Daoliang, & Kumar, 2024). Existing technologies that deliver value during wind turbine deployment include ML methods for anomaly detection, such as detecting anomalies in vibration or temperature data to enable early warning systems, and AI-driven predictive models that can improve maintenance planning, e.g., predicting the RUL of vital components. Additionally, digital twins support real-time optimization as operators simulate control response strategies, such as dynamic variation in blade pitch, under various environmental and loading conditions. (Lui, 2024).

As shown in Figure 1, the following sections discuss the following topics: Wind turbine fundamentals and classical control theory; Communication infrustrature, technologies and hardware requirements; AI-driven applications; AI in Wind Farms using Machine Learning Models, Predictive Maintenance, and Condition Monitoring; The architecture of the Digital Twin for turbines; Monitoring and Deployment Challenges; Cybersecurity considerations; Future trends and research directions; and lastly the conclusion of the chapter.

Figure 1. The structural overview of Chapter 6

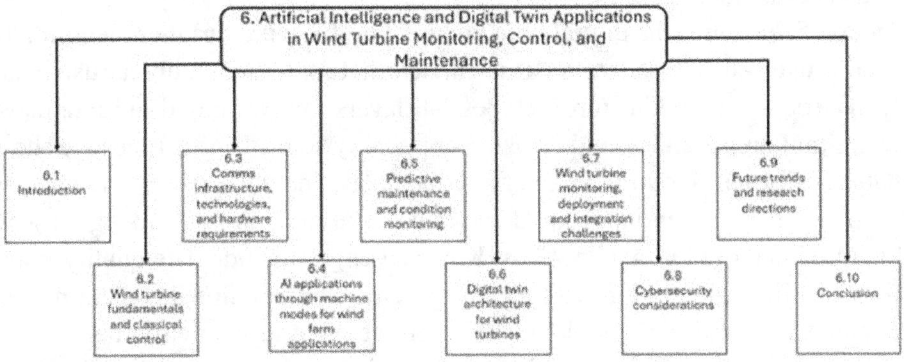

This chapter presents a comprehensive overview of the integration of artificial intelligence (AI) and digital twin technologies in modern wind turbine and wind farm systems. It begins by establishing the technical foundations of wind turbine operation and progressively advances toward intelligent, secure, and future-ready energy systems.

Section 1 introduces the fundamentals of wind turbines, covering aerodynamic principles, mechanical components, and power conversion systems. Classical control strategies, such as PID, pitch, and torque control, are discussed, highlighting their roles in maintaining system stability and efficiency under varying wind conditions. Limitations of traditional control approaches in handling nonlinear and uncertain environments are also emphasized.

Section 2 examines the communication infrastructure and hardware requirements for supporting AI-driven and digital twin applications in wind turbines. It discusses sensor networks, data acquisition systems, edge and cloud computing platforms, and communication technologies such as SCADA, IoT, 5G, and fiber-optic networks. The section highlights the importance of low-latency, high-reliability communication for real-time monitoring and decision-making.

Section 3 focuses on AI applications using machine learning models for wind farm operations. Supervised, unsupervised, and reinforcement learning techniques are explored for tasks such as power forecasting, fault detection, wind speed prediction, and optimal turbine coordination. The benefits of data-driven models in improving operational efficiency and reducing uncertainty across large-scale wind farms are discussed.

Section 4 addresses predictive maintenance and condition monitoring, emphasizing AI-based approaches for early fault detection and for estimating remaining useful life. Techniques including vibration analysis, thermal imaging, anomaly detection, and deep learning-based health monitoring are presented. This section highlights how predictive maintenance reduces unplanned downtime, maintenance costs, and component failures.

Section 5 describes the digital twin architecture for wind turbines, detailing the integration of physical assets with virtual models that continuously update using real-time sensor data. The architecture includes data layers, physics-based and data-driven models, simulation engines, and decision-support systems. Digital twins are shown to enable performance optimization, fault diagnosis, and what-if scenario analysis.

Section 6 discusses monitoring, deployment, and integration challenges associated with AI and digital twin systems. Key challenges include data quality issues, model scalability, integration with legacy systems, computational constraints, and the complexity of real-time deployment in remote wind farm environments.

Section 7 highlights cybersecurity considerations, addressing vulnerabilities in connected wind turbine systems. Threats such as data breaches, data injection attacks, and control system attacks are discussed, along with mitigation strategies including encryption, authentication, intrusion detection, and secure communication protocols.

Section 8 concludes the chapter by exploring future trends and research directions, including autonomous wind farms, self-healing control systems, edge AI, federated learning, hybrid physics–AI models, and standardized digital twin frameworks. The section emphasizes the growing role of intelligent, resilient, and sustainable wind energy systems in the global energy transition.

2. WIND TURBINE FUNDAMENTALS AND CLASSICAL CONTROL

Advanced monitoring techniques and intelligent control strategies can only be understood with a good understanding of the physical behaviour of wind turbines and their classification dynamics in classical control systems. This section opens with the basics of wind as a source of energy, explaining how well the aerodynamic forces are converted mechanically and, after that, electrically. It then describes the key elements of turbine dynamics, including rotor aerodynamic, drivetrain mechanical, and generator electric behaviour, all of which underpin any control strategy.

In light of this physical motivation, the section provides a readily comprehensible review of classical control theory, including feedback loops, proportional-integral-derivative (PID) controllers, and state-space representations. These techniques have been used for many years to control turbine speed, pitch the blades, and direct yaw to stabilize and optimize performance in changing wind fields. Although these classical control methods form the basis of turbine operation, their inability to handle nonlinear dynamics and volatile conditions calls for more adaptive, AI-informed approaches, as detailed later in this chapter.

The basic design principles for wind turbines, such as aerodynamics, mechanical loads, drivetrain dynamics, and control system interoperability, are well covered in the literature on wind energy. Detailed discussions of these principles appear in classic references such as Burton et al.'s Wind Energy Handbook, one of the seminal works on turbine design, load analysis, and control system considerations that can apply throughout the turbine's life cycle. (Burton, Jenkins, Sharpe, & Bossanyi, 2011). These concepts serve as the foundation for both classical control methods and their modern AI-augmented counterparts, meaning they set the physical limits within which advanced monitoring, control, and optimisation techniques must operate.

2.1. Wind Energy Conversion and Power Output

Wind turbine systems harness kinetic energy from wind flow. The power contained in the wind flows across a rotor swept area. $A = \pi R^2$ is given by the kinetic energy flux:

$$P_{wind} = \frac{1}{2}\rho A v^3 \tag{1}$$

where ρ is the air density, and υ is the wind speed. This wind-speed-cubed relationship is why slight increases in wind velocity have an exponential effect on energy generation.

Not all of this power can be extracted by a wind turbine. The ratio of available kinetic energy usable for mechanical output was constrained by aerodynamic factors. The Betz law states that NO wind turbine can capture more than 59.3% of the kinetic energy of the wind, and this arises physically from conservation of momentum within the rotor disc. (Coelho, 2023). This is the limiting factor represented by the power coefficient *Cp*, and is used to define turbine mechanical power:

$$P_t = \frac{1}{2}\rho A\, C_p(\lambda,\beta)v^3 \tag{2}$$

Here, *Cp*(λ,β) is the dimensionless power coefficient as a function of tip-speed ratio λ and blade pitch angle β. Open literature reviews provide comprehensive discussions of the dependence of *Cp* on these parameters, which are necessary for control design that seeks to maximize energy capture. (Castillo, Andrade, Rivas, & González, 2023).

2.2. Tip-Speed Ratio and Aerodynamic Efficiency

The tip-speed ratio (TSR) describes the relationship between the rotational speed of the rotor and the incoming wind speed and is defined as:

$$\lambda = \frac{\omega_r R}{v} \tag{3}$$

where ω_r is the rotor's angular speed and *R* is the blade radius. TSR characterises how fast the blades move relative to the wind. For a given turbine design, there exists an optimal TSR λ* at which the rotor achieves maximum *Cp* and therefore maximum energy extraction. High-fidelity performance curves for modern turbines typically show peak *Cp* values around λ between 6 and 8 under fixed pitch, as shown in Figure 6.2. below. (Järvinen, et al., 2025).

This concept is the basis of MPPT (maximum power point tracking) control strategies in region II (below rated wind speed), which operate to keep rotor speeand generator torque such that λ remains near λ* (Muñoz-Palomeque, Sierra-García, & Santos, Wind turbine maximum power point tracking control based on unsupervised neural networks, 2022).

Figure 2. Power coefficient curve of a modern wind turbine (Järvinen, et al., 2025).

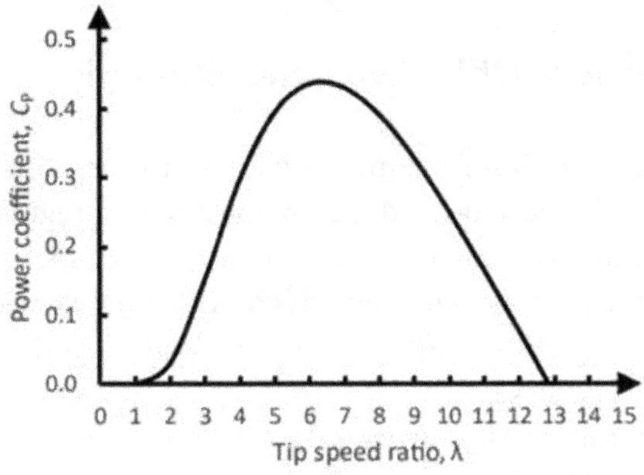

2.3. Tip-Speed Ratio and Aerodynamic Efficiency

The mechanical power generated by the turbine is associated with the aerodynamic torque acting on the rotor. This relationship is expressed as:

$$P_t = T_a \omega_r \tag{4}$$

Replacing the turbine power term gives the aerodynamic torque:

$$T_a = \frac{1}{2} \rho A R \frac{C_p(\lambda, \beta)}{\lambda} v^3 \tag{5}$$

This non-linear dependence underscores the highly interactive nature of wind speed, rotor speed, and blade angle. The dynamic representation of the rotor and drivetrain can be written as per Newton's 2nd law of rotation:

$$J \dot{\omega}_r = T_a - T_g \tag{6}$$

where J is the equivalent rotational inertia and T_g is the generator torque. This equation is the essential dynamic model for the design of wind turbine control.

The operation of a wind turbine is conventionally divided into regions based on wind speed, with each region having different control targets. (Pao, Pusch, & Zalkind, 2024):

- Region II (below rated): Optimize power capture using torque control.

- Region III (above rated): Control the amount of power produced and rotor run speed with pitch control, while protecting materials from failure.

2.4. Classical Control Objectives and Strategies

The physical relations above naturally give rise to the wind turbine's control objectives. For control in the sub-rated region, the aim is to regulate an optimal tip-speed ratio by controlling generator torque, so that the rotor speed tracks changes in wind speed. In contrast, the available aerodynamic power exceeds the generator's rated capacity at wind speeds slightly above its rated value, and active torque control (blade pitch) becomes necessary.

Reducing fatigue, minimizing structural expenditures, and maintaining system stability under turbulent wind excitation are also common objectives in any working conditions. There is a need for a suitable control design to reconcile these competing objectives.

- Generator Torque Control

In Region II, the appropriate control objective is to adjust generator torque to maintain rotor speed near the optimum TSR during turbine operation. Classical controllers often employ a quadratic law of the form:

$$T_g = K_{opt}\omega_r^2 \qquad (7)$$

This control law is derived from the aerodynamic power equation under the assumption of operation at the optimal tip-speed ratio. This simple law is widely used because it naturally increases torque with rotor speed when wind speeds are high and decreases it when wind speeds drop, thereby maximizing energy capture when wind speed measurements are unreliable or unavailable.

- Pitch Control

Above the rated wind speed, aerodynamic power exceeds the turbine's design limits. Blade pitch control is used to regulate rotor speed and limit power extraction. A typical classical feedback controller is the proportional–integral (PI) form:

$$\beta(t) = K_p e(t) + K_i \int e(t)\, dt \qquad (8)$$

where $e(t)=\omega r(t)-\omega r*$ represents the error between measured and desired rotor speed. This threshold is still the industrial benchmark for its simplicity and robustness (Muñoz-Palomeque, Sierra-García, & Santos, 2024).

2.5. Linearisation and Stability

The lack of linearity in wind turbine dynamics makes control design and analysis challenging. To handle this, the system is often linearised about operating points to obtain a state space model:

$$\dot{x} = Ax + Bu + Dw \tag{9}$$

This linearised model permits the application of traditional control analysis tools, such as frequency-domain stability analysis and robustness considerations. The stability margins should be chosen to account for unmodeled dynamics, communication delays, and stochastic wind variations.

2.6. Limitations of Classical Control and Motivation for Advanced Methods

While classical control is still prevalent in current turbine operation, it has limitations when dealing with the following, and hence drives the development of adaptive model-predictive and data-driven control methods using AI and a digital twin framework discussed later in this chapter. (Mulders, Gallo, & Rotea, 2024):

- Big nonlinearities in a wide operation range
- System uncertainties due to wakes and turbine induction effects
- Structural versatility and fatigue load resistance
- Long-term degradation in performance

To be deployable in actual wind farms, advanced control and digital twin solutions for wind turbines need to comply with globally established safety and certification guidelines. Especially the IEC 61400 series and among these the IEC 6140001 specify design requirements for turbines, which include design load cases and safety factors (cf. also operational envelopes) as well as model-based limitations on control actions such as generator torques, blade pitch rates, or yaw dynamics that can be applied at a particular time instance. (Freudenreich & Argyriadis, 2007).

Thus, AI-based controllers and digital twin–driven optimisation approaches must respect such certification constraints to avoid violating structural load assumptions or endangering turbine safety. This demand is already common practice from the wind energy control perspective, since certified load envelopes are the foundation for control system validation. (Burton, Jenkins, Sharpe, & Bossanyi, 2011).

Moreover, in the context of the digital twin, IEC Standards provide a verified baseline for physics-based turbine models, ensuring that simulated aerodynamic loads, fatigue damage, and control actions remain aligned with certified turbine performance. Hence, incorporating IEC-compliant constraints into AI and digital twin frameworks is a key element on which to build to deliver advanced control and optimisation from experimental studies to certified industrial deployment.

2.7. Summary

Chapter 6 reviewed the fundamental physics principles and classical control solutions for wind turbine operation, including aerodynamic power extraction, rotor and drive-train dynamics, generator torque control, blade pitch regulation, and system stability. Although classical controllers guarantee safe and efficient control across different wind conditions, they cannot effectively address parametric uncertainties, structural flexibility, or long-term performance degradation in WT systems. This has driven demand for more adaptive, data-driven control approaches.

These principles are essential to consider in the context of AI and other digital twin applications. A good understanding of how the turbine functions, control parameters, and operational limits is guiding the definition of a communication architecture and sensor network, as well as hardware specifications to support best real-time monitoring, feedback, situational awareness, prediction, and applied control strategies. Toward the end of Section 6.3, we turn our attention to communication infrastructure, its technologies, and the hardware capabilities required for AI-based and digital twin applications, grounded in the operational principles described in this section.

3. COMMUNICATION INFRASTRUCTURE, TECHNOLOGIES, AND HARDWARE REQUIREMENTS FOR AI AND DIGITAL TWIN APPLICATIONS IN WIND TURBINES

The deployment of artificial intelligence (AI) and digital twin (DT) technologies in wind turbines requires communication infrastructures capable of supporting high-rate data acquisition, ultra-low latency, deterministic delivery, and high reliability. Traditional SCADA-centric communication architectures were designed primarily for supervisory monitoring and low-frequency control and are therefore insufficient for the real-time data exchange and tight synchronisation demands imposed by AI inference pipelines and digital twin synchronisation. As a result, recent research highlights the need for advanced wired and wireless communication technologies, edge computing hardware, and time-synchronised network architectures to enable scalable and resilient AI- and DT-enabled wind energy systems. (Fuller, Fan, Day, & Barlow, 2020) (Kandemir, Hasan, Kvamsdal, & Abdel-Afou Alaliyat, 2024).

3.1. Communication Performance Requirements

AI-driven condition monitoring and digital twin synchronisation are critical for the timely exchange of high volumes of sensor data, model parameters, and state updates. Research on industrial AI and cyber–physical systems suggests that end-to-end latency of 5–10 ms is commonly required to maintain temporal consistency for real-time analytics and closed-loop control applications. (Saad, Bennis, & Chen, 2020). Such latency requirements directly affect the accuracy of vibration-based fault detection, blade load estimation, and real-time digital twin state matching in wind turbines.

Apart from latency and jitter, time synchronization accuracy is also important. Packet arrival time jitter can hinder feature extraction and temporal alignment in AI pipelines, just as poor packet synchronisation hinders state estimation in DT frameworks. Consequently, real-time monitoring and control of a turbine require deterministic communication. (Finn, Thubert, Varga, & Farkas, 2019) (Nasrallah, et al., 2019).

Key performance constraints commonly reported in the literature include:

$$\mathcal{T}_{e2e} < 5ms,\ \sigma_{jitter} < 1ms\ P_{loss} < 10^{-5}$$

The first parameter, $\mathcal{T}_{e2e} < 5ms$, specifies that the end-to-end latency should be less than 5ms. This is an ultra-low-latency demand and can be observed in mission-critical services(e.g., grid protection, autonomous driving, or industrial automation). The second argument, $\sigma_{jitter} < 1ms$, indicates that the jitter (the difference in time between packets) should be less than 1 millisecond. This way, the timing is deterministic, ensuring stable control loops and sync. Finally, the cap P sub l o s s less than 10 to the minus 5 end superscriptindicates that the network can have at most " out of packet loss (corresponding to less than one lost packet in 100,000 packets). This provides high reliability, which is important for protective relays or tripping signals, etc. These concepts can be associated with URLLC (Ultra-Reliable Low-Latency Communication) in 5G and IEC 61850-based smartgrid communications network architecture (Popovski, Trillingsgaard, Simeone, & Durisi, 2018).

3.2. Wired Communication Technologies

Wired communication remains the dominant solution for on-turbine and substation-level networking, particularly in offshore wind farms. Fibre-optic Ethernet backbones provide high bandwidth, low signal attenuation, and strong immunity to electromagnetic interference, making them highly suitable for high-rate sensor streams and aggregated digital twin (DT) data transfer requirements. (Gungor, et al., 2011). In harsh offshore conditions, such as an electrical-noisy environment and long transmission distances, a fiber-optic network offers significant advantages over its copper counterparts.

To support deterministic behaviour, Time-Sensitive Networking (TSN) extensions to Ethernet have been proposed for industrial energy systems. Time-Sensitive Networking (TSN) is a set of IEEE 802.1 standards that enhances standard Ethernet to support deterministic communication with bounded latency, high reliability, and precise time synchronisation. TSN introduces mechanisms such as traffic scheduling through the IEEE 802.1Qbv Time-Aware Shaper and time synchronisation through IEEE 802.1AS (generalised Precision Time Protocol), enabling Ethernet to meet stringent real-time control and monitoring requirements (Thangamuthu, Concer, Cuijpers, & Lukkien, 2015) (Finn, 2018). These capabilities allow TSN-enabled Ethernet to support time-critical monitoring and control tasks traditionally handled by proprietary fieldbus systems while maintaining interoperability and scalability.

The IEEE 802.1AS synchronization profile is especially crucial for digital twin and AI-based monitoring applications, as it provides sub-microsecond clock alignment between nodes in a distributed network. Correct time synchronisation ensures precise temporal alignment of data from multiple sensors, which is important for vibration-based fault diagnosis, blade load monitoring, and coordinated control of cyber-physical energy systems. (Thangamuthu, Concer, Cuijpers, & Lukkien, 2015).

However, the wired communication infrastructure lacks flexibility and has high installation and maintenance costs, especially for expanding existing wind farms or distributed sensor networks. Physical cabling requirements and deployment complexity can limit scalability, particularly for large or distant wind plants. As a consequence, more research is currently being devoted to wireless communication technologies as alternatives or complements to flexible turbine monitoring and control architectures.

3.3. Advanced Wireless Communication: 5G for Wind Energy Applications

Fifth-generation (5G) wireless communication introduces advanced capabilities that directly overcome the inherent limitations of conventional wind farm networks.

Ultra-Reliable Low-Latency Communication (URLLC) delivers sub-millisecond latency with exceptionally high reliability, making it ideally suited for mission-critical functions such as real-time monitoring, protection signaling, and digital twin synchronisation. (Popovski, Trillingsgaard, Simeone, & Durisi, 2018). The URLLC service class is specifically designed to support closed-loop control, fault detection, and real-time cyber-physical coordination, all of which are essential for AI-enabled turbine monitoring and digital twin state alignment.

Moreover, network slicing enables the logical partitioning of a shared physical infrastructure into multiple virtual networks, each with tailored quality-of-service guarantees. This capability allows strict prioritisation of AI-driven control and supervisory traffic while isolating non-critical flows such as maintenance logs, augmented reality maintenance support, or video surveillance streams (Foukas, Patounas, Elmokashfi, & Marina, 2017). Such traffic differentiation is particularly valuable in wind farms, where heterogeneous data streams vary significantly in terms of latency sensitivity, reliability requirements, and bandwidth demand.

Three types of services for industrial communication by 5G cellular technology are supported:

- URLLC (Ultra-Reliable Low Latency Communication): To provide a sub-millisecond latency and ultra-high reliability for mission-critical services such as real-time control, protection signalling, and monitoring.
- eMBB (enhanced Mobile Broadband): The high-capacity data streams from widespread sensor data collection, high-resolution inspection video exchange, and syncing digital-twin models.
- mMTC (massive Machine-Type Communication): It goes with the connection of a massive number of remote sensors and the whole Internet-of-Things (IoT) generation, which in turn empowers scalable condition monitoring architectures.

The integration of 5G into industrial automation and energy systems has been widely analysed as a promising approach to replace or augment traditional fieldbus and wired communication infrastructures to meet latency, reliability, and scalability requirements. (Siddiqi, Yu, & Joung, 2019). In wind energy applications, 5G enables flexible deployment of additional sensors, mobile inspection platforms such as drones, and distributed AI inference nodes without extensive cabling infrastructure.

Recent work has also considered the convergence of wired TSN with 5G wireless networks to provide deterministic end-to-end performance across both wired and wireless segments. These hybrid TSN-5G architectures provide ultra-reliable, low-latency communication and guarantee seamless time synchronisation and deterministic traffic scheduling, which are crucial for distributed industrial use cases with both mobility and reliability requirements, e.g., cooperative turbine control. (Cavalcanti, et al., 2019) (5G-ACIA, 2021).

Finally, private 5G will increase operational determinism, augment cybersecurity through network isolation, and enable smooth integration with edge computing platforms that sit close to turbines or substations. Private 5G networks minimize communication latency, enhance availability and autonomy against network failures by supporting local data processing and reducing reliance on remote cloud infrastructure. In combination, these capabilities enable 5G to become a foundation for resilient, intelligent, and secure communication infrastructures for future wind energy systems.

3.4. Emerging 6G Technologies

Next-generation (6G) wireless communication systems are currently being developed to go beyond the fundamental constraints of 5G for ultra-reliable cyber–physical systems, particularly those involving tightly coupled sensing, communication, computation, and control. Unlike past generations of wireless that have been predominantly designed to improve data bandwidth, 6G is envisioned as an AI-native, sensing-integrated network that supports data collection, real-time analytics, and distributed intelligence. This paradigm combines communication and sensing, with embedded machine learning in network management and resource allocation. (Saad, Bennis, & Chen, 2020) (Tataria, et al., 2021).

One of the core performance targets of 6G is sub-millisecond end-to-end latency, extending beyond the Ultra-Reliable Low-Latency Communication (URLLC) capabilities of 5G. This is expected to be achieved through joint optimisation of physical-layer transmission, network-layer routing, and distributed computation placement, enabling latency budgets on the order of hundreds of microseconds for mission-critical applications (Dang, Amin, Shihada, & Alouini, 2020) (Saad, Bennis, & Chen, 2020). Such ultra-low latency is essential for industrial cyber-physical systems requiring real-time closed-loop control and high-fidelity digital twin synchronisation.

For real-time wind turbine monitoring and control, communication latency directly constrains closed-loop stability and the accuracy of digital twin state estimation. The end-to-end delay can be expressed as:

$$\tau_{e2e} = \tau_{tx} + \tau_{queue} + \tau_{proc} + \tau_{rx} \tag{10}$$

where τ_{tx} and τ_{rx} are transmission and reception delays, τ_{queue} is the network-induced queuing delay, and τ_{proc} is the computation delay at the edge or fog nodes.

6G architectures aim to minimise τ_{queue} and τ_{proc} by co-locating computation with communication resources, a feature critical for high-rate vibrationmonitoring and AI-driven fault detection in wind turbines (Zhang, et al., 2019) (Saad, Bennis, & Chen, 2020). Sub-millisecond latency enables near-instantaneous state updates between physical turbines and their digital twins, reducing model drift and improving control fidelity during fast transients such as gusts or grid faults.

In wind turbine systems, low-delay components are important, particularly for high-frequency vibration monitoring, blade load estimation, and AI-based fault detection algorithms. Sub-millisecond latency enables high-speed state matching between the physical turbines and their digital twins, helping mitigate model drift in control fidelity during fast transients such as wind gusts, wake effects, and grid disturbances.

The next generation of wind farms will likely be deployed as distributed cyber–physical energy systems that leverage AI inference, sensor fusion, and digital twin updates across multiple computational nodes at different locations. 6G facilitates this by a set of key technologies, including cell-free massive MIMO, which increases the spatial diversity and coverage by deploying APs over a network, while THz communication can achieve very high data rates, which apply to realistic URC transmission, such as high resolution of sensor stream and model synchronization (Dang, Amin, Shihada, & Alouini, 2020) (Tataria, et al., 2021). Edge intelligence also supports distributed AI inference near turbines, thereby reducing communication overhead and increasing resilience to network failures.

Even though 6G is still at the research and standardization stage, its potential, e.g., for ultra-low latency, sensing-integrated communication (SIC), AI-native control, and distributed intelligence, can address the communication bottleneck experienced by AI-driven monitoring and digital twin systems in wind energy. As a result, 6G offers a visionary architecture to underpin the mature and sustainable deployment of advanced AI and digital twin technologies, as studied in the following sections of this chapter.

3.5. Sensor Infrastructure and Time Synchronisation

Sensors with high resolution (e.g., accelerometers, strain gauges, LiDAR systems, acoustic emission sensors, and temperature probes), when employed in AI-enabled condition monitoring and digital twin (DT) platforms, produce continuous, high-frequency data streams that must be transmitted under strict temporal constraints. These sensing technologies enable enhanced diagnostic capabilities, including vibration-based fault detection, structural load identification, aerodynamic health monitoring, and wake interaction assessment. But the huge data sizes and high sampling rates make intense demands on network interfaces and processing platforms.

(Maldonado-Correa, Martín-Martínez, Artigao, & Gómez-Lázaro, 2020) (Hussain, et al., 2024). Hence, state-of-the-art network interface cards (NICs) with hardware-based Precision Time Protocol (PTP) timestamping are utilised alongside commercial TSN-capable switches and bridges to ensure deterministic timing behaviour while keeping communication jitter low in industrial cyber-physical systems.

Accurate time synchronisation between units of distributed sensing and control devices is a fundamental requirement for deterministic monitoring, coordinated control, and coherent multi-sensor data fusion. In digital twin architectures, it is essential to correctly align timestamps to ensure consistency between the physical states of turbines and their virtual representations. The Precision Time Protocol (PTP), standardized as IEEE 1588, allows sub-microsecond clock synchronisation over packet-switched Ethernet networks and is thus very well suited for real-time industrial automation and intelligent energy systems. (Eidson, 2006) (Messenger, 2018).

Multiple revisions of IEEE 1588 have been developed; among them, we can cite the newer IEEE 1588-2008 and IEEE 1588-2019, which improve synchronization performance, increase robustness to failures, and support networks with a higher number of devices. These updates include new features from boundary clocks and the introduction of transparent clocks, as well as enhanced delay-measuring mechanisms, resulting in much reduced Great time errors in complex industrial networks. Fast and accurate clock synchronisation enables high-resolution event correlation, easing distributed fault localisation and the coordinated sampling of multiple turbines in a wind farm.

PTP synchronisation also extends to TSN profiles, in particular IEEE 802.1AS, which defines a generalised Precision Time Protocol (gPTP) framework to re-time all network elements of Ethernet devices on a standard time base. IEEE 802.1AS enables deterministic communication scheduling and synchronized network traffic shaping, both of which are required for time-sensitive AI inference pipelines or closed-loop turbine control. The integration of TSN and PTP-based synchronisation ensures the deterministic transfer of high-rate sensor data with guaranteed timing consistency among distributed wind energy measurement and control systems. (Thangamuthu, Concer, Cuijpers, & Lukkien, 2015) (Messenger, 2018).

Advanced sensor infrastructure and high-precision time synchronisation technologies together provide the foundational layer for scalable, deterministic digital twin implementations with high fidelity in contemporary wind energy systems.

3.6. Hardware Constraints and Computational Requirements

The deployment of artificial intelligence (AI) and digital twin applications in wind turbine systems is significantly constrained by the computational and hardware capabilities of embedded controllers, supervisory control systems, and communication infrastructures. These limitations directly affect the fidelity of monitoring, the responsiveness of control actions, and the scalability of integration across geographically dispersed assets.

- **Embedded Controller Limitations:** Programmable logic controllers (PLCs) and remote terminal units (RTUs) typically have limited CPU cycles and support only a few floating-point operations; they cannot execute advanced AI models at the edge. Most of the time, digital twin simulations rely on synthetic, high-resolution vibration, temperature, and load data that common turbine controllers cannot process. (Pandit, Astolfi, Hong, Infield, & Santos, 2023).
- **Memory and Storage Bottlenecks:** Anomaly detection and predictive maintenance via AI depend on large-scale training data and inference. Light local devices with less than 1 GB of RAM and a small amount of NV storage cannot maintain a long history. This requires hybrid architectures in which raw data is preprocessed at the edge and dispatched to cloud servers for more comprehensive analysis. (Castellani, Pandit, Natili, Belcastro, & Astolfi, 2023).
- **Computational Load Distribution:** Edge computing is increasingly used to overcome hardware limitations by performing primary filtering, feature extraction, and real-time fault detection. Cloud infrastructure, however, runs massively parallel, compute-intensive workloads such as digital twin simulations, machine learning model training, and fleet-wide optimization. Time-critical applications , such as blade pitch control and gearbox fault detection, require sub-second response times, necessitating the partitioning of workloads with some thought. (Arroba, Buyya, Cárdenas, Risco-Martín, & Moya, 2024).
- **Energy and Thermal Constraints:** The monitoring hardware needs to be power-limited to around 50 W and must survive harsh environmental conditions. High-performance processors are at risk of overheating or exceeding power budgets; hence, rugged, low-power designs. (Huang, Zhou, Li, & Shen, 2025).
- **Scalability and Integration Challenges:** For a 200-turbine wind farm, the digital twin ecosystem can produce sensor data at 10+ TB/day. Centralized systems cannot operate in real time without distributed preprocessing, as the computational and storage loads would be too high. The protocol interoperability (IEC 61400-25, OPC UA, and MODBUS) also introduces additional computational overhead due to harmonization across diverse vendor platforms. (Kestel, et al., 2025).

- **Cybersecurity Implications:** Resource-constrained devices limit the use of advanced IDS systems and encryption. Light-weight cryptographic solutions are commonly used, but these compromises between security and efficiency render the digital twin infrastructures vulnerable to cyber-attacks (Murray, et al., 2025).

3.7. Summary

In section 6.3, the essential communication and hardware structures for AI and digital twin deployment in wind turbines were discussed. The key takeaways are the requirements for ultra-low-latency, deterministic networks, high-fidelity sensor infrastructure, and accurate time synchronization; moreover, distributed computation architectures are used to process high-rate data streams. Wired (e.g., TSN-enabled Ethernet) and advanced wireless solutions (e.g., 5G/next-gen 6G) were debated regarding their benefits and drawbacks for real-time monitoring and control. The section also underscored hardware and computational constraints, such as embedded controller capabilities, memory/storage limits, and power/thermal budgets, as well as requirements for scalable architectures to manage turbine fleets while upholding cybersecurity.

These hardware and infrastructure requirements are critical to enabling AI-based analysis and predictive modeling. A sharp-winding system for a wind turbine can provide accurate, instantaneous synchronization data for advanced machine learning algorithms, real-time digital twin simulation, and intelligent control strategies. Section 6.4 extends this theme by implementing AI applications and machine learning models to improve turbine monitoring, fault detection, and operational optimization.

4. ARTIFICIAL INTELLIGENCE APPLICATIONS THROUGH MACHINE MODELS FOR WIND FARM APPLICATIONS

In this section, we present an artificial intelligence application in wind farm operations using machine learning models. It focuses on how such models leverage large datasets of SCADA and sensor data to enhance turbine performance, support early fault detection, and enable predictive maintenance. Supervised and unsupervised learning, anomaly detection, and time-series forecasting are the techniques emphasized, which are necessary for fine-tuning asset management and reducing operational costs.

4.1. Overview of AI and ML Techniques

ML and AI methodologies are now playing a more significant role in optimizing wind turbine systems. To predict the fault category and wind power output, machine learning methods, including neural networks and decision trees (developed algorithms), use a training set with labelled input data (fault logs and healthy operation) (Soe & Htet, 2024), which can, for example, be SCADA. Here, we extract the operational behaviour without labels (i.e., unsupervised learning) and then use that to detect underlying patterns or anomalies from unlabelled sensor data (e.g., temperature, vibration signals) (Dibaj, Valavi, & Nejad, 2024). Reinforcement learning differs because systems can adapt to environmental changes and learn optimal control strategies, such as pitch and yaw angles, by repeatedly driving a wind turbine. (Liu, Wu, He, & Gupta, 2025). These methods build strong, data-driven models using multiple data sources, such as sensor records, SCADA logs, and weather monitoring data. Wind energy operations become more reliable and efficient through an AI/ML-based system that enhances the effectiveness of problem detection, condition monitoring, predictive maintenance, and related processes. In the context of input data, popular application domains, popular algorithms, and the sources of data on which they are based, we compare supervised, unsupervised, and reinforcement learning in Table 6.1. It also provides examples of how each could be implemented in wind power.

Table 1. Comparison of Machine Learning Techniques Applied to Wind Turbine Systems (Soe & Htet, 2024) (Dibaj, Valavi, & Nejad, 2024) (Lui, 2024)

	Supervised Learning	**Unsupervised Learning**	**Reinforcement Learning**
Input Data	• Labeled data.	• Unlabeled data.	• State, action, reward feedback loop from a real or simulated environment.
Common Algorithms	• Decision Trees. • Support Vector Machines. • Neural Networks.	• Clustering (K-means). • Autoencoders. • Principal Component Analysis (PCA).	• Q-learning. • Deep Q-Networks (DQN). • Policy Gradients.
Typical Applications	• Fault classification. • Power output prediction. • Wear estimation.	• Anomaly detection. • Performance pattern clustering	• Blade pitch/yaw optimization. • Adaptive energy control. • Load balancing.
Data Dependency	• Requires large amounts of accurately labeled historical data.	• Effective when labeled data is scarce or unavailable.	• Learns from ongoing or simulated turbine operation.
Example in Wind Energy	Predicting gearbox failure using SCADA fault logs	Identifying abnormal vibration patterns without prior labels	Optimizing rotor speed under varying wind conditions

AI and machine learning models in wind turbine systems rely on various data sources for prediction and analysis. Other operational variables (system status, power generation, rotor speed) that underpin the majority of predictive models (Soe & Htet, 2024). Sensor signals (vibration, temperature, pressure, acoustics) provide detailed information on mechanical conditions, enabling early fault detection. (Dibaj, Valavi, & Nejad, 2024). Weather conditions, such as wind speed, direction, and temperature, increase forecasting by providing environmental data regarding turbine operation. (Sun & Han, 2025). This includes high-quality preprocessing, such as denoising and feature extraction, to ensure robust learning and accurate prediction. (Backhus, et al., 2024).

4.2. AI-Based Monitoring Applications in Wind Turbine Systems

AI-based monitoring underpins predictive maintenance, control optimisation, and digital twin synchronisation. Unlike traditional threshold-based SCADA monitoring, AI-driven systems continuously learn normal operational behaviour and detect deviations early, often before conventional systems trigger alarms. Recent reviews highlight that such approaches significantly reduce false alarms while extending fault detection horizons in large-scale wind farms. (Kostecka, Miller, Durlik, & Nerć, 2025).

In wind turbine applications, AI-based monitoring typically operates across three layers: signal-level monitoring, component-level health assessment, and system-level performance supervision. At the signal level, machine-learning models analyse high-frequency SCADA and sensor streams, including vibration, temperature, rotational speed, and power output, to identify subtle statistical deviations. Unsupervised techniques such as autoencoders, principal component analysis (PCA), and clustering methods are widely used to establish baseline operational envelopes and flag anomalies through reconstruction error or distance metrics. (Milani, Leoni, Cacciola, Croce, & Tanelli, 2025).

At the component level, supervised and semi-supervised methods are used to assess the condition of key components, such as gearboxes, bearings, generators, and pitch systems. The literature shows that SCADA-introduced data-driven prognostics algorithms can predict drivetrain abnormalities several days or weeks before the machine fails, and these methods achieve over 90% classification accuracy when applied to real field data. (Habbouche, Amirat, & Benbouzid, 2025). Security and scalability. More decentralized methods, such as federated learning, also have the potential to enable co-training across wind farms, so that sensitive operational data does not need to be centrally aggregated, thereby contributing to scaling. (Bawane, et al., 2025).

At a system scale, performance variation and energy yield monitoring are implemented using AI. Machine learning power curve models update the expected power output based on current wind and environmental conditions. Remainders of this type between predicted and actual power are commonly used to detect blade degradation, aerodynamic imbalance, wake-induced aerodynamic interaction, or control ineffectiveness. Monitoring based on residuals has been successful for online performance monitoring without any labeled fault data. (Kandemir, Hasan, Kvamsdal, & Abdel-Afou Alaliyat, 2024).

AI-based monitoring algorithms enable earlier fault detection (compared with conventional rule-based approaches), adaptable performance monitoring, and greater robustness under varying environmental conditions. Such functionalities are today considered a must for the new wind farms, in particular offshore and floating ones, where anold-school inspection system and alarm thresholds do not satisfy anymore. (Kostecka, Miller, Durlik, & Nerć, 2025).

4.3. Failure Prediction and Fault Diagnosis

Early failure detection and fault diagnosis are essential in ensuring the reliability of wind turbines, especially as wind farms grow in size and complexity. An unexpected equipment outage can lead to unscheduled production loss and increased O&M cost. (Sun & Han, 2025). Classical maintenance policies are not always accurate enough to early identify such problems. On the other hand, the use of data-driven approaches enables us to exploit real-time data from the SCADA system, sensors, and weather inputs to achieve more precise, early fault detection. (Soe & Htet, 2024). Machine learning and advanced analytics are possibilities that improve condition monitoring, allow predictive maintenance and contribute immensely to the operational resiliency of wind energy systems (Dibaj, Valavi, & Nejad, 2024).

ML is at the heart of this, as it allows us to “teach” machines how to predict whether wind turbines will fail or not by 'looking' at historical and real-time data. Unlike a rule-based approach driven by static thresholds, which are unable to detect early signs of faults, ML models can learn non-linear and complex correlations between variables. Based on network logs, sensor data, and environmental conditions, these models seek minor anomalies and degradation behaviours that are early signs of faults. (Alagha, Khairuddin, Haitaamar, Al-Khatib, & Kanesan, 2025). Proactive maintenance decisions enabled by ML-based predictive analytics reduce unplanned downtime and lengthen the equipment life, and increase operational efficiency. (Lee & Maceren, 2024).

Various machine learning techniques have been utilized for fault detection in the wind turbine generator industry, providing distinct advantages in terms of data type when applied to specific faults based on their complexity. Table 6.2 lists a selection of state-of-the-art techniques, including neural networks (NN), support vector machines (SVM), and random forests (RF), along with their main characteristics, advantages/disadvantages, and common areas of application in wind energy systems.

Table 2. Comparison of Machine Learning Algorithms for Fault Detection and Classification in Wind Turbines

Algorithm	Key Features	Advantages	Typical Applications
Support Vector Machines (SVM)	• Handles binary and multiclass classification. • Works well with high-dimensional data.	• Robust to overfitting in high-dimensional spaces. • Effective with limited training data.	• Classifying SCADA-based fault events. • Bearing fault detection.
Random Forests	• Ensemble of decision trees. • Uses bagging and random feature selection.	• High accuracy and generalization • Provides feature importance insights.	• Multi-class fault detection. • Identifying root causes of performance loss
Artificial Neural Networks (ANNs)	• Captures complex nonlinear relationships. • Composed of interconnected processing nodes.	• Highly flexible and powerful. • Performs well in large-scale, multi-variable datasets.	• Diagnosing generator and gearbox faults. • Condition monitoring based on sensor fusion.
k-Nearest Neighbors (k-NN)	• Instance-based learner. • Classifies based on proximity in feature space.	• Simple to implement. • No training phase required.	• Detecting anomalies from sensor data. • Early fault stage identification.
Gradient Boosted Trees (e.g., XGBoost)	• Sequential ensemble of weak learners (trees).	• Excellent for imbalanced datasets. • Handles missing data and nonlinearity well.	• Power output prediction with fault indicators. • Fault severity classification.

However, the introduction of machine learning into fault detection for wind turbines raises several practical issues. A preponderance of one side of data distribution (number of failure operations versus number of normal operations) can cause an equalization point for a priori false negatives in the models (Alagha, Khairuddin, Haitaamar, Al-Khatib, & Kanesan, 2025). Sensor noise, missing data, and high-dimensional features also affect the model's usefulness and generalization. Therefore, preprocessing the data as a pre-requisite step, including normalization, noise reduction, and feature extraction, is essential (Backhus, et al., 2024).

4.4. Trend Identification and Power Output Estimation

In the context of wind turbines, trend lines are used to analyze historical SCADA and sensor data to monitor slow-moving performance changes and wear degradation, which is crucial for early fault prevention. Typical techniques are moving averages and regression models to display trends of the smooth kind, as well as ML-based anomaly detection (like autoencoders, clustering) to find hidden, subtle exceptions (Backhus, et al., 2024). These techniques enable trend-based use cases, such as blade efficiency erosion, seasonal performance deviation, and bearing or gearbox wear, to drive Proactive maintenance and optimised turbine life.

Scientific methods for trend analysis vary from statistical filters to sophisticated machine learning approaches. Moving average techniques can, for example, remove the short-term noise to expose longer-term patterns as shown in Equation (11):

$$SMA_t = \frac{1}{n}\sum_{i-0}^{n-1} x_{t-i} \tag{11}$$

where x_t is the performance signal at time t, and n is the smoothing window. Exponential moving averages (EMAs) place greater emphasis on recent data, improving responsiveness to rapid operational changes as shown in Equation (12), where α represents the smoothing factor ($0 < \alpha \leq 1$):

$$EMA_t = \alpha x_t + (1 - \alpha)EMA_{t-1} \tag{12}$$

To model long-term degradation or efficiency loss, polynomial regression can be applied as shown in equation 13:

$$P_{trend}(t) = \beta_0 + \beta_1 t + \beta_2 t^2 + \ldots + \beta_k t^k + \in_t \tag{13}$$

Here, $P_{trend}(t)$ represents the predicted trend of power output or other health indicators, $\in_t$ is the error term and $\beta_0, \beta_1, \ldots, \beta_k$ are the polynomial coefficients.

In parallel, unsupervised machine learning techniques such as autoencoders and clustering are employed to uncover complex, nonlinear anomalies in multi-sensor data. Autoencoders reconstruct normal system behavior, and deviations from reconstruction serve as anomaly indicators. The anomaly score can be calculated using equation 14, where x is the original observed data vector and $\hat{x}$is the reconstructed or predicted vector:

$$Anomaly\ Score = \|x - \hat{x}\|^2 \tag{14}$$

Clustering approaches, e.g., k-means, discover normal operation states. Significant deviations from cluster centers at different timestamps could indicate abnormal operation or deterioration in blades, gearboxes, bearings, etc.

Predicting the nominal power production of a wind turbine and comparing it with actual power production is an essential step in early identification of underperformance or concealed failures. The theoretical baseline is based on the wind power equation as shown in Equation (15):

$$P = \frac{1}{2} \cdot \rho \cdot A \cdot C_p(\lambda,\beta) \cdot V^3 \tag{15}$$

where:

- ρ is the air density (kg/m^3),
- $A = \pi R^2$ is the rotor swept area (m^2),
- Cp is the power coefficient, dependent on the tip-speed ratio λ and pitch angle β,
- V is the wind speed (m/s).

While useful as a first approximation, this model does not account for environmental disturbances, aging components, or interactions with the control system.

To overcome these limitations, machine learning models—such as Random Forests, XGBoost, and Artificial Neural Networks (ANNs) - are used to map multiple inputs to a powerful output more accurately, as shown in equation 16:

$$P_{pred} = f\left(V, \theta, T_{amb}, \omega, Yaw, Pitch\right) \tag{16}$$

where:

- V: Wind speed,
- θ: Wind direction,
- Tamb: Ambient temperature,
- ω: Rotor speed,
- Yaw and Pitch: Control settings.

These models capture nonlinear and context-dependent relationships, thereby enhancing prediction accuracy.

Once the predicted output P_{pred} is obtained, residual analysis is performed by comparing it with the actual SCADA-recorded power output as shown in equation 17:

$$\in_t = P_{actual}(t) - P_{pred}(t) \tag{17}$$

Persistent harmful residuals (i.e., $\in_t < 0$) indicate underperformance and may suggest mechanical degradation, sensor drift, blade contamination, or wake interference. Conversely, positive residuals could point to sensor calibration errors or data quality issues.

4.5. AI-Enhanced Turbine Control

AI-enhanced turbine control is projected to be a game-changer in the wind farm industry, enabling real-time wind farm operations to underpin adaptive decisions that optimize energy capture and minimize mechanical wear and unplanned downtime. Despite the robust chance of the shift, multiple existential challenges resist safe deployment. Safety-critical conditions are challenging to operate in due to the intrinsic complexity, opacity, and explainability of many AI systems, including deep neural networks, which are currently the most promising known architecture. It is challenging to operate as an operator who needs decisions to be auditable. Furthermore, the AI integration into legacy turbine control systems generally requires substantial renewal investments, vital for structural compatibility, and increases cost complexity. (Maldonado-Correa, Torres-Cabrera, Martín-Martínez, Artigao, & Gómez-Lázaro, 2023).

This is in direct contrast to the fact that, according to evidence, "AI-based controllers have been shown to dominate conventional PID or rule-based systems in dynamic wind gusts, adapting control strategies to the underlying control policies." (Liu, Wu, He, & Gupta, 2025). The performance increase is mainly due to the AI's ability to learn an optimal control policy in a rapidly changing environment. Thus, AI-controlled systems offer enhanced fault tolerance and performance compared to established options, and there are technical, economic, and regulatory barriers to adoption. One of the most promising paradigms is for AI to formulate the turbine control problem as an MDP Reinforcement Learning.

Learning-based control approaches, particularly reinforcement learning (RL), provide a principled framework for sequential decision-making under uncertainty and have gained increasing attention in wind turbine and wind farm control applications. In contrast to classical model-based control strategies, RL formulates the control problem as a Markov decision process, where control policies are learned through interaction with the environment to optimise long-term performance objectives such as energy capture, load mitigation, or maintenance cost reduction. The theoretical foundations of RL, including value functions, policy iteration, and temporal-difference learning, are well established in the control and artificial intelligence literature (Sutton & Barto, 1998).

In the context of wind energy, RL-based controllers can leverage high-fidelity digital twins as safe training environments, enabling extensive exploration of control strategies without risking physical turbine assets. Digital twins, therefore, play a critical role in addressing one of the key challenges of RL: the need for large volumes of interaction data, by providing a physics-informed simulation platform for policy learning and validation. This integration of reinforcement learning with digital twin architectures bridges classical control theory and modern data-driven optimisation, while maintaining consistency with operational constraints and safety requirements.

The aim is to find an optimal policy $\pi*$ as shown in Equation (18), which is a mapping of a state to a control action associated with the state, such that it maximizes the long-term cumulative rewards:

Formal RL Objective:

$$\pi^* = \arg\max_{x} \mathbb{E}\left[\sum_{t-0}^{\infty} \gamma^t R(S_t, a_t)\right] \tag{18}$$

Where

- S_t: State at time (e.g., wind speed, yaw angle, rotor speed),
- a_t: Action at time (e.g., blade pitch, yaw command, torque),
- $R(S_t, a_t)$: Immediate reward (e.g., power coefficient, load penalty),
- $\gamma \in (0,1]$: Discount factor,
- π: Policy function mapping states to actions.

An example of a reward function used in wind control could be calculated using equation 19:

$$R(S_t, a_t) = \eta \cdot \frac{P_t}{P_{rated}} - \lambda \cdot \sigma_L(t) \tag{19}$$

Where:

- P_t: Actual power output,
- P_{rated}: Rated turbine capacity,
- $\sigma_L(t)$: Load variance (to penalize mechanical stress),
- η, λ: Reward scaling coefficients.

Traditional turbine control strategies use fixed setpoints or linear feedback to govern blade pitch (β), generator torque (Tg), and yaw angle (θ). These are often insufficient in unsteady wind conditions.

Torque Control Region II (Optimal Power Tracking):

In Region II (partial load), the generator torque Tg is typically proportional to the square of the rotor speed ω and can be calculated as shown in equation 20:

$$T_g = K_{opt} \cdot \omega^2, \; K_{opt} = \frac{1}{2}\rho A R^3 \frac{C_p}{\lambda^3} \tag{20}$$

AI-based control can learn to adapt this relationship to varying turbulence intensities or aerodynamic wear, achieving better energy yield than static curves.

Pitch/Yaw Optimization:

The pitch control problem is to track the maximum power point while ensuring load constraints, and can be calculated using equation 21:

$$\min_{\beta(t)} \mathcal{J} = -P(t) + \mu \cdot L_{fatigue}(\beta(t)) \tag{21}$$

Where:

- $\mathcal{J}$: Cost function to minimize,
- $P(t)$: Instantaneous power output,
- $L_{fatigue}$: Load-based fatigue damage estimator (e.g., rainflow counting),
- μ: Trade-off coefficient.

AI-enhanced turbine control presents a radical departure from static, model-based regulation to data-driven, adaptive optimization. Moreover, RL and other AI models can enhance control decision-making throughout their lifecycle. This can include sensor inputs, uncertainty awareness, sensor data utilization, responsive wear and drift in turbine dynamics, fault endurance, and operative profit. While solving the illustrious pregame and optimum regulation problems might sound alluring, several factors, such as explainability, system integration, and real-time deployment issues, must be addressed to enable safe, cost-effective implementation in massive wind turbine installations.

4.6. Case Study: Machine-Learning-Based Fault Prediction for Wind Turbine Generator Bearings

A representative peer-reviewed case study demonstrating the practical application of artificial intelligence in wind farm operations is presented by (Williams, Phillips, Sheng, Dobos, & Wei, 2020) through a collaboration supported by the U.S. Department of Energy (DOE) and the National Renewable Energy Laboratory (NREL). The study focuses on the use of machine-learning models for generator bearing-fault prediction using absolute Supervisory Control and Data Acquisition (SCADA) data from an operational wind power plant.

In this work, multiple machine learning algorithms are trained and evaluated to identify early indicators of bearing degradation in wind turbine generators. The models leverage high-frequency operational data, including temperature, power output, rotational speed, and other SCADA-derived features, to distinguish normal operating behaviour from incipient fault conditions. Unlike purely simulation-based studies, this case study is grounded in field data from utility-scale turbines, providing a realistic assessment of AI model performance under operational variability and noise.

Scalability and deployability. The study emphasizes scalability and deployment feasibility. The authors address challenges associated with their application across large turbine fleets, including data disparity, model generalisation, and computational costs. Overall, the performance parameters presented in this study indicate that computational intelligence-based fault prognosis can forecast bearing failure notably earlier than the conventional threshold-by-threshold strategy and that it provides a conceptual tool for proactive maintenance and minimizing unplanned downtime.

From the wind farm perspective, this case study demonstrates how models can be incorporated directly into existing monitoring infrastructures without the need for additional sensing hardware. Since the approach processes SCADA data streams directly, it is compatible with industry concerns about costs and interface standards, as well as with the retrofit of existing systems. This makes the approach especially suitable for large, geographically distributed wind farms with less ubiquitous instrumentation across turbines.

Notably, the work provides numerical threshold values for AI-based fault detection performance and underscores relevant practical issues for model implementation, including data preprocessing, feature selection, and false-alarm strategy. As a consequence, it serves as a tangible illustration of how machine-learning approaches introduced in this section can be transferred from algorithmic development into the context of actual wind farm applications.

Overall, this DOE- and NREL-supported case study demonstrates that artificial intelligence–based machine-learning models can deliver measurable operational value in wind farm monitoring and maintenance. Its inclusion provides empirical validation of the AI methodologies discussed in this chapter and establishes a practical reference point for deploying data-driven intelligence in modern wind energy systems.

4.7. Summary

In Section 6.4, the use of artificial intelligence (AI) and machine learning (ML) in wind farm operations was presented, with a focus on their contributions to improving turbine performance, early fault detection, and predictive maintenance. The priority strategies were supervised, unsupervised, and reinforcement learning algorithms, as well as artificial intelligence (AI)-based monitoring in a signal, component, or system perspective. The section described high-resolution SCADA and sensor data, trend analysis, and residual-based performance computation for anomaly detection, fault prediction, and adaptive turbine control. Success stories – such as an ML model to predict bearing faults on generators – demonstrated how AI models can be deployed at scale and integrated with real-world operations across wind farms.

They build directly on the communication infrastructures, sensor networks, and computational architectures described in Section 6.3. Real-time learning, fault detection, and adaptive turbine control are supported through timely synchronized data streams. Digital twin frameworks provide safety when emulating advanced reinforcement learning and policy optimization.

These AI-driven insights from Sections 6.3 and 6.4 can then be applied to predictive maintenance and condition monitoring, the subject of Section 6.5, where the focus shifts from model development and control strategies to their operational implementation. Here, AI outcomes are used for predictive maintenance scheduling to reduce unplanned downtime and increase turbine lifetimes, bringing intelligence and analytics into viable operational decisions.

5. PREDICTIVE MAINTENANCE AND CONDITION MONITORING

Data-driven techniques are employed in predictive maintenance for wind turbines to predict faults before they occur. This allows maintenance crews to schedule their workload rather than reactively. It uses online information from SCADA systems, sensors, and ecological models to assess the health and remaining life of mechanical components in turbines. The goal is to prevent unplanned downtime and increase turbine availability, as ram-air preventive maintenance interventions are performed only when needed. Perfective maintenance enhances operational performance by rectifying failures at an embryonic stage, in parallel with a phased or reactive tactic. (Alagha, Khairuddin, Haitaamar, Al-Khatib, & Kanesan, 2025).

5.1. Data Sources for Predictive Maintenance

Predictive Maintenance of wind turbines is a process in which data from multiple sources is combined to assess component health and predict failures. The data consist of SCADA, sensors, and weather data. High-frequency operational parameters, including power output, rotor speed, and air temperature, are represented by data points in the SCADA system. (Gigoni, Betti, Tucci, & Crisostomi, 2019). Similarly, sensor data is collected using vibration, temperature, acoustic, and pressure sensors to detect mechanical and electrical issues early. (Granados, et al., 2023). The environmental data records wind speeds, directions, temperatures, and turbulence intensity, which puts the performance causes into perspective. (Gonzalez, Stephen,

Infield, & Melero, 2019). This fusion enables fault diagnosis and Remaining Useful Life (RUL) prognosis.

5.2. Machine Learning Techniques for Early Fault Detection

In this context, machine learning (ML) methodologies are key for early identification of mechanical, electrical, and control-related faults in wind turbine systems. Using historical and live SCADA and sensor data, ML algorithms recognize subtle, multivariable patterns that signal leading indicators of catastrophic failures – patterns that are frequently imperceptible to traditional threshold-based monitoring. (Zheng, et al., 2023). These techniques are broadly categorized into supervised, unsupervised, and residual-based approaches:

5.2.1. Supervised Learning for Fault Classification

In supervised learning, the labeled dataset is fed into correlated input features with specific fault types. The most successful application of the supervised/unsupervised mix has been the classification of gearbox wear stages, bearing spalling, and generator overheating. Supervised systems have achieved remarkable predictive accuracy, with some exceeding 90% on SCADA-labeled fault logs. For instance, Haroon et al. validated the performance of several ML algorithms on SCADA-fed gearbox temperature data and reported early warnings from the analysis. (Haroon, Erfan, Jawad, & Canras, 2020). These models learn a mapping $f: X \rightarrow Y$, where:

- $X \in \mathbb{R}^n$: Input features (e.g., rotor speed, generator temperature, vibration amplitude),
- $Y \in \{0, 1, 2, ...\}$: Fault categories (e.g., normal, bearing failure, generator fault).

Popular supervised methods include:

a. Support Vector Machines (SVM):

SVM aims to find the optimal separating hyperplane between fault classes as shown in equation 22:

$$f(x) = sign(w^T x + b) \tag{22}$$

Where w and b define the decision boundary, SVMs are effective for high-dimensional sensor data, particularly with kernel methods for non-linear faults.

b. Random Forests:

An ensemble of decision trees where each tree votes on the class label. The model estimates the most probable fault type using equation 23:

$$\hat{y} = mode(h_1(x), (h_2(x), \dots, \left(h_T(x)\right) \tag{23}$$

where $h_i(x)$ is the prediction from tree i. Random Forests are robust to noise and class imbalance, making them suitable for SCADA-based fault detection.

c. Artificial Neural Networks (ANNs):

ANNs approximate complex, nonlinear mappings using layered activation functions. For classification, equation 24 can be used:

$$y = \sigma\left(W^{(2)} \cdot \phi(W^{(1)}x + b^{(1)}) + b^{(2)}\right) \tag{24}$$

Where:

- ϕ: Activation function (e.g., ReLU),
- $W^{(i)}, b^{(i)}$: Weights and biases for the layer
- σ: Softmax function for multi-class output

5.2.2. Unsupervised Learning for Anomaly Detection:

Unsupervised methods apply when little or no labeled fault data is available to detect anomalies from the baseline solely by analyzing the input data structure. Such models are used to detect early degradation, misalignment, and other rare fault modes that were previously undefined. These unsupervised models are suitable because accurately classifying high-dimensional machinery data does not require extensive labeled data. (Campoverde-Vilela, del Cisne Feijóo, Vidal, Sampietro, & Tutivén, 2023).

a. Principal Component Analysis (PCA):

PCA reduces dimensionality while preserving variance, helping to detect abnormal patterns as shown in equation 25:

$$Z = XW \tag{25}$$

where:

- $X \in \mathbb{R}^{m \times n}$: Data matrix,
- W: Matrix of eigenvectors of the covariance matrix $\Sigma = X^T X$

Anomalies are detected by examining the reconstruction error or Hotelling's T^2 A statistic which suggests abnormal system behavior when values are high, as shown in equation 26:

$$T^2 = x^T \sum^{-1} x \tag{26}$$

b. Clustering:

Techniques such as k-means cluster normal operational data and flag outliers based on distance from cluster centers. The anomaly score can then be calculated using equation 27:

$$AnomalyScore = \min_k \|x - \mu_k\|_2 \tag{27}$$

where μ_k is the centroid of cluster k. Anomalous observations fall outside the confidence radius of any cluster.

6.5.2.3. Residual-Based Methods for Performance Deviation

Residual-based approaches detect faults by comparing the measured turbine performance with predictions from physics-based or machine learning models (Xu, Lu, Zhai, & Jiang, 2020).

Let:

- $P_{pred}(t)$: Expected power output from a predictive model,
- $P_{actual}(t)$: Measured power from SCADA,
- Residual: $\in_t = P_{actual}(t) - P_{pred}(t)$

Persistent negative residuals (i.e., $\epsilon(t) < -\delta$) may indicate:

- Blade soiling or damage,
- Generator derating,
- Control system failure,
- Sensor drift.

To monitor for sustained underperformance, Equation (28) can be used:

$$\epsilon(t) \in [-3\sigma, +3\sigma] \Rightarrow \textit{Normal Behaviour} \quad (28)$$

Where σ is the residual's standard deviation under healthy conditions.

Alternatively, CUSUM (Cumulative Sum Control) methods can track small but accumulating residual shifts as per equation 29:

$$S_t = \max(0, S_{t-1} + \in_t - k) \quad (29)$$

Fault alerts are triggered when S_t exceeds a threshold, enabling early interventions.

5.3. Remaining Useful Life (RUL) Estimation

Remaining Useful Life (RUL) prediction is the process of estimating the time, cycles, or operational intervals remaining before a turbine component fails or must be replaced. RUL prediction is the base of Condition-Based Maintenance CBM strategies, which have the goal of replacing components before failure, thus reducing unplanned downtime, decreasing operating costs, and lengthening asset lifetime. RUL prediction is most effective for high-value components like gearboxes, bearings, generators, and pitch systems because these tend to demonstrate constant degradation in the SCADA and high-resolution sensor data (Yang & Zhang, 2021).

5.3.1. Regression-Based RUL Modeling

Regression techniques estimate degradation trends by fitting models to condition-monitoring features (e.g., vibration, oil temperature, or acoustic emissions) over time. (Lei, et al., 2018).

a. **Linear Regression Model:** This simple model assumes monotonic degradation and is useful when degradation follows a consistent linear trend and is represented by Equation (30).

$$RUL(t) = \frac{X_{fail} - X(t)}{\frac{dX}{dt}} \quad (30)$$

Where:

- $X(t)$: Degradation signal (e.g., bearing RMS vibration),
- X_{fail}: Defined failure threshold,

- $\frac{dX}{dt}$: Estimated degradation rate from historical data.

b. **Support Vector Regression (SVR):** SVR is adequate for capturing non-linear wear signatures in components like gear teeth or pitch actuators. SVR models capture nonlinear degradation patterns using a kernel-based approach as shown in Equation (31):

$$RUL(t) = \sum_{i-1}^{n}(\alpha_i - \alpha_i^*)K(x_i, x) + b \tag{31}$$

Where:

- $K(x_i,x)$: Kernel function (e.g., radial basis function),
- α_i, α_i^*: Lagrange multipliers,
- x: Input condition data at time t.

5.3.2. Survival Analysis

Survival analysis estimates the probability of survival of a component beyond a given time horizon, incorporating censored data (instances where failure has not yet occurred) (Ozturk, Fthenakis, & Faulstich, 2018).

A Kaplan–Meier Estimator can be modelled as shown in equation 32:

$$\hat{S}(t) = \prod_{i:t_i \leq t}\left(1 - \frac{d_i}{n_i}\right) \tag{32}$$

Where:

- d_i: Number of failures at time
- n_i: Number of components at risk just prior to
- $\hat{S}(t)$: Estimated survival probability at time t.

A Cox Proportional Hazards Model can be modelled as shown in equation 33:

$$h(t|x) = h_0(t) \cdot \exp(\beta^T x) \tag{33}$$

Where:

- $h(t|x)$: Hazard function (instantaneous failure rate),
- $h_0(t)$: Baseline hazard rate,

- x: Covariates (e.g., operational conditions),
- β: Coefficients estimated from data.

5.3.3. Deep Learning for RUL Forecasting

Deep learning models can learn long-term temporal dependencies and complex degradation signatures from raw SCADA and sensor data. (Xiang, Wang, Yang, Hu, & Su, 2021).

a. **Long Short-Term Memory (LSTM) Networks**: LSTMs are effective in modeling delayed or oscillatory degradation trends typical in rotating machinery. LSTM models are a type of recurrent neural network (RNN) designed to capture sequential patterns in time-series data. Given sensor sequence $x_1, x_2, \ldots, x_t$, LSTM outputs a forecasted degradation state $\hat{x}_{t+k}$, from which RUL can be inferred as shown in equation 34:

$$RUL(t) = \min\left\{k \middle| \hat{x}_{t+k} \geq X_{fail}\right\} \quad (34)$$

b. **Convolutional Neural Networks (CNNs):** CNNs can process multi-sensor SCADA images or temporal patterns as 2D arrays and extract spatially correlated degradation features across time and sensor type.

Both models can be trained using historical run-to-failure datasets or simulated degradation profiles generated from digital twins.

5.3.4. Applications

RUL estimation has direct applications in (Vieira, et al., 2024):

a. Proactive Maintenance Scheduling: Maintenance is performed just-in-time, avoiding unnecessary interventions.
b. Asset Life Extension: Components are replaced based on actual condition, not calendar-based intervals.
c. Downtime Reduction: Early detection of imminent failure enables advanced preparation of parts, labor, and tools.
d. Spares Inventory Optimization: Accurate RUL reduces the need for overstocking critical spare parts.

5.4. Benefits of Preventive Maintenance Over Scheduled Maintenance

Maintenance carried out in this way offers several significant advantages over traditional scheduled interventions. Firstly, as maintenance actions are predicated on actual equipment health, rather than pre-determined time-based intervals (Santiago, et al., 2024). Another benefit of this approach is a reduction in seemingly unnecessary servicing and spare part costs. Secondly, as there is a direct correlation between availability and maintenance, operators have to perform fewer maintenance tasks, both the ones completed too early and the ones deemed unnecessary, which cuts down on labor, materials, and inventory management costs. (Frederiksen, Bocewicz, Radzki, Banaszak, & Nielsen, 2024). Thirdly, Turbines that are maintained with a predictive approach experience significantly higher availability and reliability (Frederiksen, Bocewicz, Radzki, Banaszak, & Nielsen, 2024). For one, predictive maintenance activities detect failures early enough to prevent them and improve uptime once the maintenance event is complete. Lastly, with more accurate foresight of a failure, operators will be able to schedule them better. By creating work plans that prioritize faults predicted to occur during periods of low demand, or by conducting multiple planned interventions at different, uncoordinated locations simultaneously, they will save several hours of logistical planning and overhead. (Kandemir, Hasan, Kvamsdal, & Abdel-Afou Alaliyat, 2024).

5.5. Challenges and Practical Considerations

However, despite the potential of these technologies to improve wind turbine reliability, predictive maintenance, and Remaining Useful Life (RUL) estimation, RUL still faces several challenges in the field. The first major pitfall is the quality and availability of well-labeled historical data, which are necessary for the efficient training of machine learning-based models. In most wind farms, SCADA and condition-monitoring historical data are partial, inconsistent, contain missing data points, and lack sufficient examples of rare but critically important failure modes. The latter circumstance prevents the models' development from being generalizable and robust. Moreover, the problem of class imbalance complicates the situation since the frequency of turbine failures is much lower than that of regular operation. This influences a dataset, in which normal states can be severely overrepresented, leading to overfitting or high false-alarm error rates. The solution can be achieved through artificial data synthesis, such as synthetic data generation or weighted learning techniques (Oliveira-Filho, et al., 2024) (Maldonado-Correa, Valdiviezo-Condolo, Artigao, Martín-Martínez, & Gómez-Lázaro, 2024).

Another substantial challenge is model interpretability. In the critical safety category of wind energy systems, operators must be able to integrate the model's predictions into their decision-making processes. While complex deep learning models achieve high performance, they are relatively hard to interpret and appear opaque to an average operator, making them "black boxes." To address this issue, researchers suggest introducing explainable AI approaches, including artificial intelligence, such as the SHAP feature attribution method, rule-based approximators, attention mechanisms, boosting interpretability, and supporting informed decision-making (Astolfi, De Caro, & Vaccaro, 2023) (Letzgus & Müller, 2024).

5.6. Case Study: Predictive Maintenance and Condition Monitoring Using NREL's Gearbox Reliability Collaborative Data

A basic predictive maintenance and condition-monitoring research program for wind energy was conducted at the National Renewable Energy Laboratory (NREL) through the Gearbox Reliability Collaborative (GRC), sponsored by the United States Department of Energy (DOE). A key publication of the GRC is the technical report titled Wind Turbine Drivetrain Condition Monitoring During GRC Phase 1 and Phase 2 Testing, which provides comprehensive field measurements and test stand test results for drivetrain components (i.e., gearboxes and bearings) under realistic operational loads. In this work, the context, motivation for testing in GRC, instrumentation system, and condition monitoring systems layout are discussed alongside the performance results that were obtained throughout phase 1 and phase 2 of the GRC testing campaign, covering also some practical challenges faced during field deployment and sharing some findings regarding signal processing and algorithm performance for anomaly detection. (Sheng, et al., 2011).

As an addition to this work, the Wind Turbine Gearbox Condition Monitoring Round Robin Study – Vibration Analysis offers a technical comparative study of different condition-monitoring concepts based on vibration and sensor data from both dynamometer and field tests. The Round Robin study assesses the capability of typical vibration analysis methods at detecting response measurements from realistic gearbox faults and allows comparison of monitoring performance using different analytical tools. (Sheng, 2012).

These technical reports provide a practical evidence base for predictive maintenance systems by grounding condition-monitoring frameworks in operational measurements and controlled physical tests rather than synthetic datasets. By documenting specific monitoring approaches, comparative performance of algorithms, and absolute failure or degradation events in drivetrain hardware, they provide benchmarks for model accuracy, sensor selection, and early-fault-detection lead times that are directly applicable to operational wind farms and digital twin–assisted maintenance systems.

5.7. Summary

In section 6.5, an overview of predictive maintenance and condition monitoring in wind turbine systems was provided, focusing on data-driven methodologies for predicting equipment failures and scheduling maintenance. Operators can analyze component health and identify anomalies, as well as estimate Remaining Useful Life (RUL) for crucial assets such as gearboxes, bearings, and generators by combining SCADA, sensor, and environmental data. In that section, we presented machine learning methods (i.e., supervised, unsupervised, and residual-based) for early fault detection, as well as for RUL forecasting regression analysis and survival analysis, followed by deep learning techniques (LSTM, CNN) for studying RUL. Practical applications were demonstrated through real-world use cases, including the NREL Gearbox Reliability Collaborative, with a focus on operational gains and challenges, such as data quality, class imbalance, and model interpretability.

Predictive maintenance offers clear advantages over non-data-driven scheduled programmes - less downtime, optimal spares holding, extended asset life, and proactive fault management. Such advantages depend on reliable monitoring and resilient modeling, which are built directly upon our AI and ML methods described in section 6.4.

Section 6.6 moves from prediction to digital twin architectures, and computational models of physical turbines reflect their real-time behaviour during operation. With their high level of integration, digital twins provide a uniform foundation for monitoring, fault detection, predictive maintenance, and optimized control, bridging the gap between AI-based analytics and real wind turbine operation. This facilitates not only safer testing of control strategies and degradation scenarios, but also better decision-making for wind farm asset management.

6. DIGITAL TWIN ARCHITECTURE FOR WIND TURBINES

The idea of the digital twin was introduced in product lifecycle management. It was first formally defined by Grieves as the convergence of a physical system, its virtual model, and a data stream that connects them over the life cycle of an asset. (Grieves, 2014). This basic definition emphasizes the importance of continuous data sharing and life-cycle collaboration, distinguishing digital twins from static simulation models. Applied to the domain of wind energy, Grieves' framework for digital twins establishes the conceptual structure for integrating physics-based turbine models with operational data and AI analytics to enable monitoring, control, and maintenance decision-making.

In the context of wind turbines, a digital twin is an instantaneous digital copy of a turbine or an entire wind farm that fully reproduces its physical state, structure, and motion. This digital twin connects real-time data feeds from SCADA systems, sensors, and weather information providers with physics-based and machine-learning models. Its primary objective will be to facilitate real-time monitoring, enhanced diagnostics, and prognostics by computing turbine performance over an operational range. By establishing a reliable correlation between the physical turbine and its virtual twin, it is easier for operators to understand how their system behaves sub-optimally and how they can optimise control and maintenance actions. (Haghshenas, Hasan, Osen, & Mikalsen, 2023).

The diagram below (Figure 3) shows a classification of models based on the level of integration between the digital object and its physical counterpart. The chapter focuses on the digital twin model. In addition, Figure 6.4. shows the digital twin schematic of a wind turbine, including the physical system, its virtual replica, and the supervising entity. The components of digital twins and their appplications are explained in detail in this section.

Figure 3. Classification of models depending on the level of integration between the digital object and its physical counterpart (Leon-Medina, Tibaduiza, Parés, & Pozo, 2025).

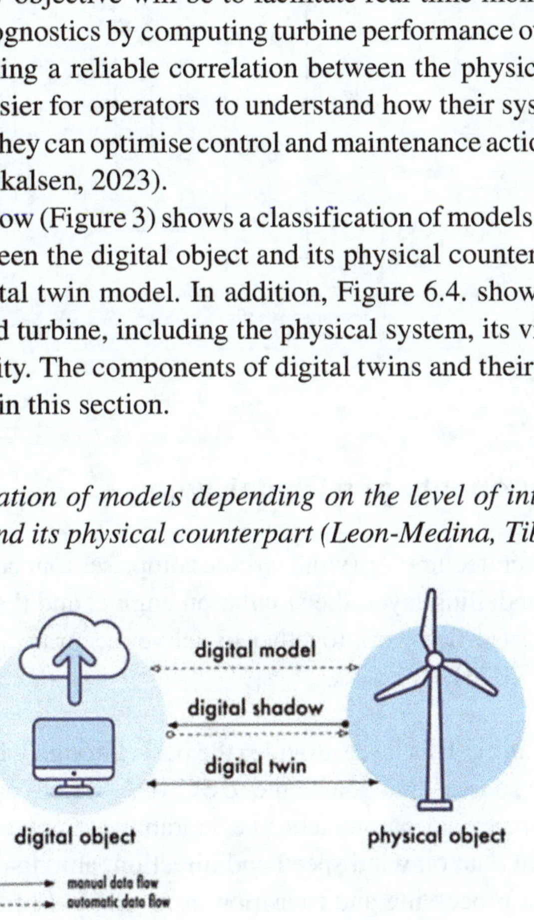

Figure 4. Digital twin schematic of a wind turbine, including the physical system (left), its virtual replica (right), and the supervising entity (De Kooning, Stockman, De Maeyer, Jarquin-Laguna, & Vandevelde, 2021).

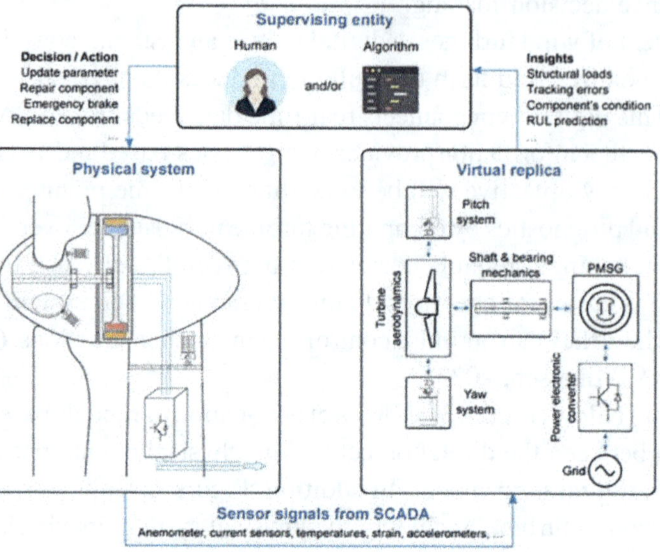

6.1. Core Components of a Digital Twin

A digital twin architecture for a wind turbine comprises four components: the data intake layer, the modelling layer, the simulation engine, and the virtualization and interface layer, all of which work together to deliver accurate, actionable insights. (Stadtmann, et al., 2023):

- Data Intake Layer: This layer provides the basis through which to integrate the multiple data sources that contribute to SCADA system outputs, sensor measurements for critical parameters like vibration, temperature, pressure, and environmental data on wind speed and direction, among others. Continuous, seamless data processing and transitioning in and out of the digital twin ensure the twin continues to represent the turbine's actual state in real time.
- Modeling Layer: This layer is concerned with the use of physics-based models and machine learning algorithms to represent the mechanical and aerodynamic behavior of the turbine. Here, physics-based, detailed models perform mathematically rigorous simulations of the turbine's essential dynamics. In contrast, machine learning models perform large-scale analytics

and pattern recognition to accurately predict conditions and support condition monitoring.

- Simulation Engine: This layer permits the digital twin to simulate "what-if scenarios of various natural disasters like hurricanes and component degradation, and more, in addition to simulation and testing of maintenance strategies and their outcomes.
- Visualization and Interface: It is an interactive layer that allows system operators to monitor system performance results and get predictive alerts on performance on real-time simulation and testing plans. This facilitates informed decision-making and efficient management of wind farm operations.

In the diagram below (Figure 5), the components required to estimate remaining useful life for a wind turbine are shown. Figure 6 shows a conceptual framework of a digital twin for a wind turbine. The physical asset consists of sensors and IoT devices. The digital twin platform consists of three main fronts: big data and analytics, simulation and property modeling, and visualization. Data is sent from the physical assets to the digital twin platform, where information and processes are then sent back to the physical assets.

Figure 5. Digital Twin framework for continuous remaining functional life estimation in wind turbine drivetrain components (Mehlan, Keller, & Nejad, 2023).

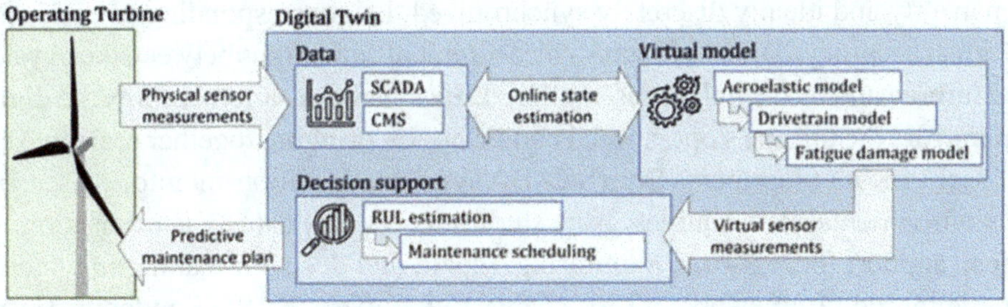

Figure 6. Conceptual framework of digital twin for a wind turbine (Kandemir, Hasan, Kvamsdal, & Abdel-Afou Alaliyat, 2024).

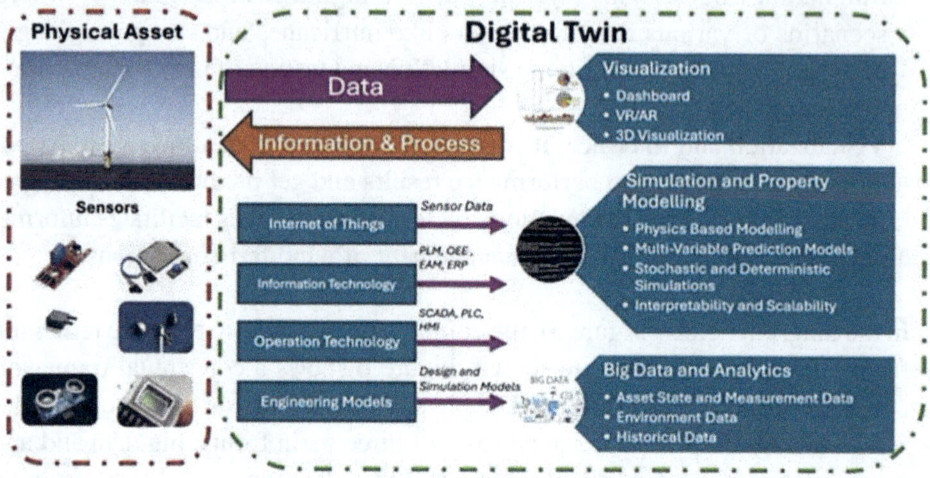

6.2. Data Synchronisation and Model Updating

One of the basic preconditions for a useful digital twin is that both should be permanently and ideally discretely synchronized, i.e., corresponding to each other with great accuracy, data-wise, across measurement campaigns between the physical wind turbine and its digital representation. Data synchronisation ensures the digital twin mirrors the turbine's operational conditions by bringing together real-time and near-real-time measurements from SCADA systems, condition-monitoring sensors, and environmental data sources. New studies assert that data synchronisation is a critical support for real-time monitoring, repair, and decision-making in wind energy digital twin frameworks. (Tao, Zhang, Liu, & Nee, 2019) (Kandemir, Hasan, Kvamsdal, & Abdel-Afou Alaliyat, 2024).

In practice, wind turbine systems receive data streams from different sources with diverse sampling rates, latencies, and signal reliabilities. SCADA data are usually obtained at low sample rates (e.g., 1–10 min), whereas condition monitoring systems provide high-frequency measurements, such as vibration, acoustic emission, and electrical signals. Optimal data synchronisation, therefore, necessitates temporal alignment, resampling, and filtering of the different sources so that they are consistent before being injected into the digital twin environment. It is well known that these challenges are observed in the implementation of digital twins for a large wind farm, depending on their roles. (Stadtmann, et al., 2023) (Leon-Medina, Tibaduiza, Parés, & Pozo, 2025).

Model operation is essential for keeping the model updated and accurate over time. As wind turbines age and components wear, existing discrepancies between model predictions and observations are further exacerbated by changes in environmental conditions. To mitigate this, the digital twins implement an adaptive model-updating algorithm that periodically recalibrates physics-based model parameters and data-driven model weights using online operational data. Standard approaches, such as parameter estimation, state observers' design, Bayesian updating, and retraining based on machine learning, are used to decrease the prediction error while maintaining the fidelity of the model. (Rasheed, San, & Kvamsdal, 2020).

In the context of monitoring and diagnosis, synchronized data and online re-estimated models enable residual-based analysis of temporal trends in differences between measurements and predictions. Continued residuals may indicate abnormal operation, sensor drift, or component degradation. Transient deviations are generally caused by short-term disturbances (e.g., remote transfer centers closing or noise). An adaptive model-updating-based residual monitoring scheme has been reported to enhance early anomaly detection performance in wind turbine drivetrain components and structural elements. (Kandemir, Hasan, Kvamsdal, & Abdel-Afou Alaliyat, 2024).

By combining continuous data synchronisation with adaptive model updating, digital twin architectures maintain a high degree of correspondence between physical and virtual systems. This capability is essential for enabling accurate condition monitoring, control optimisation, and predictive maintenance, ensuring that digital twins remain actionable and reliable decision-support tools throughout the operational lifecycle of wind turbines. (Tao, Zhang, Liu, & Nee, 2019) (Leon-Medina, Tibaduiza, Parés, & Pozo, 2025).

6.3. Digital Twin–Based Condition Monitoring and Control Optimisation

Digital twins enable a tightly integrated framework in which condition monitoring and control optimisation are treated as interconnected processes rather than independent functions. By continuously synchronising real-time operational data with physics-based and data-driven models, the digital twin provides a comprehensive representation of turbine health, performance, and dynamic behaviour. This comprehensive picture serves as the foundation for intelligent decision-making in wind turbine operation and maintenance throughout its lifetime, from early fault detection to adaptive control strategy optimization. (Zhang & Zhao, 2023).

With a focus on condition monitoring, the digital twin fuses diverse data streams from SCADA systems, vibration and strain sensors, temperature measurements, and environmental conditions to provide a real-time assessment of critical turbine components, including rotor blades, drivetrain elements, and support structures. Such deviations could be so-called "unclear" - partial repetitive responses that may indicate abnormal functioning or progressive deterioration at an early stage. Recent developments in digital twin technology have shown the potential of this approach for health monitoring and aerodynamic performance, which can be effectively realized when physics-informed or reduced-order modeling enables real-time operation. (Marykovskiy, Clark, Deparday, Chatzi, & Barber, 2024).

Beyond fault detection, digital twin–based condition monitoring supports advanced diagnostics and prognostics by capturing the evolving relationship between operational conditions and component degradation. Predictive digital twin platforms for offshore wind farms have shown how continuous data assimilation and model updating can be used to estimate degradation trends and remaining useful life under variable loading and environmental conditions. These capabilities significantly enhance situational awareness and enable maintenance planning to transition from reactive to predictive strategies. (Branlard, Jonkman, Brown, & Zhang, 2024).

Importantly, the outputs of the condition monitoring layer are used directly to control optimisation in the digital twin environment. Health measurements, degradation metrics, and state estimates produced by the digital twin, feeding directly into control decision-making, mean operational strategies can evolve according to the present and expected states of turbine components. For instance, control setpoints for pitch, torque, or yaw can be tuned to alleviate overloading of aging drivetrain components or minimize fatigue due to high turbulence, while simultaneously compromising energy capture targets with respect to asset life. (Zhang & Zhao, 2023) (Branlard, Jonkman, Brown, & Zhang, 2024).

Digital twins also enrich control optimisation through virtual commissioning and scenario-based assessment of control solutions. In the virtual setting, different control algorithms can be simulated across an extreme range of operating conditions, including high-wind events, wake interference, and component damage, without risking real-world assets. Recent research has shown that diagnostic digital twins with real-time state estimation and an anomaly detector can also be used to validate adaptive and predictive control algorithms before deployment in the field.

Further, in more sophisticated applications, digital twins are embedded in a closed-loop optimization scheme, with monitoring-generated outputs iteratively updating the internal models and affecting the control tactics in near real time. This enables predictive control strategies that account for the life cycle and adapt to variations in conditions, system dynamics, and component states. By explicitly integrating condition monitoring with control optimisation, digital twins enable a shift from static, rule-based turbine operation to intelligent, data-driven control strategies in modern wind energy systems. (Zhang & Zhao, 2023).

6.4. Applications of Digital Twins

Digital twins are dynamic virtual replicas of turbines that enable a range of essential applications to advance operational safety, efficiency, and reliability. They include, first, simulation of extreme scenarios to safely test one's turbine's resilience to severe weather events or to analyze the behavior of a critical component. Digital twins allow operators to test the response of their infrastructure to distressing conditions, such as gusts, storms, or a vital component failure, without physically damaging the turbines. (Kandemir, Hasan, Kvamsdal, & Abdel-Afou Alaliyat, 2024). They also provide operators with a contingency plan for such events and increase the turbine's overall readiness by understanding how it responds in critical scenarios.

Secondly, virtual commissioning: a digital twin provides an environment for developing, testing, and validating control logic or control strategies. Control algorithms and logic can be validated on a digital twin before being applied to physical turbines. The opportunity for virtual commissioning accelerates the development of new control logics, reduces the risk of malfunctions during rollout, and increases expertise in controlling the turbine. (Stadtmann, et al., 2023).

Third, real-time health monitoring and predictive analytics: a digital twin is constantly fed with real-time sensor data streams from SCADA and assesses the current health status of all components. When coupled with machine learning and physics-based models, DTs help predict the remaining valid lifetime of an element and faults, enabling operators to act in advance to minimize unforeseen downtime and extend a turbine's lifespan. (Leon-Medina, Tibaduiza, Parés, & Pozo, 2025).

6.5. Implementation Challenges and Practical Considerations

While digital twins hold great promise for advancing wind turbine monitoring, control, and maintenance, there are also several technical and organizational challenges in realizing them in practice. These challenges need to be addressed for digital twin platforms to remain robust, scalable, and economically feasible solutions for real-world wind farms.

One of the key barriers to deploying digital twins is the availability and quality of data. The development of wind turbine digital twins is based on heterogeneous data feeds from SCADA systems, condition-monitoring sensors, and environmental observations, which are often plagued by missing values, sensor drift, asynchronous communication, and uneven sampling rates. Bad data quality can degrade model accuracy and lead to unreliable condition assessments if not well handled through preprocessing, filtering, and validation techniques. (Stadtmann, et al., 2023) (Leon-Medina, Tibaduiza, Parés, & Pozo, 2025).

Further challenges arise from computational complexity and scalability, particularly for large wind farms or high-fidelity digital twins that can run close to real time. Physics-based simulations that involve ongoing data assimilation and machine-learning inference can be computationally intensive. When it comes to scaling, practical implementations require stringent tradeoffs between computational efficiency and model fidelity as well as the use of edge computing, model reduction techniques or hierarchical digital twin architectures (Rasheed, San, & Kvamsdal, 2020) (Kandemir, Hasan, Kvamsdal, & Abdel-Afou Alaliyat, 2024).

Interoperability with incumbent systems is another key driver. In many operational wind farms, existing SCADA systems and proprietary control solutions were not even intended to be combined with digital twins. To attain frictionless interoperability, standardised data interfaces and middleware solutions are indispensable, as is thoughtful adaptation of available operational processes. Without this combination, digital twins may become stand-alone analysis tools rather than becoming fully integrated decision-support systems. (Fuller, Fan, Day, & Barlow, 2020).

As DTs enable bidirectional data flow from physical assets to virtual models, cybersecurity and other data governance concerns become more relevant. Because wind turbines continuously generate operational data and control information, they are at risk of cyberattacks; therefore, a robust security architecture, access control mechanisms, and adherence to industrial cybersecurity standards should be implemented. This becomes of paramount importance , especially for offshore wind farms and geographical spread generators. (Alhamam, Hafizur Rahman, & Aljughaiman, 2025).

Last but not least, organisational and economic factors impact the successful deployment of digital twin technologies. The development and operation of digital twins require an interdisciplinary skill set spanning control engineering, data science, and domain-specific knowledge in wind turbines. In addition, the digital twin cost-benefit argument should consider performance benefits and factor in implementation, data infrastructure costs, and long-term maintenance. Therefore, phased roll-out approaches -beginning by using component-level digital twins or monitoring-focused digital twins – are commonly advised to reduce risks, and to prove the value incrementally (Leon-Medina, Tibaduiza, Parés, & Pozo, 2025).

The solution to these implementation challenges is crucial for advancing digital twins from early prototypes to ready-to-use, industrial-grade technologies that support the reliable and sustainable operation of wind turbines. The following subsection provides depth on the challenges of wind turbine monitoring, deployment, and integration.

6.6. Case Study: Digital Twin Architecture for Floating Offshore Wind Turbines Validated Using a Full-Scale Prototype

A well-established and rigorously validated example of a digital twin architecture for wind turbine systems is presented by (Branlard, Jonkman, Brown, & Zhang, 2024), who developed and experimentally validated a digital twin for a floating offshore wind turbine using data from a full-scale prototype. This study represents one of the most comprehensive demonstrations of how physics-based digital twin frameworks can be deployed and verified under real operational conditions, thereby providing an important benchmark for digital twin architectures in wind energy applications.

The proposed digital twin integrates high-fidelity aero-hydro-servo-elastic models with measured operational data to maintain synchronisation between the physical turbine and its virtual counterpart. Environmental excitations, including wind and wave conditions, are combined with SCADA measurements and structural response data to enable continuous state estimation. Through this hybrid modelling approach, the digital twin can estimate internal system states, such as structural loads, platform motions, and fatigue damage accumulation, that are not directly measurable during regular turbine operation. (Branlard, Jonkman, Brown, & Zhang, 2024).

One of the distinctive architectural features of this work is verification at full scale, which, unlike many digital twin-related works that centre solely on computational modeling or laboratory testing, is carried out. The authors report a strong correlation between digital twin predictions and measured responses across different operational conditions, providing a quantitative assessment of the model's accuracy and robustness. This validation stage is crucial for safety-critical wind scenarios, where trust in the digital twin's outputs is a key requirement for operational decisions.

From a system design perspective, the work demonstrates several principles for deploying digital twins on wind turbine systems. This includes modular physical model integration, ongoing data assimilation for state updating, and a clear separation of sensing, modelling, and decision-support layers. These architectural aspects promote the scalability and flexibility of the digital twin framework, enabling its application across different turbine designs and operating conditions.

Crucially, the verified digital twin enables numerous lifecycle operations, such as structural health monitoring (SHM), fatigue life prediction, control optimization, and maintenance planning, for offshore structures where inspection and maintenance tasks are expensive or logistically intensive. Grounding digital twin predictions in actual measurement data, the research offers a viable approach to integrating AI-based analytics with physics-based models to improve turbine reliability and performance further.

Generally, this case study demonstrates that digital twin architectures for wind turbines are no longer theoretical concepts but can be implemented and validated at full scale. Therefore, it serves as a solid reference to support the architectural ideas presented in this chapter and further validate that digital twin–based monitoring and control integrated with wind energy systems is practically feasible.

6.7. Summary

In Section 6.6, it has also been noted that the DT provides a unifying foundation for WT monitoring, predictive maintenance, and control optimisation. By integrating SCADA and sensor data in real time with physics-based/dataset-driven models, digital twins enable continuous condition monitoring, early fault warning, RUL prediction, and user-adjustable operational tactics. Applications range from virtual commissioning and scenario simulation to predictive maintenance and control optimisation, paving the way for increased reliability, efficiency, and asset service life.

Nevertheless, despite these potential benefits, the real-world deployment of digital twins in wind farms remains challenging. The quality and heterogeneity of the data, complexity, compatibility with legacy systems, security, and preparedness all contribute to the success of a pragmatic deployment in a digital twin. These challenges are critical to overcome if the digital twins' capacity potential is to be realised through operational benefits.

These deployment and integration challenges are further discussed in the next section (6.7), which offers a critical review of practical considerations to ensure the successful implementation of advanced monitoring and digital twin frameworks for modern wind energy systems.

7. WIND TURBINE MONITORING, DEPLOYMENT, AND INTEGRATION CHALLENGES

Despite the tremendous potential of artificial intelligence and digital twin technologies to revolutionize the routine monitoring and maintenance of wind turbines, these concepts remain far from deployment at a practical scale in real wind farms. Many solutions have been proven successful in small-scale deployments or pilot studies. However, their implementation is unfeasible for turbine fleets numbering in the thousands and spread across diverse geographic territories. The purpose of this section is to identify the significant practical challenges of the integration of AI and digital twin systems, such as data standardization, communication infrastructure, and operational alignment, and discuss the critical steps of overcoming these hurdles to reach the deployment stage (Clifton, et al., 2023).

7.1. Data Quality and Accessibility

DTs and AI systems for wind turbines rely heavily on high-quality, consistent, and accessible operational data. In reality, uncertainties in sensor accuracy and calibration procedures, long-term exposure to other environmental stressors, and variability exist and can degrade the reliability of models. (Stadtmann, et al., 2023). Incomplete, noisy, and inconsistent SCADA data may arise from communication failures, sensor failures, and harsh environments; a new nomenclature identifies missingness as band miss and feature miss, and prescribes GAN-based imputation architectures tailored to spatial-temporal manifestations when handling multimodal data. (Yang, Huang, Mo, & Luo, 2025). Preprocessing and validation steps, including Dynamic ICA, DEWMA-based fault detection, noise filtering, and anomaly screening, are critical; they must ensure data integrity before the model learning phase. (Kini, Harrou, Madakyaru, & Sun, 2023). Significant data challenges, including data volume, heterogeneity, veracity, standardized schemas, and metadata schema and metadata standardization, are more complex to overcome as AC/digital twin systems scale across multiple wind farms. (Clifton, et al., 2023).

7.2. Communication and Latency

Onshore wind is becoming increasingly important as a renewable energy source and has been installed in numerous countries. However, the industrialisation of AI and digital twin-based solutions in wind farms would be impossible without a corresponding communication infrastructure. In fact, the extensive geographical coverage of wind turbines that provide SCADA readings, along with a variety of sensor measurements and environmental details, implies frequent monitoring of different wind properties at each location. This data needs to be uploaded to a cloud or a centralized control data center. It is expected that data transmission latency is determined by available bandwidth. However, such latency, stemming from geological differences in wind farms or the presence of nearby cellular towers, can compromise real-time control or data-based prediction. (Kandemir, Hasan, Kvamsdal, & Abdel-Afou Alaliyat, 2024).

Even modest delays (e.g., >100 ms) can negatively impact:

- Dynamic load balancing,
- Turbine curtailment decisions,
- Condition monitoring response times.

This aspect of the problem is vital to the urgency of the situation; for example, it may involve responding to grid disturbances or natural disasters by changing turbine settings. In that case, without timely, high-quality data transmission, the full potential of accurate predictions and adaptive responses that advanced analytics provides, such as predictive maintenance and control, may not be realized. (Mohamed & Chul-Hwan, 2017).

To tackle the issues mentioned above, edge computing-based approaches are becoming more popular. Edge computing performs local processing near or on the turbine, reducing reliance on high-bandwidth communication channels and lowering latency, resulting in faster preliminary analysis and response times. Additionally, employing optimized communication protocols, such as lightweight messaging systems and priority-based data routing, can achieve even greater data transport efficiency and trustworthiness across the wind farm. (Ma, Yuan, & Chen, 2023).

In addition, optimized communication protocols improve efficiency:

- Lightweight message formats (e.g., MQTT, CoAP) reduce overhead,
- Priority-based routing ensures time-sensitive alerts (e.g., overspeed warnings) are transmitted before routine data,
- Quality-of-Service (quality of service) mechanisms enhance reliability under congested conditions.

The diagram below (Figure 6.7.) shows a proposed architecture for a cyber-physical wind energy system. The architecture includes a wind farm power system, a data acquisition and monitoring layer, a communication and network layer, and the application layer. The diagram also contains some abbreviations, including SCADA (supervisory control and data acquisition), CCTV (closed-circuit television), VOIP (voice over internet protocol), IED (intelligent electronic device), WTG (wind turbine generator), WTC (wind turbine controller), HV (high voltage), and Met. Mast: meteorological mast.

Figure 7. Proposed architecture of the cyber-physical wind energy system (Ma, Yuan, & Chen, 2023).

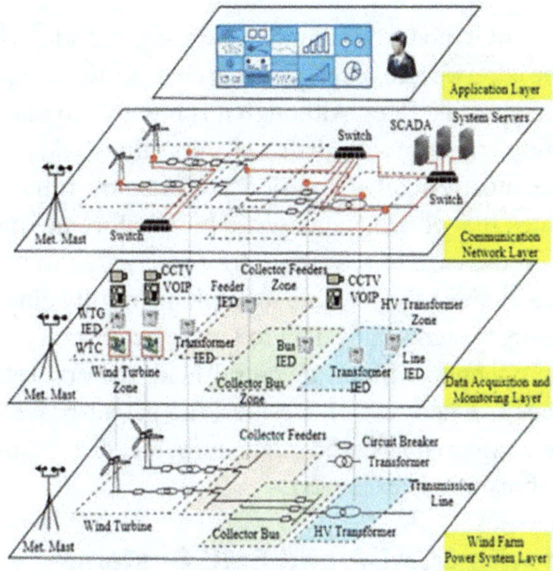

7.3. Model Drift and Updating

Model drift is the gradual decrease in model quality over time as the turbine, environment, or operational context evolves. For example, natural sounds, wear on components, sensor aging, seasonal effects, or changes in grid regulation can create new patterns not present in the historical data used to train AI models. If unattended, model drift decreases the performance of failure detection, power production prediction, and maintenance planning. (Zhu, et al., 2024).

- To detect drift, operators can monitor residual trends whereby persistent error indicates model mismatch as shown in equation 35:

$$\epsilon_t = y_t - \hat{y}_t \tag{35}$$

- The drift can also be detected using the Population Stability Index (PSI), whereby a PSI > 0.25 signals significant feature drift by using equation 36:

$$PSI = \sum_i (P_i - Q_i) \cdot \ln\left(\frac{P_i}{Q_i}\right) \tag{36}$$

- Prediction confidence: A drop in classifier certainty or rise in output entropy can reveal concept drift.

To stay reliable, regular model retraining, validation, and adaptive learning should be implemented. The latter ensures explicitly that changes to the model closely match the currently operated turbine fleet. Although it is possible to monitor performance metrics and retraining triggers with drift detection thresholds based on a model update's results, the most effective solution is to deploy fully automated model management pipelines that conduct such validation before the model is deployed. (Zhu, et al., 2024). In the case of retraining frameworks, as in most cases, even at the edge, it would be technically possible to implement fully autonomous systems. For instance, retraining systems can adapt to local circumstances, such as seasonal cycles or aging components, without a central load, thereby retaining prediction accuracy. (Ma, Yuan, & Chen, 2023). Besides, the patent-based industry systems have logic to update comparisons based on the most recent conditions and to retrain when a predefined threshold is exceeded.

7.4. Integration with SCADA and EMS Systems

Integrating AI and digital twins with current Supervisory Control and Data Acquisition and Energy Management System platforms poses significant challenges. In most cases, the SCADA/EMS platforms are developed on antiquated infrastructure with predetermined data formats, codified communication protocols, and limited capacity to incorporate external analytics products. Therefore, aligning real-time insights, predictions, and recommendations generated by AI and digital twin products can be complex and require significant customization.

A primary concern is compatibility and interoperability – ensuring that AI-generated data, alarms, and control recommendations can be understood and executed by SCADA/EMS without introducing errors or delays. Variations in data structures, naming conventions, and protocol standards may otherwise create integration bottlenecks. (Jørgensen & Ma, 2025).

Given these challenges, many common strategies include modular system design, the adoption of standardized APIs, and the use of middleware solutions to convert between AI/digital twin outputs and SCADA/EMS inputs. Given that these strategies scale and cut down on integration expenses, wind farm operators may begin utilizing levels of advanced analytics even without completely reworking their current control systems (Mchirgui, Quadar, Kraiem, & Lakhssassi, 2024).

The diagram below (Figure 6.8.) shows the SCADA and EMS architecture for a wind farm. The architecture shows that two wind farms can be controlled from a single remote control center via a wide-area network to their respective on-site control rooms.

Figure 8. Wind farm SCADA and EMS Architecture (Singh, Sharma, & Govindarasu, 2020).

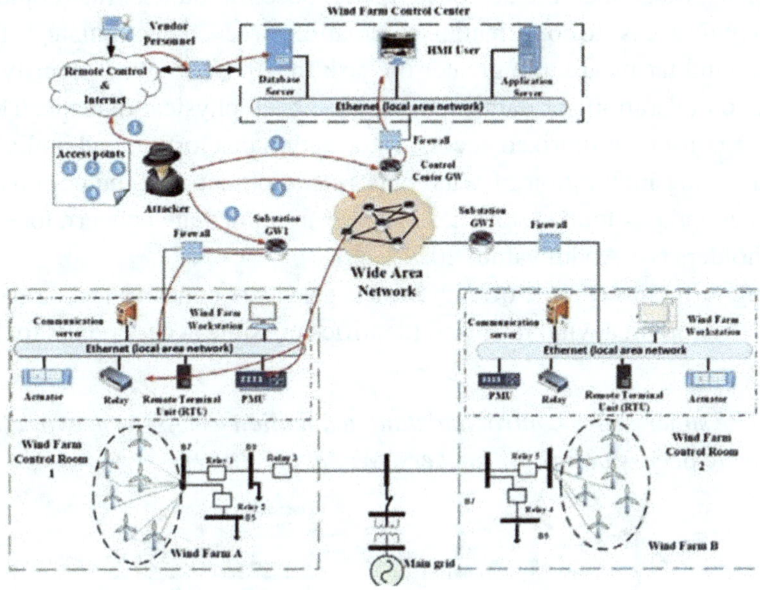

7.5. Summary

Section 6.7 outlines the real-world obstacles to implementing AI and digital twin technologies in wind farm operations. These tools offer the potential for improved monitoring, predictive maintenance, and control optimization, but how to address key challenges such as data quality and availability, communication latency, model drift, and integration with legacy SCADA/EMS platforms remains an open question. Overcoming these challenges will require the development of robust preprocessing, edge computing, adaptive model updating, and modular integration approaches to enable real-time, reliable, and scalable operations across wind turbine fleets worldwide.

The following section (6.8) elaborates on this theme in the context of cybersecurity, considering how the bidirectional data flows and the systems-of-systems nature of digital twin-enabled wind farms give rise to new risks. Securing operational integrity and confidence in the collected data is crucial to fully exploit advanced monitoring and predictive control in current wind energy systems.

8. CYBERSECURITY CONSIDERATIONS

The increasing complexity and digital sophistication of wind turbine systems, partly driven by AI, IoT, and digital twins, have already significantly improved efficiency and predictability. However, increased connectivity also means higher cybersecurity risks, since the sector is now exposed to attacks that could disrupt operations, gain access to confidential information, or even manipulate it. The more connected wind farms are, the greater the risk they pose to cybersecurity, including attacks on operations or data exposure, and even physical damage. Therefore, protection against unauthorized access, information disclosure, and malicious acts in this expanding integration of wind energy equipment into the communication sphere or networked cloud system has become an immediate concern for operators and stakeholders (Al Abdulwahid, 2024).

The diagram below (Figure 6.9.) shows a supervisory control and data acquisition (SCADA) network in a wind farm and identifies possible access points for hackers.

Figure 9. The supervisory control and data acquisition (SCADA) network in a wind farm, showing access points for hackers (Wu, et al., 2019).

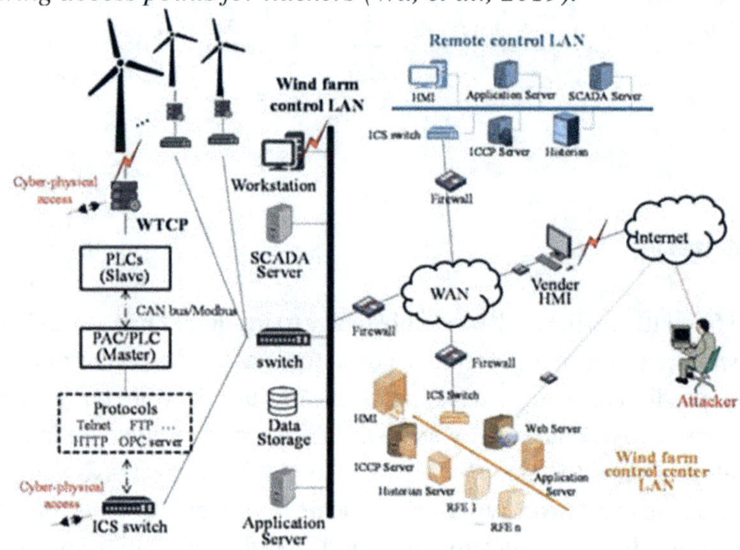

8.1. Common Cyber Threats

The ever-increasing cyber threat landscape poses a growing threat to the digital infrastructure of next-generation wind turbines and wind farms, potentially affecting safe operation, data integrity, and decision support. In such AI- and digital twin–enabled systems, cybersecurity threats should not be limited to data confidentiality but must also encompass data integrity and control integrity needed for effective monitoring, prediction, and control.

From a cyber threat modelling perspective, many of the risks faced by wind turbine systems can be classified using well-established models such as STRIDE (Spoofing, Tampering, Repudiation, Information disclosure, Denial of service, and Elevation of privilege) and cyber–physical attack taxonomies developed for industrial control systems. These models are particularly relevant for wind energy systems, where cyber intrusions can directly lead to physical consequences.

One of the most significant threat classes is attacks related to data tampering and integrity, respectively corresponding to the Tampering and Spoofing classes under the STRIDE model. In these attacks, malicious actors intentionally provide modified sensor readings, SCADA logs, or input data to AI algorithms and digital models. By tampering with operational data, integrity attackers can degrade the performance of machine learning models and digital twin simulations, leading to incorrect fault detection, misleading maintenance scheduling, or less efficient, possibly unsafe control actions in cyber–physical systems (CPS) such as wind turbines. These integrity attacks are particularly harmful because misguiding the process output can directly influence physical operation of the device, turbine's life cycle loading and availability (Yang, et al., 2024).

Unauthorized access and control overhead is the second significant class of threads. These classes of threats are Elevation of Privilege and Unauthorised Command Injection, which are predominantly found in the ICS threat model. An attacker might be able to exploit a turbine by sending malicious control commands, disrupt coordination amongst turbines, and gain access to sensitive operating data if they can break into the Supervisory Control and Data Acquisition (SCADA) systems or energy management nodes. In the rare case that an attack command overcomes all defense mechanisms and operational constraints, such attacks could lead to physical damage to hardware, continued losses due to unavailability, financial damage, or even hazardous conditions. (Datta & Rahman, 2017).

An additional sophisticated attack vector , explicitly mentioned in both STRIDE and ICS threat taxonomies, is spoofing, a concern for wind turbine applications. In such conditions, intruders forge trustworthy sensors, controllers, or communication nodes to introduce fake data flows to disguise themselves as trusted nodes. For example, if wind speed and related sensors, vibration sensors, or temperature sensors are spoofed, AI-based monitoring systems could trigger unnecessary shutdowns, bury real fault alarms, or cause unsafe operating conditions. These attacks undermine the trustworthiness of sensor data and can cause significant reliability degradation in digital twin–based diagnosis and control applications. (Zhang, Lin, & Wang, 2020).

Apart from targeted attacks, denial-of-service (DoS) attacks and communication breakdowns also endanger vulnerable wind farm infrastructure with extensive interconnections. Such attacks can affect data availability, real-time monitoring, and coordinated control at turbine- or wind-farm levels, mainly when digital twins depend on continuous information exchange for state estimation and prediction.

Given the increasing digitisation, connectivity, and reliance on AI-driven decision-making in wind energy systems, systematically recognising and mitigating these cyber threats through established threat modelling frameworks is essential. Such an approach enables the identification of attack surfaces, prioritisation of risks, and design of resilient AI and digital twin architectures that preserve system integrity, operational continuity, and safety.

8.2. Cybersecurity Safeguarding Strategies

The further spread of these advanced winds of change makes the development of strong cybersecurity a necessity rather than an option, as exploits in wind turbine and farm control systems are becoming increasingly common. In industrial control system (ICS) environments, security must focus not only on confidentiality but also on operational safety, availability, and integrity. A key to these mitigations is the use of fail-safe and fail-secure design principles, in which the turbines automatically move to a safe, stable state when cyber anomalies, data integrity attacks, or control system failures are detected. Preventing safe operation and physical destruction is achieved when a turbine control system detects manipulation or tampering by separating affected subsystems, turning off non-functional control functions, or returning to the certified default control state. (Cid Montoya, Rubio-Medrano, & Kareem, 2024).

Apart from fail-safe operation, network segmentation and zoning, as recommended for ICS environments, are a crucial line of defence. Isolating a wide range of turbine controllers, SCADA networks, digital twin platforms, and enterprise IT systems with security zones and DMZs slows down attackers' lateral movement and the potential spread of cyber contamination. Such segmentation is especially vital for AI and cloud-based digital twins, where data exchange must be strictly controlled to prevent unauthorized access to real-time operational systems.

Encryption of the collected data is necessary to protect it against unauthorized access during and after the operation of a given health record system. By using secure communication protocols like Transport Layer Security and high-level encryption standards such as AES-256, data transmitted between control centers or the cloud and turbines is shielded from eavesdropping and tampering with messages exchanged. What is more, in encrypted storage or storage chip level, our risk-sensitive ops logs or AI training data set privacy is guaranteed from unauthorized access or leakage (Alshammari, 2025).

Access to the system should be restricted to authorized entities, which can be achieved only with robust authentication systems. Multi-factor authentication is often recommended as passwords working in tandem with secondary factors, such as hardware tokens or biometric verification, demonstrate a 76% improvement in odds against credential theft. Moreover, digital certificates and Public Key Infrastructure enable device-to-device authentication, which will be necessary for identifying sensor IDs and preventing spoofing attacks. Role-based access control restricts roles to permissions, reducing the likelihood of insider actions and erroneous use. (Shu, Yuan, Ke, & Kou, 2022).

Continuous monitoring and ICS-aware anomaly detection provide an additional layer of defence. Unlike conventional IT systems, industrial control systems exhibit highly structured and predictable operational behaviour, making them well-suited to behaviour-based intrusion detection. AI-driven security analytics can identify subtle deviations in command sequences, communication timing, or sensor correlations that may indicate cyber intrusion or data manipulation. Such monitoring is especially valuable in digital twin–enabled environments, where discrepancies between predicted and observed system behaviour can serve as early indicators of cyber compromise. Regular security audits, vulnerability assessments, and penetration testing tailored to ICS environments are also essential for identifying weaknesses and maintaining an up-to-date defensive posture. (Alshammari, 2025).

To characterize the security of digital wind farm management, the sanctioned layers of ICS-specific mitigations — including fail-safe control, network segmentation, secure channel communication, strong authentication, and continuous monitoring — will create a strong cybersecurity posture that enables safe and reliable operation for AI- and digitally twined–enabled wind energy systems. Securing physical and digital assets is vital to unlocking the power of advanced analytics and automation. Indeed, in the absence of such guarantees, data integrity, system availability, and control authority may be at risk -and operational safety and user trust may ultimately suffer. Therefore, resilient system design, secure data management, and controlled access are indispensable for the implementation of practical, reliable, and safe digital wind farm operations.

8.3. Regulatory and Standardization Frameworks for Cybersecure Wind Energy Systems

The adoption of artificial intelligence and digital twins for wind turbine monitoring, control, and maintenance will have to comply with the existing regulatory and standardization frameworks governing turbine safety, cybersecurity, and operational resilience. These are the guardrails that allow advanced analytics, automated decision-making, and real-time control to be safely and effectively rolled out.

From a turbine safety and control perspective, the IEC 61400 series, particularly IEC 61400-1, defines internationally accepted design requirements, safety factors, and load case assumptions for wind turbines (Freudenreich & Argyriadis, 2007). Although primarily focused on structural and mechanical integrity, these standards implicitly constrain digital control actions and supervisory commands generated by AI-based controllers and digital twin optimisation layers. Consequently, digital twins used for control validation or operational optimisation must ensure that simulated load responses and control interventions remain consistent with IEC-certified operating envelopes.

As wind farms become more instrumental in electricity generation, cyber needs go beyond turbine-level defenses and encompass plant-wide and system-level operational technology (OT) systems. From this point of view, NERC CIP (North American Electric Reliability Corporation's Critical Infrastructure Protection) standards are a common set of regulations intended to ensure security measures across different transmission lines and bulk electric system assets against cybersecurity threats. (Dolezilek & Hussey, 2011). Although the foundation of NERC CIP standards was established within the North American regulatory regime, their basic tenets (e.g., asset identification + access control, system monitoring, incident response planning, and recovery planning) are akin to AI-enabled wind farm SCADA systems and digital twin-driven platforms that integrate real-time operational data with supervisory control signals.

Complementary to compliance-driven standards, the NIST Cybersecurity Framework (CSF) provides a risk-centric prescription for cybersecurity across entire operational technology systems from cradle to grave. (Shen, 2014) (Pascoe, 2023). The five fundamental functions of the framework – Identify, Protect, Detect, Respond, and Recover – are also a perfect match for AI-based wind turbine monitoring and digital twin systems. AI-driven anomaly detection, for instance, can support the Detect function, and the availability of digital twins can assist with the Respond and Recover functions, such as system reconfiguration and preparation following cyber incidents, or scenario analysis. These principles are subsequently supported by OT-oriented guidance, including NIST SP 800-82, which outlines cybersecurity considerations for industrial control systems (ICS) (Stouffer, Falco, & Scarfone, 2011).

By explicitly mapping AI and digital twin use cases to existing regulatory and standardization domains, trustworthiness among system users, regulators, and operators is assured. This alignment is necessary to move advanced, research-based, data-driven methods toward industrial-scale, safety-critical wind energy applications.

8.4. Case Studies: Cyber-Attacks on Wind Turbine Manufacturers and Service Providers

In recent years, major wind turbine manufacturers and service organisations have been targeted by high-profile cyber-attacks, underscoring the real-world vulnerabilities of the wind energy supply chain and the importance of robust industrial-grade cybersecurity measures.

One of the most notable incidents occurred in March–April 2022, when the Conti ransomware group claimed responsibility for a ransomware attack on Nordex SE, a leading global wind turbine manufacturer. The attack forced Nordex to shut down its IT systems and restrict remote access to turbines under contract as a precautionary measure. However, the turbines themselves continued to operate normally, and there was no indication of direct operational damage. Forensic investigations coordinated with internal and external experts confirmed that the intrusion was limited to internal IT infrastructure and did not spread to customer assets. (Nordex, 2022).

The Nordex breach would be followed by one at Deutsche Windtechnik AG, an extensive German wind turbine maintenance and service provider operating thousands of turbines, in April 2022. The business said it was forced to temporarily disconnect remote data monitoring connections for about 2,000 turbines due to security issues. However, full service had been resumed after just a few days. Deutsche Windtechnik described the attack as "targeted and professional" and implemented additional security measures to strengthen its systems (Deutsche Windtechnik, 2022).

Moreover, in November 2021, Vestas Wind Systems A/S, the largest maker of wind turbines and a global service provider, disclosed that it was hit by a cybersecurity attack following unauthorized access to a part of its IT systems. In response, Vestas shut down IT systems at various plants this week to contain the incident and launched forensic investigations with external partners. There had been no signs that customer systems or turbine operations were affected. However, the attackers made off with internal data, and some of it was later published online, prompting alerts urging employees and suppliers to be on guard against possible exploitation of the stolen information. (Vestas, 2021).

These incidents collectively illustrate that wind energy OEMs and service organisations are not immune to sophisticated cyber threats — including ransomware and targeted intrusion campaigns — that can affect remote monitoring, data integrity, and operational resilience. Even where physical turbine operations were not directly harmed, the need to isolate IT systems, interrupt remote connectivity, and undertake extensive forensic investigation highlights vulnerabilities in supply-chain and maintenance ecosystems that must be mitigated as part of any cybersecurity strategy for AI-enabled and digital twin-supported wind infrastructure.

8.5. Summary

Section 6.8 also emphasizes that, in wind turbine systems that are becoming digital and interconnected via SCADA networks, cloud platforms, or DTs, cybersecurity is an important dimension of operational reliability. Although advanced monitoring, AI-powered analytics, and digital twin-based architectures offer significant operational advantages, they also expand the attack surface, making wind farms susceptible to threats such as data tampering, spoofing, unauthorized access, and service disruption. Only a multi-layer strategy can truly address them, layering fail-safe design, secure communications, authentication protocols, anomaly detection, and compliance with regulatory frameworks such as IEC, NERC, CIP, or NST CSF. As you will see in real cases, such as one that failed and another that succeeded, allowing attackers to take control of a turbine, you need to face it.

The following section (Section 6.9) extends this discussion to future trends and research directions that investigate how new technologies, better architectures, and evolving regulatory and operational guidance can be leveraged to overcome current limitations while also enhancing the resilience of these turbines as well as unlocking the full potential for such wind energy systems aided by AI- and digital twin-enabled wind energy systems.

9. FUTURE TRENDS AND RESEARCH DIRECTIONS

This section outlines key emerging trends and research directions that are expected to define the next generation of intelligent, scalable, and resilient wind turbine monitoring and control systems. The main areas include integrated DT with model-based health monitoring and risk-informed inspection planning. Others are proof-of-concept predictive DT platforms installed on offshore turbines equipped with real-time SCADA, visualization, and anomaly forecasting. Similarly, diagnostic DTs employ unsupervised learning for early fault detection and immersive monitoring interfaces. These works provide a research path toward a realistic, affordable, and usable DT-AI system for the future operational wind farm.

9.1. Key Trends

9.1.1. Federated Learning

Federated learning has emerged as a promising paradigm for collaborative model development across distributed wind turbines and wind farms while preserving data privacy. In this approach, machine-learning models are trained locally on turbine-level data, and only model parameters or updates are shared with a central aggregator, avoiding the need for centralised storage of sensitive SCADA or condition-monitoring data and aligning with commercial and regulatory constraints in the wind energy sector.

Recent work has shown that federated learning leads to performance matching, or even exceeding, that of a centralised heuristic when the data distributions across turbines are heterogeneous. For instance, federated digital twin models for wind turbine performance prediction that split the training across edge units and aggregate model updates without sharing raw data achieve good prediction accuracy in both homogeneous and heterogeneous fleets of wind turbines. (Erturk, Al-Dubai, Gursu, & Canberk, 2025), confirming the importance of privacy-preserving learning in digital twin architectures for wind energy systems. Likewise, federated learning for condition monitoring can enable turbines with only a small amount, or even no, representative local data to benefit from collaborative model training and significantly increase fault detection accuracy compared with purely local models, particularly under more diverse operating conditions.vv (Jenkel, Jonas, & Meyer, 2023). Further efforts on intra- and inter-farm federated learning show that collaborative models trained across turbines outperform isolated ones in settings with little to no data, since the more diverse global training set reduces the amount of local data required for robust normal behavior modelling. However, they bring the challenge of statistical heterogeneity between farms and turbine types. (Grataloup, Jonas, & Meyer, 2024).

However, federated learning faces challenges of communication overhead, non-IID (not independently and identically distributed) data convergence, and robustness to intermittent connectivity/heterogeneous datasets. Addressing these challenges via adaptive aggregation methods, clustered learning schemes, and hybrid edge–cloud architectures is an ongoing research topic in the context of federated digital twin systems for wind energy. (Zhou, Li, & Lin, 2025).

9.1.2. Edge-Based AI

Edge-based AI involves deploying machine-learning models directly on turbine controllers or embedded computing platforms. The primary advantage of this approach is the ability to process data close to its source, thereby reducing communication latency and enabling near real-time detection of abnormal operating conditions. This capability is significant for safety-critical functions, such as fault detection, alarm triggering, and protective control actions.

Recent research demonstrates that lightweight neural networks can be adapted for edge deployment in wind turbine condition monitoring without significant loss of diagnostic accuracy. Compact deep learning models can perform vibration-based fault diagnosis on edge devices with acceptable inference latency. (Huang, et al.,

2024). In addition, edge-based vision systems have been demonstrated for real-time blade inspection, enabling on-site detection of surface damage using embedded hardware. (Wang, Wang, Wang, & Fang, 2025).

Nevertheless, hardware limitations, energy consumption, and model update mechanisms remain important constraints. Future research should focus on optimising model compression, adaptive inference, and synchronisation between edge-level intelligence and higher-fidelity digital twins.

9.1.3. Explainable AI

As AI models are increasingly used to guide maintenance planning and operational decision-making in wind energy systems, explainability becomes crucial for trust building, safety assurance, regulatory acceptance, and engineering understanding. Explainable Artificial Intelligence (XAI) methods aim to provide insight into why a model behaves as it does by showing which input features, signals, or operating conditions drive particular predictions, thereby helping manage the potential downsides of black-box decision-making.

Recent wind turbine fault-diagnosis studies using XAI have shown that explainable models can achieve high predictive performance and provide decisions whose interpretations align with known mechanical breakdown phenomena. For instance, ensemble learning-based models, along with post-hoc explainability methods such as SHAP and LIME, have been shown to estimate the importance of SCADA features such as temperature, rotational speed, and load characteristics for fault detection tasks. (Rengasamy & Ramachandran, 2025). Furthermore, vibration-based diagnostics using XAI methods highlight interpretable associations between frequency-domain features and gearbox or bearing wear and tear, thereby enabling physically meaningful fault diagnosis across a variety of XAI models. (Gbashi, Olatunji, Adedeji, & Madushele, 2024).

Explainable deep learning models have also demonstrated the ability to extract such degradation trends from high-dimensional sensor data, aligning data-driven outputs with engineering understanding. (Letzgus & Müller, 2024). Such explainable AI is of key importance to engineers in digital twin architectures by allowing them to validate AI-predictions against physical expectations, thereby increasing confidence in AI-informed diagnostics, prognostics, and control decisions through the lifecycle of the turbine. (Astolfi, De Caro, & Vaccaro, 2023).

9.1.4. Blockchain for Data Security

Blockchain technology offers a promising option for securing and validating operational and maintenance data for wind energy systems. Using a distributed ledger system, all information related to the system, including sensor logs, maintenance logs, maintenance records, operational events, and others, is stored in a largely tamper-proof and immutable way. It is especially relevant in the multi-stakeholder setup, when turbine operators, manufacturers, maintenance providers, and regulators may need to access trustworthy data for auditing, compliance, and contract legality. Recent publications have already revealed potential blockchain frameworks for securely logging wind turbine maintenance and operational data, with considerable improvements in trust, provenance, and coordination among the parties involved. (Faheem, Raza, Bhutta, & Madni, 2024).

9.2. Opportunities for Research

9.2.1. Lightweight Models for Edge Deployment

One pressing challenge is developing lightweight, resource-efficient machine learning models for edge deployment in wind turbines. Combined compression techniques like knowledge distillation and neural architecture search can effectively reduce model size and inference latency while retaining diagnostic accuracy, as demonstrated in recent work on gearbox fault detection. (Wu, Tang, Deng, & Li, 2022).

9.2.2. Scalability of Federated Learning Frameworks

Federated learning frameworks have demonstrated potential for privacy-preserving, collaborative model training across distributed turbines. However, scaling these frameworks to large wind farms comprising hundreds or thousands of turbines introduces challenges related to communication overhead, heterogeneity of local data distributions, and system robustness. To address these, recent research explores efficient aggregation algorithms and partial-update schemes to reduce bandwidth consumption. (Wang, Fan, Jiang, & Xie, 2023). Comprehensive surveys across energy domains underscore the importance of adaptive aggregation, heterogeneity-aware model adaptation, and fault-tolerant protocols to support resilient, fleet-wide federated learning deployment. (Grataloup, Jonas, & Meyer, 2024).

9.2.3. Integration of Blockchain with SCADA and Digital Twin Platforms

Blockchain technology offers promising solutions for securing and validating operational and maintenance data in wind energy systems. However, integrating blockchain with existing SCADA infrastructures and emerging digital twin platforms remains an open research area. Investigations into interoperability standards, latency reduction, consensus mechanisms suitable for industrial settings, and the design of smart contracts to automate maintenance workflows will be essential. This integration can enhance data provenance, transparency, and trustworthiness across the entire wind farm lifecycle. (Faheem, Raza, Bhutta, & Madni, 2024).

Emerging technologies such as federated learning, edge-based AI, and blockchain hold significant promise for advancing the intelligence, security, and efficiency of wind energy systems. These innovations offer new pathways to optimize turbine operations, safeguard critical data, and enable real-time decision-making. To fully realize their potential, interdisciplinary collaboration among researchers, engineers, and industry stakeholders is essential to address the complex technical and operational challenges ahead.

9.2.4. Sustainability and Computational Efficiency

Although wind energy helps with decarbonisation, the computational footprint of massive AI models is now a matter of concern. New sustainability AI research focuses on the trade-off between prediction performance and computational energy use and emissions. Sustainable AI-based wind forecasting systems can reduce computational costs while maintaining prediction accuracy and aligning with the significant environmental goals of deploying artificial intelligence. (Elmousalami, Alnaser, & Hui, 2025).

The integration of sustainability metrics into model-based monitoring and control concepts (i.e., digital twins) might be a fruitful area for future investigation in lifecycle-aware WES. By going beyond conventional performance and reliability metrics to also capture computational energy consumption and emissions, they can enable environmentally conscious trade-offs between model complexity, control performance, and operational efficiency over the life cycle of an asset.

9.3. Summary

Based on the challenges and opportunities outlined in Sections 6.1–6.8, Section 6.9 departs from the state of the art and looks ahead to trends and research directions in wind turbine monitoring & control. Sections 6.6–6.8 have shown the promise of digital twins, AI-analytic edge processing, and secure deployment frameworks. However, they also revealed practical, operational, and cyber limitations that need to be addressed for scalable/implementable/reliable deployment. Lessons from those studies could form the basis for detecting emerging technology trends (e.g., federated learning, edge-based AI, and blockchain) that would enable data-driven monitoring, predictive maintenance, and security-aware operational decision support.

This chapter thus sets the stage for the roadmap ahead for the future generation of wind turbine systems. It presents research directions, innovative technology development, and approaches that must be pursued to create intelligent, reliable, and readily deployable digital twin-enabled wind farms.

10. CONCLUSION

This chapter has provided insights into the disruptive nature of AI, ML, and DT-based technologies in contemporary wind power systems. It has been shown that conventional maintenance and control methods are becoming insufficient to handle the complexity, variability, and challenges that arise from the operation of modern wind turbines, and how data-driven intelligent solutions are changing the landscape in terms of monitoring turbines, fault detection and diagnosis (FDD), and prognostic health management. Leveraging multiple data sources, including SCADA systems, condition-monitoring sensors, and environmental and meteorological readings, advanced analytics provide a more comprehensive assessment of performance, the ability to make better decisions in real-time, as well as improved overall operational reliability.

The chapter also emphasized ongoing challenges, including data quality and access, system interoperability, cybersecurity threats, and the computational burden of real-time digital twin simulations. Overcoming these bottlenecks is essential to translating digital twin and AI frameworks from research prototypes to industry-scale implementations. In the future, technologies like federated learning, edge AI, and blockchain may offer greater opportunities for secure, scalable, and real-time wind farm control. However, reaching their potential will continue to depend on cross-disciplinary teamwork among engineers, data scientists, operators, and regulators.

Ultimately, as the wind industry transitions to a greener, safer, and more efficient embodiment, continued innovation from AI, digital twins, and related technologies will be necessary to address operational, technical, and sustainability challenges for next-generation wind systems.

REFERENCES

5G-ACIA. (2021). Integration of 5G with Time-Sensitive Networking for Industrial Communications. *White Paper, 5G ACIA, Frankfurt Germany*.

Abdullahi, I., Longo, S., & Samie, M. (2024). Towards a Distributed Digital Twin Framework for Predictive Maintenance in Industrial Internet of Things (IIoT). *Sensors (Basel)*, *24*(8), 1–30. DOI: 10.3390/s24082663 PMID: 38676279

Al Abdulwahid, A. (2024). Cyber-physical threat mitigation in wind energy systems: A novel secure architecture for industry 4.0 power grids. *Energy Informatics*, *7*(1), 141. DOI: 10.1186/s42162-024-00449-6

Alagha, N., Khairuddin, A., Haitaamar, Z., Al-Khatib, O., & Kanesan, J. (2025). Artificial Intelligence in Wind Turbine Fault Detection and Diagnosis: Advances and Perspectives. *Energies*, *18*(7), 1–23. DOI: 10.3390/en18071680

Alhamam, N., Hafizur Rahman, M., & Aljughaiman, A. (2025). A Comprehensive Review on Cybersecurity of Digital Twins Issues, Challenges, and Future Research Directions. *IEEE Access : Practical Innovations, Open Solutions*, *13*, 45106–45124. DOI: 10.1109/ACCESS.2025.3545004

Alshammari, A. (2025). Securing smart microgrids with a novel multi-layer cybersecurity framework for Industry 4.0 renewable energy systems. *Discover Computing*, *28*(1), 80. DOI: 10.1007/s10791-025-09600-7

Alves, D., Mendonça, F., Mostafa, S., & Morgado-Dias, F. (2024). A comprehensive IoT cloud-based wind station ready for real-time measurements and artificial intelligence integration. *e-Prime - Advances in Electrical Engineering. Electronics and Energy*, *10*, 100862. DOI: 10.1016/j.prime.2024.100862

Arroba, P., Buyya, R., Cárdenas, R., Risco-Martín, J., & Moya, J. (2024). Sustainable edge computing: Challenges and future directions. *Software, Practice & Experience*, *54*(11), 2272–2296. DOI: 10.1002/spe.3340

Astolfi, D., De Caro, F., & Vaccaro, A. (2023). Condition Monitoring of Wind Turbine Systems by Explainable Artificial Intelligence Techniques. *Sensors (Basel)*, *23*(12), 1–24. DOI: 10.3390/s23125376 PMID: 37420542

Backhus, J., Rao, A., Venkatraman, C., Padmanabhan, A., Kumar, A., & Gupta, C. (2024). Equipment Health Assessment: Time Series Analysis for Wind Turbine Performance. *Applied Sciences (Basel, Switzerland)*, *14*(8), 1–20. DOI: 10.3390/app14083270

Bawane, S., Matange, G., Shrivastava, A., Qureshi, A., Sultan, S., & Shrivastava, D. (2025). Smart Wind Farm Management Using IoT and Predictive in Analytics. *International Journal of Environmental Sciences*, *11*(3), 1079–1095.

Branlard, E., Jonkman, J., Brown, C., & Zhang, J. (2024). A digital twin solution for floating offshore wind turbines validated using a full-scale prototype. *Wind Energy Science*, *9*(1), 1–24. DOI: 10.5194/wes-9-1-2024

. Burton, T., Jenkins, N., Sharpe, D., & Bossanyi, E. (2011). *Wind energy handbook*. John Wiley \& Sons.

Campoverde-Vilela, L., del Cisne Feijóo, M., Vidal, Y., Sampietro, J., & Tutivén, C. (2023). Anomaly-based fault detection in wind turbine main bearings. *Wind Energy Science*, *8*(4), 557–574. DOI: 10.5194/wes-8-557-2023

Castellani, F., Pandit, R., Natili, F., Belcastro, F., & Astolfi, D. (2023). Advanced Methods for Wind Turbine Performance Analysis Based on SCADA Data and CFD Simulations. *Energies*, *16*(3), 1081. Advance online publication. DOI: 10.3390/en16031081

Castillo, O., Andrade, V., Rivas, J., & González, R. (2023). Comparison of Power Coefficients in Wind Turbines Considering the Tip Speed Ratio and Blade Pitch Angle. *Energies*, *16*(6), 2774. Advance online publication. DOI: 10.3390/en16062774

Cavalcanti, D., Perez-Ramirez, J., Rashid, M., Fang, J., Galeev, M., & Stanton, K. (2019). Extending Accurate Time Distribution and Timeliness Capabilities Over the Air to Enable Future Wireless Industrial Automation Systems. *Proceedings of the IEEE*, *107*(6), 1132–1152. DOI: 10.1109/JPROC.2019.2903414

Cid Montoya, M., Rubio-Medrano, C., & Kareem, A. (2024). On the cybersecurity of smart structures under wind. *Journal of Wind Engineering and Industrial Aerodynamics*, *251*, 105777. DOI: 10.1016/j.jweia.2024.105777

Clifton, A., Barber, S., Bray, A., Enevoldsen, P., Fields, J., Sempreviva, A., Williams, L., Quick, J., Purdue, M., Totaro, P., & Ding, Y. (2023). Grand challenges in the digitalisation of wind energy. *Wind Energy Science*, *8*(6), 947–974. DOI: 10.5194/wes-8-947-2023

Coelho, P. (2023). The Betz limit and the corresponding thermodynamic limit. *Wind Engineering*, *47*(2), 491–496. DOI: 10.1177/0309524X221130109

Dang, S., Amin, O., Shihada, B., & Alouini, M.-S. (2020). What should 6G be? *Nature Electronics*, *3*(1), 20–29. DOI: 10.1038/s41928-019-0355-6

Datta, A., & Rahman, M. (2017). Cyber Threat Analysis Framework for the Wind Energy Based Power System. *Proceedings of the 2017 Workshop on Cyber-Physical Systems Security and PrivaCy*, pp. 81–92. DOI: 10.1145/3140241.3140247

De Kooning, J., Stockman, K., De Maeyer, J., Jarquin-Laguna, A., & Vandevelde, L. (2021). Digital Twins for Wind Energy Conversion Systems: A Literature Review of Potential Modelling Techniques Focused on Model Fidelity and Computational Load. *Processes (Basel, Switzerland)*, *9*(12), 1–26. DOI: 10.3390/pr9122224

Deutsche Windtechnik, T. (2022). *Cyber attack on Deutsche Windtechnik.* Retrieved January 29, 2026, from https://www.deutsche-windtechnik.com/en/news/press-releases/details/cyber-attack-on-deutsche-windtechnik/?

Dibaj, A., Valavi, M., & Nejad, A. (2024). Unsupervised anomaly detection of permanent-magnet offshore wind generators through electrical and electromagnetic measurements. *Wind Energy Science*, *9*(11), 2063–2086. DOI: 10.5194/wes-9-2063-2024

Dolezilek, D., & Hussey, L. (2011). Requirements or recommendations? Sorting out NERC CIP, NIST, and DOE cybersecurity. In *2011 64th Annual Conference for Protective Relay Engineers* (pp. 328-333).

Eidson, J. (2006). A Detailed Analysis of IEEE 1588. In *Measurement, Control, and Communication Using IEEE 1588* (pp. 61–131). Springer London.

Elmousalami, H., Alnaser, A., & Hui, F. (2025). Sustainable AI-driven wind energy forecasting: Advancing zero-carbon cities and environmental computation. *Artificial Intelligence Review*, *58*(6), 191. DOI: 10.1007/s10462-025-11191-0

Erturk, M., Al-Dubai, A., Gursu, K., & Canberk, B. (2025). A Digital Twin model for predicting wind turbine performance using federated learning. *Energy*, *337*, 138644. DOI: 10.1016/j.energy.2025.138644

Faheem, M., Raza, B., Bhutta, M., & Madni, S. (2024). A blockchain-based resilient and secure framework for events monitoring and control in distributed renewable energy systems. *IET Blockchain*, *4*(S1), 644–658. DOI: 10.1049/blc2.12081

Finn, N. (2018). Introduction to Time-Sensitive Networking. *IEEE Communications Standards Magazine*, 2(2), 22–28. DOI: 10.1109/MCOMSTD.2018.1700076

Finn, N., Thubert, P., Varga, B., & Farkas, J. (2019). Deterministic networking architecture. In *RFC 8655*. IETF. DOI: 10.17487/RFC8655

Foukas, X., Patounas, G., Elmokashfi, A., & Marina, M. (2017). Network Slicing in 5G: Survey and Challenges. *IEEE Communications Magazine, 55*(5), 94–100. DOI: 10.1109/MCOM.2017.1600951

Frederiksen, R., Bocewicz, G., Radzki, G., Banaszak, Z., & Nielsen, P. (2024). Cost-Effectiveness of Predictive Maintenance for Offshore Wind Farms: A Case Study. *Energies, 17*(13), 1–24. DOI: 10.3390/en17133147

Freudenreich, K., & Argyriadis, K. (2007). The Load Level of Modern Wind Turbines according to IEC 61400-1. *Journal of Physics: Conference Series, 75*(1), 012075. DOI: 10.1088/1742-6596/75/1/012075

Fuller, A., Fan, Z., Day, C., & Barlow, C. (2020). Digital Twin: Enabling Technologies, Challenges and Open Research. *IEEE Access : Practical Innovations, Open Solutions, 8*, 108952–108971. DOI: 10.1109/ACCESS.2020.2998358

Gbashi, S., Olatunji, O., Adedeji, P., & Madushele, N. (2024). An Explainable AI Approach Using SHapley Additive exPlanations for Feature Selection in Vibration-Based Fault Diagnostics of Wind Turbine Gearbox. *2024 IEEE 5th International Conference on Electro-Computing Technologies for Humanity (NIGERCON)*, pp. 1-5.

Gigoni, L., Betti, A., Tucci, M., & Crisostomi, E. (2019). *A Scalable Predictive Maintenance Model for Detecting Wind Turbine Component Failures Based on SCADA Data. 2019 IEEE Power & Energy Society General Meeting*. PESGM., DOI: 10.1109/PESGM40551.2019.8973898

Gonzalez, E., Stephen, B., Infield, D., & Melero, J. (2019). Using high-frequency SCADA data for wind turbine performance monitoring: A sensitivity study. *Renewable Energy, 131*, 841–853. DOI: 10.1016/j.renene.2018.07.068

Granados, D., Ruiz, M., Acosta, J., Lara, S., Domínguez, R., & Kañetas, P. (2023). A Wind Turbine Vibration Monitoring System for Predictive Maintenance Based on Machine Learning Methods Developed under Safely Controlled Laboratory Conditions. *Energies, 16*(5), 1–17. DOI: 10.3390/en16052290

Grataloup, A., Jonas, S., & Meyer, A. (2024). A review of federated learning in renewable energy applications: Potential, challenges, and future directions. *Energy and AI, 17*, 100375. DOI: 10.1016/j.egyai.2024.100375

Grieves, M. (2014). Digital Twin: Manufacturing Excellence through Virtual Factory Replication. *White Paper*.

Gungor, V., Sahin, D., Kocak, T., Ergut, S., Buccella, C., Cecati, C., & Hancke, G. (2011). Smart Grid Technologies: Communication Technologies and Standards. *IEEE Transactions on Industrial Informatics*, *7*(4), 529–539. DOI: 10.1109/TII.2011.2166794

GWEC. (2024). *Global wind report 2024.* Global Wind Energy Council. Retrieved from https://img.saurenergy.com/2024/05/gwr-2024_digital-version_final-1-compressed.pdf

Habbouche, H., Amirat, Y., & Benbouzid, M. (2025). Leveraging Digital Twins and AI for Enhanced Gearbox Condition Monitoring in Wind Turbines: A Review. *Applied Sciences (Basel, Switzerland)*, *15*(10), 5725. Advance online publication. DOI: 10.3390/app15105725

Haghshenas, A., Hasan, A., Osen, O., & Mikalsen, E. (2023). Predictive digital twin for offshore wind farms. *Energy Informatics*, *6*(1), 1–26. DOI: 10.1186/s42162-023-00257-4

Haroon, R., Erfan, K., Jawad, R., & Canras, B. (2020). Fault Prediction of Wind Turbine Gearbox Based on SCADA Data and Machine Learning. *2020 10th International Conference on Advanced Computer Information Technologies (ACIT)*, pp. 391-395. DOI: 10.1109/ACIT49673.2020.9208884

Huang, J., Zhou, S., Li, G., & Shen, Q. (2025). Real-time monitoring and optimization methods for user-side energy management based on edge computing. *Scientific Reports*, *15*(1), 24890. DOI: 10.1038/s41598-025-07592-4 PMID: 40640293

Huang, Y., Liang, S., Cui, T., Mu, X., Luo, T., Wang, S., & Wu, G. (2024). Edge Computing and Fault Diagnosis of Rotating Machinery Based on MobileNet in Wireless Sensor Networks for Mechanical Vibration. *Sensors (Basel)*, *24*(16), 1–14. DOI: 10.3390/s24165156 PMID: 39204852

Hussain, M., Hussain Mirjat, N., Shaikh, F., Luxmi Dhirani, L., Kumar, L., & Sleiti, A. (2024). Condition Monitoring and Fault Diagnosis of Wind Turbine: A Systematic Literature Review. *IEEE Access : Practical Innovations, Open Solutions*, *12*, 190220–190239. DOI: 10.1109/ACCESS.2024.3514747

IEA. (2024). *Renewables 2024 – Market update.* International Energy Agency. Retrieved from https://iea.blob.core.windows.net/assets/17033b62-07a5-4144-8dd0-651cdb6caa24/Renewables2024.pdf

Järvinen, M., Paulomäki, H., Floors, R., Pansch, C., Thompson, P., Apajalahti, E.-L., & Kangasoja, J. (2025). Wind Energy. In Järvinen, M., & Paulomäki, H. (Eds.), *Designing Renewable Energy Systems within Planetary Boundaries: A Textbook for Energy Engineers* (pp. 345–460). Springer Nature Switzerland., DOI: 10.1007/978-3-031-69856-9_5

Jenkel, L., Jonas, S., & Meyer, A. (2023). Privacy-Preserving Fleet-Wide Learning of Wind Turbine Conditions with Federated Learning. *Energies*, *16*(17), 6377. Advance online publication. DOI: 10.3390/en16176377

Jørgensen, B., & Ma, Z. (2025). Digital Twin of the European Electricity Grid: A Review of Regulatory Barriers, Technological Challenges, and Economic Opportunities. *Applied Sciences (Basel, Switzerland)*, *15*(12), 1–32. DOI: 10.3390/app15126475

Kandemir, E., Hasan, A., Kvamsdal, T., & Abdel-Afou Alaliyat, S. (2024). Predictive digital twin for wind energy systems: A literature review. *Energy Informatics*, *7*(1), 68. DOI: 10.1186/s42162-024-00373-9

Kestel, K., & Chesterman, X. Zappal\'a, D., Watson, S., Li, M., Hart, E., . . . Helsen, J. (2025). Condition monitoring of wind turbine drivetrains: State-of-the-art technologies, recent trends, and future outlook. *Wind Energy Science Discussions*, 1--85. DOI: 10.5194/wes-2025-168

Kini, K., Harrou, F., Madakyaru, M., & Sun, Y. (2023). Enhancing Wind Turbine Performance: Statistical Detection of Sensor Faults Based on Improved Dynamic Independent Component Analysis. *Energies*, *16*(15), 1–25. DOI: 10.3390/en16155793

Kostecka, E., Miller, T., Durlik, I., & Nerć, A. (2025). Artificial Intelligence in Floating Offshore Wind Turbines: A Critical Review of Applications in Design, Monitoring, Control, and Digital Twins. *Energies*, *18*(22), 5937. Advance online publication. DOI: 10.3390/en18225937

Lee, C.-Y., & Maceren, E. (2024). Wind energy system fault classification and detection using deep convolutional neural network and particle swarm optimization-extreme gradient boosting. *IET Energy Systems Integration*, *6*(4), 479–497. DOI: 10.1049/esi2.12144

Lei, Y., Li, N., Guo, L., Li, N., Yan, T., & Lin, J. (2018). Machinery health prognostics: A systematic review from data acquisition to RUL prediction. *Mechanical Systems and Signal Processing*, *104*, 799–834. DOI: 10.1016/j.ymssp.2017.11.016

Leon-Medina, J., Tibaduiza, D., Parés, N., & Pozo, F. (2025). Digital twin technology in wind turbine components: A review. *Intelligent Systems with Applications*, *26*, 200535. DOI: 10.1016/j.iswa.2025.200535

Letzgus, S., & Müller, K.-R. (2024). An explainable AI framework for robust and transparent data-driven wind turbine power curve models. *Energy and AI, 100328*, 100328. Advance online publication. DOI: 10.1016/j.egyai.2023.100328

Liu, X., Wu, X., He, J., & Gupta, R. (2025). Intelligent Operation and Maintenance and Prediction Model Optimization for Improving Wind Power Generation Efficiency. *2025 7th International Congress on Human-Computer Interaction, Optimization and Robotic Applications (ICHORA)*, pp. 1-7. DOI: 10.1109/ICHORA65333.2025.11017307

Lui, S. (2024). Wind power short-term prediction based on digital twin technology. *Frontiers in Energy Research, 12*, 1–9. DOI: 10.3389/fenrg.2024.1365237

Ma, J., Yuan, Y., & Chen, P. (2023). A fault prediction framework for Doubly-fed induction generator under time-varying operating conditions driven by digital twin. *IET Electric Power Applications, 17*(4), 499–521. DOI: 10.1049/elp2.12280

Maldonado-Correa, J., Martín-Martínez, S., Artigao, E., & Gómez-Lázaro, E. (2020). Using SCADA Data for Wind Turbine Condition Monitoring: A Systematic Literature Review. *Energies, 13*(12), 3132. DOI: 10.3390/en13123132

Maldonado-Correa, J., Torres-Cabrera, J., Martín-Martínez, S., Artigao, E., & Gómez-Lázaro, E. (2023). Wind turbine fault detection based on the transformer model using SCADA data. *Engineering Failure Analysis, 162*, 108354. DOI: 10.1016/j.engfailanal.2024.108354

Maldonado-Correa, J., Valdiviezo-Condolo, M., Artigao, E., Martín-Martínez, S., & Gómez-Lázaro, E. (2024). Classification of Highly Imbalanced Supervisory Control and Data Acquisition Data for Fault Detection of Wind Turbine Generators. *Energies, 17*(1), 1–20. DOI: 10.3390/en17071590

Marykovskiy, Y., Clark, T., Deparday, J., Chatzi, E., & Barber, S. (2024). Architecting a digital twin for wind turbine rotor blade aerodynamic monitoring. *Frontiers in Energy Research, 12*, 1428387. Advance online publication. DOI: 10.3389/fenrg.2024.1428387

Mchirgui, N., Quadar, N., Kraiem, H., & Lakhssassi, A. (2024). The Applications and Challenges of Digital Twin Technology in Smart Grids: A Comprehensive Review. *Applied Sciences (Basel, Switzerland), 14*(23), 1–32. DOI: 10.3390/app142310933

Mehlan, F., Keller, J., & Nejad, A. (2023). Virtual sensing of wind turbine hub loads and drivetrain fatigue damage. *Forschung im Ingenieurwesen, 87*(1), 207–218. DOI: 10.1007/s10010-023-00627-0

Messenger, J. (2018). Time-Sensitive Networking: An Introduction. *IEEE Communications Standards Magazine*, *2*(2), 29–33. DOI: 10.1109/MCOMSTD.2018.1700047

Milani, S., Leoni, J., Cacciola, S., Croce, A., & Tanelli, M. (2025). A machine-learning-based approach for active monitoring of blade pitch misalignment in wind turbines. *Wind Energy Science*, *10*(3), 497–510. DOI: 10.5194/wes-10-497-2025

Mohamed, A., & Chul-Hwan, K. (2017). Communication Architecture for Grid Integration of Cyber Physical Wind Energy Systems. *Applied Sciences (Basel, Switzerland)*, *7*(10), 1–20. DOI: 10.3390/app7101034

Mulders, S., Gallo, A., & Rotea, M. (2024). Analysis of Extremum Seeking Control for Wind Turbine Torque Controller Optimization by Aerodynamic and Generator Power Objectives. In *2024 American Control Conference (ACC)* (pp. 1030-1037). Toronto: IEEE. DOI: 10.23919/ACC60939.2024.10644899

Muñoz-Palomeque, E., Sierra-García, J., & Santos, M. (2022). Wind turbine maximum power point tracking control based on unsupervised neural networks. *Journal of Computational Design and Engineering*, *10*(1), 108–121. DOI: 10.1093/jcde/qwac132

Muñoz-Palomeque, E., Sierra-García, J., & Santos, M. (2024). Enhancing offshore wind turbines performance with hybrid control strategies using neural networks and conventional controllers. *Journal of Computational Design and Engineering*, *12*(3), 80–97. DOI: 10.1093/jcde/qwae103

Murray, J., Huang, C., Parkinson, S., Bolton, R., Gledhill, D., Parkinson, H., Basher, D., Vulovic, V., Yang, W., & Ball, A. (2025). Unified Digital Twin Approach for Energy Grid Condition and Cybersecurity Monitoring. *IEEE Access : Practical Innovations, Open Solutions*, *13*, 180710–180724. DOI: 10.1109/ACCESS.2025.3622468

Nasrallah, A., Thyagaturu, A., Alharbi, Z., Wang, C., Shao, X., Reisslein, M., & ElBakoury, H. (2019). Ultra-Low Latency (ULL) Networks: The IEEE TSN and IETF DetNet Standards and Related 5G ULL Research. *IEEE Communications Surveys and Tutorials*, *21*(1), 88–145. DOI: 10.1109/COMST.2018.2869350

Nordex. (2022). *Update on cyber security incident*. Retrieved January 29, 2026, from https://www.nordex-online.com/en/2022/04/update-on-cyber-security-incident/?

Oliveira-Filho, A., Comeau, M., Cave, J., Nasr, C., Côté, P., & Tahan, A. (2024). Wind Turbine SCADA Data Imbalance: A Review of Its Impact on Health Condition Analyses and Mitigation Strategies. *Energies*, *18*(1), 1–23. DOI: 10.3390/en18010059

Ozturk, S., Fthenakis, V., & Faulstich, S. (2018). Assessing the Factors Impacting on the Reliability of Wind Turbines via Survival Analysis—A Case Study. *Energies*, *11*(11), 1–20. DOI: 10.3390/en11113034

Pandit, R., Astolfi, D., Hong, J., Infield, D., & Santos, M. (2023). SCADA data for wind turbine data-driven condition/performance monitoring: A review on state-of-art, challenges and future trends. *Wind Engineering*, *47*(2), 422–441. DOI: 10.1177/0309524X221124031

Pao, L., Pusch, M., & Zalkind, D. (2024). Control Co-Design of Wind Turbines. *Annual Review of Control, Robotics, and Autonomous Systems*, *7*(1), 201–226. DOI: 10.1146/annurev-control-061423-101708

Pascoe, C. (2023). *Public draft: The NIST cybersecurity framework 2.0*. National Institute of Standards and Technology.

Pérez, E. (2022). A simulation-driven online scheduling algorithm for the maintenance and operation of wind farm systems. *Simulation*, *98*(1), 47–61. DOI: 10.1177/00375497211028605

Popovski, P., Trillingsgaard, K., Simeone, O., & Durisi, G. (2018). 5G Wireless Network Slicing for eMBB, URLLC, and mMTC: A Communication-Theoretic View. *IEEE Access : Practical Innovations, Open Solutions*, *6*, 55765–55779. DOI: 10.1109/ACCESS.2018.2872781

Rasheed, A., San, O., & Kvamsdal, T. (2020). Digital Twin: Values, Challenges and Enablers From a Modeling Perspective. *IEEE Access : Practical Innovations, Open Solutions*, *8*, 21980–22012. DOI: 10.1109/ACCESS.2020.2970143

Rengasamy, P., & Ramachandran, R. (2025). Explainable artificial intelligence framework for wind turbine fault detection using random forest – Extreme gradient boosting hybrid model. *Results in Engineering*, *28*, 107446. DOI: 10.1016/j.rineng.2025.107446

Saad, W., Bennis, M., & Chen, M. (2020). A Vision of 6G Wireless Systems: Applications, Trends, Technologies, and Open Research Problems. *IEEE Network*, *34*(3), 134–142. DOI: 10.1109/MNET.001.1900287

Santiago, R., Barbosa, N., Mergulhão, H., Carvalho, T., Santos, A., Medrado, R., Filho, J. B. M., Pinheiro, O. R., & Nascimento, E. (2024). Data-Driven Models Applied to Predictive and Prescriptive Maintenance of Wind Turbine: A Systematic Review of Approaches Based on Failure Detection, Diagnosis, and Prognosis. *Energies*, *17*(5), 1–24. DOI: 10.3390/en17051010

Shah, S., Daoliang, T., & Kumar, S. (2024). RUL forecasting for wind turbine predictive maintenance based on deep learning. *Heliyon, 10*(20), e39268. DOI: 10.1016/j.heliyon.2024.e39268 PMID: 39678167

Shen, L. (2014). The NIST cybersecurity framework: Overview and potential impacts. *Scitech Lawyer, 10*(4), 16.

Sheng, S. (2012). *Wind Turbine Gearbox Condition Monitoring Round Robin Study – Vibration Analysis.* United States: National Renewable Energy Laboratory (NREL), Golden, CO (United States). DOI: 10.2172/1048981

Sheng, S., Link, H., LaCava, W., van Dam, J., McNiff, B., Veers, P., . . . Oyague, F. (2011). Wind Turbine Drivetrain Condition Monitoring During GRC Phase 1 and Phase 2 Testing. United States: National Renewable Energy Laboratory (NREL), Golden, CO (United States). DOI: 10.2172/1027157

Shu, Y., Yuan, Q., Ke, W., & Kou, L. (2022). Security Access Control Method for Wind-Power-Monitoring System Based on Agile Authentication Mechanism. *Electronics (Basel), 11*(23), 1–12. DOI: 10.3390/electronics11233938

Siddiqi, M., Yu, H., & Joung, J. (2019). 5G Ultra-Reliable Low-Latency Communication Implementation Challenges and Operational Issues with IoT Devices. *Electronics (Basel), 8*(9), 981. DOI: 10.3390/electronics8090981

Singh, V., Sharma, R., & Govindarasu, M. (2020). *Testbed-based Performance Evaluation of Attack Resilient Control for Wind Farm SCADA System. 2020 IEEE Power & Energy Society General Meeting*. PESGM., DOI: 10.1109/PESGM41954.2020.9281431

. Soe, H., & Htet, A. (2024). A Comprehensive Review of SCADA-Based Wind Turbine Performance and Reliability Modeling with Machine Learning Approaches. *Journal of Technology Innovations and Energy*, 68–92. DOI: 10.56556/jtie.v3i3.1028

Stadtmann, F., Rasheed, A., Kvamsdal, T., Johannessen, K., San, O., Kölle, K., Tande, J. O., Barstad, I., Benhamou, A., Brathaug, T., Christiansen, T., Firle, A.-L., Fjeldly, A., Frøyd, L., Gleim, A., Høiberget, A., Meissner, C., Nygård, G., Olsen, J., & Skogås, J. (2023). Digital Twins in Wind Energy: Emerging Technologies and Industry-Informed Future Directions. *IEEE Access : Practical Innovations, Open Solutions, 11*, 110762–110795. DOI: 10.1109/ACCESS.2023.3321320

. Stouffer, K., Falco, J., & Scarfone, K. (2011). Guide to industrial control systems (ICS) security. *NIST special publication, 800*(82), 16--16.

. Sun, Y., & Han, W. (2025). A review of enhancing wind power with AI: applications, economic implications, and green innovations. *Digital Economy and Sustainable Development*, 1-11. DOI: 10.1007/s44265-025-00059-4

. Sutton, R., & Barto, A. (1998). *Reinforcement learning: An introduction.* MIT press Cambridge.

Tao, F., Zhang, H., Liu, A., & Nee, A. (2019). Digital Twin in Industry: State-of-the-Art. *IEEE Transactions on Industrial Informatics, 15*(4), 2405–2415. DOI: 10.1109/TII.2018.2873186

Tataria, H., Shafi, M., Molisch, A., Dohler, M., Sjöland, H., & Tufvesson, F. (2021). 6G Wireless Systems: Vision, Requirements, Challenges, Insights, and Opportunities. *Proceedings of the IEEE, 109*(7), 1166–1199. DOI: 10.1109/JPROC.2021.3061701

Thangamuthu, S., Concer, N., Cuijpers, P., & Lukkien, J. (2015). *Analysis of Ethernet-switch traffic shapers for in-vehicle networking applications. 2015 Design, Automation & Test in Europe Conference & Exhibition*. DATE.

Vestas. (2021). *Third update on cyber incident*. Retrieved January 29, 2026, from https://www.vestas.com/en/media/company-news/2021/third-update-on-cyber-incident-c3466518?

Vieira, J., Farias, F., Ochoa, A., de Menezes, F., Costa, A., da Costa, J., de Novaes Pires Leite, G., Vilela, O. C., de Souza, M. G. G., & Michima, P. (2024). Remaining Useful Life Estimation Framework for the Main Bearing of Wind Turbines Operating in Real Time. *Energies, 17*(6), 1–17. DOI: 10.3390/en17061430

Wang, L., Fan, W., Jiang, G., & Xie, P. (2023). An efficient federated transfer learning framework for collaborative monitoring of wind turbines in IoE-enabled wind farms. *Energy, 284*, 128518. DOI: 10.1016/j.energy.2023.128518

Wang, Y., Wang, X., Wang, Y., & Fang, S. (2025). Edge–Cloud Intelligence for Sustainable Wind Turbine Blade Transportation: Machine-Vision-Driven Safety Monitoring in Renewable Energy Systems. *Energies, 18*, 1–22. DOI: 10.3390/en19010001

Williams, L., Phillips, C., Sheng, S., Dobos, A., & Wei, X. (2020). Scalable Wind Turbine Generator Bearing Fault Prediction Using Machine Learning: A Case Study. In *2020 IEEE International Conference on Prognostics and Health Management (ICPHM)* (pp. 1-9). Detroit: IEEE. DOI: 10.1109/ICPHM49022.2020.9187050

Wu, H., Liu, J., Liu, J., Cui, M., Liu, X., & Gao, H. (2019). Power Grid Reliability Evaluation Considering Wind Farm Cyber Security and Ramping Events. *Applied Sciences (Basel, Switzerland), 9*(15), 1–18. DOI: 10.3390/app9153003

Wu, Y., Tang, B., Deng, L., & Li, Q. (2022). Distillation-enhanced fast neural architecture search method for edge-side fault diagnosis of wind turbine gearboxes. *Expert Systems with Applications, 118049*, 118049. Advance online publication. DOI: 10.1016/j.eswa.2022.118049

Xiang, L., Wang, P., Yang, X., Hu, A., & Su, H. (2021). Fault detection of wind turbine based on SCADA data analysis using CNN and LSTM with attention mechanism. *Measurement, 175*, 109094. DOI: 10.1016/j.measurement.2021.109094

Xu, Q., Lu, S., Zhai, Z., & Jiang, C. (2020). Adaptive fault detection in wind turbine via RF and CUSUM. *IET Renewable Power Generation, 14*(10), 1789–1796. DOI: 10.1049/iet-rpg.2019.0913

Yang, L., Huang, Z., Mo, X., & Luo, T. (2025). Enhanced GAIN-Based Missing Data Imputation for a Wind Energy Farm SCADA System. *Electronics (Basel), 14*(8), 1–26. DOI: 10.3390/electronics14081590

Yang, L., Liang, G., Yang, Y., Ruan, J., Yu, P., & Yang, C. (2024). Adversarial false data injection attacks on deep learning-based short-term wind speed forecasting. *IET Renewable Power Generation, 18*(7), 1370–1379. DOI: 10.1049/rpg2.12853

Yang, L., & Zhang, Z. (2021). Wind Turbine Gearbox Failure Detection Based on SCADA Data: A Deep Learning-Based Approach. *IEEE Transactions on Instrumentation and Measurement, 70*, 1–11. DOI: 10.1109/TIM.2020.3045800

Yu, Q., Patriksson, M., & Sagitov, S. (2021). Optimal scheduling of the next preventive maintenance activity for a wind farm. *Wind Energy Science, 6*(3), 949–959. DOI: 10.5194/wes-6-949-2021

Zhang, J., & Zhao, X. (2023). Digital twin of wind farms via physics-informed deep learning. *Energy Conversion and Management, 293*, 117507. DOI: 10.1016/j.enconman.2023.117507

Zhang, Y., Lin, F., & Wang, K. (2020). Robustness of Short-Term Wind Power Forecasting against False Data Injection Attacks. *Energies, 13*(15), 1–22. DOI: 10.3390/en13153780

Zhang, Z., Xiao, Y., Ma, Z., Xiao, M., Ding, Z., Lei, X., Karagiannidis, G. K., & Fan, P. (2019). 6G Wireless Networks: Vision, Requirements, Architecture, and Key Technologies. *IEEE Vehicular Technology Magazine, 14*(3), 28–41. DOI: 10.1109/MVT.2019.2921208

Zheng, M., Man, J., Wang, D., Chen, Y., Li, Q., & Liu, Y. (2023). Semi-supervised multivariate time series anomaly detection for wind turbines using generator SCADA data. *Reliability Engineering & System Safety*, *235*, 109235. DOI: 10.1016/j.ress.2023.109235

Zhou, R., Li, Y., & Lin, X. (2025). A clustered federated learning framework for collaborative fault diagnosis of wind turbines. *Applied Energy*, *377*, 124532. DOI: 10.1016/j.apenergy.2024.124532

Zhu, X., Wu, Y., Zhao, X., Yang, Y., Liu, S., Shi, L., & Wu, Y. (2024). Overview of Wind and Photovoltaic Data Stream Classification and Data Drift Issues. *Energies*, *17*(17), 1–24. DOI: 10.3390/en17174371

Chapter 4
AI-Driven Twin Framework for Intelligent Condition Monitoring and Predictive Maintenance in Grid-Integrated Wind Turbines:
Advanced ML Models for Real Time Asset Optimization

T. Manikandan
http://orcid.org/0000-0001-5613-1163
Vivekanandha College of Engineering for Women, India

G. Madhu Mitha
http://orcid.org/0009-0005-3176-3005
Vivekanandha College of Engineering for Women, India

ABSTRACT

The exponential development of wind power requires a high level of operational efficiency, reliability, and grid integration. The chapter presents a four-layer AI-powered digital twin architecture of grid-integrated wind turbines, which includes multi-sensors networks, IoT-based data acquisition (MQTT, LoRaWAN, 5G), hybrid physics and deep learning modeling, and autonomous decision-making. Machine learning models such as CNNs, RNNs, attention-based models and ensemble genetic models allow real-time condition monitoring and predictive maintenance

DOI: 10.4018/979-8-3373-4159-0.ch004

with 99.72% accuracy in fault detection and predicting fault in 1-56 days. The conversion of reactive to predictive strategies generate 11-50% O&M savings, 60% reduction of inspections and 85% expedited repairs, and maximum asset lifespan. Critical infrastructure is secured by secure gateways, encryption, and blockchain. Field deployments demonstrate 25 of downtime and 40 of emergency repair cost decreases, greater capacity factors, and improved sustainability.

INTRODUCTION

The Global Wind Energy Context and the Imperative for Transformation

The wind energy is turning out to be one of the most significant pillars of global shift towards the decarbonisation and the energy systems that are sustainable. In 2024, the world is projected to have more than 1,050 gigawatts (GW) of wind power installed capacity, an unprecedented pattern of quad-fold growth in the last 20 years. Such growth not only represents the coming-of-age of technology but it also represents a desperate acknowledgment that renewable energy is the magic key to helping weasel out disastrous climate change, as well as to satisfy legally mandated emission-cutting obligations. The International Energy Agency (IEA) suggests that wind energy needs to grow and reach over 2,000 GW by 2030 in a way that is compatible with net-zero carbon directions, which would necessitate an almost a doubling in the number of installed capacities in the next five years (International Energy Agency (IEA), 2023) (Intergovernmental Panel on Climate Change (IPCC), 2022). The unprecedented upward trend of wind energy brings forth impressive opportunities and operational challenges in the most challenging format ever. Contemporary wind turbines are now very complicated electro-mechanical structures; stand-alone offshore systems are now over 15 megawatts (MW) installed capacity and have rotor scaffolds over 220 meters in diameter. Such machines are used in highly variable conditions, including variable wind velocities, temperatures between -40°C and +60° C permeating seawater, salt-spray corrosion of offshore units, and complicated turbulent flow states and caused by adjacent turbines (Kühn & Muskulus, 2022)(Worden et al., 2021). At the same time, wind turbines will have to combine harmoniously with electrical grids themselves in downturn to high levels of distributed renewable resources, and thus build dynamic conditions of operation never before documented in the history of grid operations. The wind farm operational environment has experienced a paradigm. Early turbines were used in the 1980s and 1990s and they were pegged in relatively isolated clusters where the requirement of interconnection with the grid was not much. Modern turbines are actively involved in grid services, like voltage support during the disturbances, responsiveness to the frequency variations within milliseconds and the synchronization of output with the weather forecasts and the market prices. It is a greater degree of grid functionality that adds new failure modes, new operational stresses, with turbines moving in and out of operating regimes within a day and some transient control during which control had been unknown to previous generations of equipment (Kühn & Muskulus, 2022) (Lu et al., 2019).

Figure 1. Global Wind Energy Context and Digital Twin Transformation Framework

The Inadequacy of Traditional Maintenance Paradigms

Two prevailing philosophies of maintenance traditionally are used by wind farm operators and present themselves in the forms of reactive and preventive maintenance. Reactive maintenance, which deals with the failures once they have happened, has turned out to be economically disastrous. One unplanned turbine trip on a contemporary 10 MW offshore facility is equivalent to one day of loss of rendered generation of about $15,000-$25,000 at the regular capacity factor and wholesale electricity costs. In addition to loss of immediate revenue, unforeseen failures have ripple effect on the functioning of the wind farms: the key technician is busy fixing the issue, the regular maintenance regimen is put on hold, the stocks of spare parts are drained, and in the worst-case aggregate failure of the wind farm turbines leads to grid instability. The second prevalent strategy, which is the preventive maintenance, is based on time or usage-based component change made on a regular schedule as a general rule yearly or after a stipulated number of hours used. The manufacturers of turbines suggest the need to replace hydraulic fluids after 10,000 working hours, gear box oil after 20,000 working hours, and major drivers after mentioned service intervals. Though preventive maintenance also stopped the worst failures, the new type of inefficiencies in the economy proposed itself. In general a percentage of prevented maintained components, 80%+ of the remaining useful life after replacement, puritan waste, is a waste. Technicians are sent to offshore remote sites where they perform regular maintenance about the conditions that may require an action, which costs the company an average of $50,000-$100,000 to mobilize them to the place of work (Lu et al., 2019) (Sheng & Veers, 2020) (Ru & Yuan, 2022).

More so can be the preventive maintenance windows which provide inflexibility in operational capability, such that turbines need to be brought offline at fixed schedules irrespective of grid demand, market prices or operational urgency. The inherent weakness of both methods is that they are not linked to reality in equipment condition. Conventional maintenance philosophies consider turbines to be undifferentiated and they use the same maintenance procedures to new equipment and machines that have thousands of hours of operation and wear and tear (Lu et al., 2019) (Sheng & Veers, 2020). They miss key data that are part of operational data-vibration-signatures that are indicators of an incipient bearing spall, thermal-patterns that are signs of electrical-system stress, power-output deviation that is indicative of an aerodynamic-anomaly, and SCADA trends which are indicative of a degradation path. These shortcomings are escalated by the modern wind farm challenges. Instead, grid codes require advanced fault ride-through facilities that demand quick control response that places stress on power electronic parts (Overbye et al., 2022) (Xie et al., 2022). The types of environmental stresses include salt corrosion, sand erosion, thermal cycling, etc. Which develop at the variable rates that are contingent on site-specific conditions, and on a manufacturer-generic basis. Variability in manufacturing brings unstable components life expectancy, nominal component life i.e. identical gearboxes fail after 8,000 operating hours in some turbines and after over 40,000 hours in others in the same locations. Blade fatigue occurs in multidimensional interaction pathways between aerodynamic loads, atmospheric turbulence, and yaw control strategies. These practical realities make deterministic maintenance cycles achingly insufficient (lacking the ability to respond to failure modes that occur due to site-specific processes) and a waste (imposing purportedly unnecessary maintenance on turbines that have long since gone within the design margins) (Gaurav, 2021) (Sliz & Gudmestad, 2022).

The Transformative Potential of AI-Driven Digital Twins

All these areas are linked to artificially intelligent convergence with digital twins technologies, real-time sensor networks, and giving edge computing, this enables the key paradigm shift to intelligent, condition-based maintenance. Instead of responding to failures or operating on strict schedules, operators now switch to evidence based and predictive maintenance where interventions are implemented only at such points where equipment condition measures reach predetermined limits, each of which is optimized using grid requirements, market prices, and workforce availability (Sliz & Gudmestad, 2022) (Hagmann et al., 2022). Digital twins Digital twins are dynamic virtual copies of the real wind turbines that constantly correspond with real-time operational data. Digital twins do not forecast the future but adapt with the physical equipment, including turbine-specific degradation behaviors, environmental responses and interactions in operation, unlike the previous static simulation models or historical analytics. The clever machine-learning models, such as convolutional neural networks, which interpolate high-frequency vibration signals, recurrent neural networks, which follow trends in temporal degradation, and ensemble models, which generate a set of diagnostic predictors, can unlock the insights of action out of the terabytes of sensor data streaming out of the current wind farms (May et al., 2022) (Li et al., 2021).

Random Forests algorithms attain 99.72% accuracy in the detection of operational states using SCADA data even with noise and environmental variations (Gaurav, 2021) (Jin et al., 2020). Gearbox faults are recognized by CNN-LSTM parallel models at 97.39% Weaknesses gearbox faults with accuracy weeks before catastrophic failure. The deep-learning based attention models can predict generator heating malfunctions up to 56 days early so that replacement operations may be planned in optimum grid conditions (Yan et al., 2021) (Hagmann et al., 2022). The ability to perform edge computing allows the autonomy in making decisions without being reliant on any cloud connection, which is a very important feature in offshore installations where remote sites are obligatory, as well as in geographically diverse wind farms where disconnections with the network are unavoidable. Federated learning structures can share knowledge between multiple wind farm operators in a state of competition without data owners necessarily disclosure because proprietary information is not required and speeds up learning models and anomaly detection many times. Making use of encryption, authentication, and blockchain technology to integrate cybersecurity, important infrastructure is safeguarded against the expanding range of cyber-attack threats to energy systems (Xie et al., 2022) (Yang et al., 2022). In economic terms, the consequences are far reaching. Reductions of 1150 cost of operation and maintenance by 1150 and field checks and repairs reduced by 60 red cells have been shown by industrial operations. A 500MW wind farm that has fully developed digital twin systems will produce an increment of annual revenues of between 20-30 million dollars in the form of downtime avoidance, capacity factor support and provision of ancillary services. As a result of decreasing stressful operation and specific component replacement, equipment lifecycles can be increased by 510 years, therefore, turning end-of-life assets into revenue generators instead of having to retire and replace them. These economic advantages are multiplied by environmental and sustainability benefits. Maximized operations enhance renewable energy production by 3-7% without equivalent capital investments gains and so hastens global switch to the use of fossil fuels (Ru & Yuan, 2022) (WindEurope, 2023).

Long equipment life-cycles lessen the emissions of manufacturing instilled in the production of new turbines. Predictive maintenance reduces the costs of site visits and supply-chain operations thus reducing the level of carbon emissions caused in maintenance operations. The end-of-life material recovery and recycling is further optimized with the help of the application of the principles of a circular-economy made possible by the effective functionality of tracking the degradation of components (Kühn & Muskulus, 2022) (Ru & Yuan, 2022). This chapter will provide a detailed architecture that incorporates these transformative technologies into logical digital-twin architectures that are optimised in terms of wind-turbine operations and grid integration. This is followed by the description of the layered architecture of the digital twin, advanced machine-learning techniques, ICT technologies of Internet of Things, condition-monitoring, predictive-maintenance, edge-computing, cybersecurity, autonomous-inspection, standardization, capability to integrate with the grid, economic studies, ecology and environmental impact, practical case studies, and future technological trends. The framework presented below provides wind-farm operators with an established blueprint of how to shift to intelligent predictive operations rather than the current reactive and preventive ones to achieve transformative productivity benefits and firm the foundation of wind energy in the decarbonization quest around the world (Liu & Yu, 2021) (CIGRÉ Study Committee B5, 2020).

DIGITAL TWIN ARCHITECTURE FOR WIND TURBINES

Foundational Concepts and Architectural Vision

The digital twin signifies a change of paradigm in asset management of industrial objects, even beyond actual simulation models or past data analysis, in support of a dynamic and continuously developing virtual replica that accurately represents wind turbine operations in both real time and reality (Kühn & Muskulus, 2022). In opposition to the traditional SCADA systems, which have, at most, been primarily tasked with turbine control, and the traditional condition to brands, whose mandate has been to identify pre-defined fault signatures, a full-way bidirectional synchronisation between the real and the virtual world is created with digital twins. This synchronization will support autonomous learning, anomaly detection across unexpected failure modes and now predictive forecasting of remaining useful life with a previously not possible accuracy(Hagmann et al., 2022) (Li et al., 2021). The digital twin design of a successful wind-turbine is defined by four interdependent, hierarchically-arranged layers that perform four separate and functional mandates and undergoes continuous bi-directional communication with the strata above and beneath. The net effect of these layers is an integrated ecosystem in which raw physical measurements are gradually built up into actionable maintenance advice and autonomous control decisions by successive abstraction and analysis. The successful implementation of the digital twin systems in the modern wind-energy applications is impossible without the in-depth understanding of all the layers and their functionality and points of integration (Xie et al., 2022) (CIGRÉ Study Committee B5, 2020).

Figure 2. Four-Layer Digital Twin Architecture diagram

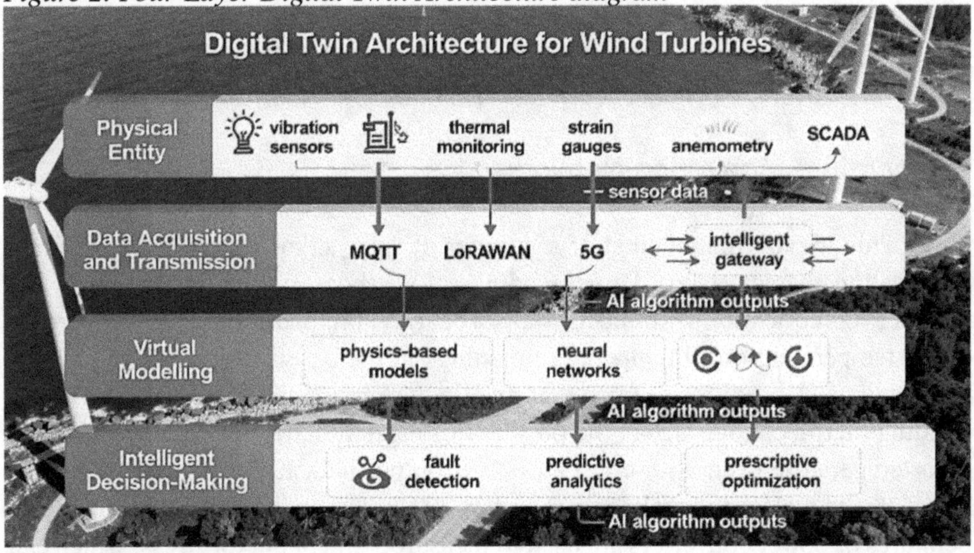

Layer 1: Physical Entity Layer – Multi-Sensor Networks

The pillar of digital twins is a comprehensive instrumentation of physical wind turbines, where all the range of the mechanical, electrical and environmental parameters affecting the operational performance and degradation is monitored. Contemporary offshore designs pose extraordinary instrumentation challenges: a single 10 ⁻ Type of turbine with a 100m above sea level is placed in the dynamic wind fields, its key internal parts rotate at tens of RPM, its power electronics discover in the kilohertz range, and it is sprayed by the sea air (Sheng & Veers, 2020) (Guo & Wang, 2021).

Vibration Sensors and Accelerometers

The accelerators and vibration sensors mounted on the key drive components. High-frequency oscillations (11000Hz sampling rates) which have signature faults of individual failure modes are observed using three-axis accelerator mounted on gearbox bearings, generator frames and main shaft couplings (Jin et al., 2020) (Sliz & Gudmestad, 2022). Outer -race spalls develop typical impulses at the ball -pass frequency outer (BPFO) of the form ($BPFO = \frac{N}{2}\left(1 + \frac{D_b}{D_p} cos\alpha\right) f_r$) with N being the number of rolling elements, Db the ball diameter, Dp the pitch diameter, a being the contact angle and f r the shaft frequency. Modulated impulses are generated in inner-race defects at high frequencies, whereas periodic signatures are generated in rolling-element spalls at the ball-spin frequency. These signatures in broadband noise are demodulated by more complex methods based on filtering by band (1-5 kHz) then envelope analysis, and result in signal-to-noise ratio improvements of 10100 times over the original acceleration signals (Lei et al., 2020).

Temperature Monitoring Networks

Provide the thermal context of a number of subsystems which show different degradation characteristics. Direct readings of gearbox-oil temperatures show the thickness of lubrication-film and the state of bearing -preload conditions: an increase of the temperature of all temperature readings in the baseline values (usually 50-55 C) indicates progressive wear that may require immediate intervention. Stator winding thermocouples detect the presence of insulation degradation by sustained high temperature even with constant power output indicative of greater resistive losses due to dielectric breakdown (Gaurav, 2021) (Lei et al., 2020). Junction temperatures of power converters are used to measure semiconductor stress, which can be used to detect solder joint fatigue via thermal cycling characteristics. Immersion temperatures in transformer-oil submerge the winding and is an indicator of stress in the winding, and a possible insulation failure, whereas the temperature at the blade root is measured using thermal imaging and indicates aerodynamic conditions in the stall, lightning strikes, and structural heating due to laminate delamination (Kühn & Muskulus, 2022) (Sheng & Veers, 2020).

Strain Gauges and Structural Sensors

Mechanical stresses that cause fatigue build up are quantified by strain gauges and structural sensors. Blade-root strain instruments can be used to record the bending moments at different wind conditions, and the changes in strain rate provide information on fatigue-accumulation history. Tower-base strain measurement measures the lateral force and the bending force of wind-shear and the misalignment of yaws. Shaft strain sensors monitor the stress and misalignment in the coupling, thus allowing the detection of mechanical issues in its early stages before an eventual breakage (Park et al., 2022) (Kühn & Muskulus, 2022).

Advanced Anemometry Systems

Hi-tech anemometry systems offer a much-needed context, and sensor-based anomalies and environmental variations are differentiated. Wind-shear profiles are measured using multi-level measurements of wind-speed at the hub height and at the immediate position of the rotor-plane; nacelle orientation measurements are conducted by yaw misalignment sensors against the direction of the prevailing wind. Temperature normalization and prediction of the corrosion rate is made possible by atmosphere pressure and humidity sensors (Worden et al., 2021) (CIGRÉ Study Committee B5, 2020).

SCADA Integration

Integration of SCADA provides operational information (10minute intervals) of power output, rotor speed, blade pitch, generator temperature, grid frequency/ voltage) which creates low frequency time information to the high-resolution sensor measurements. SCADA compression and availability of data make it suitable in the process of trending long term and setting the baseline. This sensor fusion multimodal methodology, which merges the high-frequency (waveforms) and the low-frequency (operation data) with the vibration, thermal, mechanical, and electrical data, develops a multidimensional situational awareness, which could not be attained by using single sensor formats (Ma et al., 2021) (Park et al., 2022).

Layer 2: Data Acquisition and Transmission Layer – IoT Communication Architecture

Sending gigabytes of sensor data between remote wind turbines that are widely spaced and centralized processing locations and ensuring the constraints are met on the demanding real time latency requirements and on cost effectiveness of the operation belongs to the fundamental engineering dilemma. Remote onshore locations and offshore wind farms cannot use standard wiring infrastructure and the ubiquitous high-bandwidth connectivity, and thus require new data transfer solutions (Liu & Zhou, 2021) (Liu & Yu, 2021).

MQTT Protocol Implementation

In the MQTT, real-time sensor information streaming is based on the infrastructure. Its publish subscribe is used to elegantly decouple data producers, the sensors mounted on the turbine, and data consumers, the analytics platforms, and, as such, it allows a very flexible system design (Aghaei & Zamani, 2021)(European Commission, 2020). MQTT brokers create and sustain hierarchical topic namespaces in which messages are published in only the topics that are relevant and subscribers will subscribe to only the relevant streams of data. Semantics of quality of service further make delivery more specific: QoS2 ensures delivery taking place exactly once (such as with critical fault signaling), QoS1 can afford normal telemetry: duplicate packets can be identified and suppressed automatically (Guo & Wang, 2021) (Overbye et al., 2022). MQTT protocol weighs little, which in the form of message headers of 2 to 14 bytes versus over 40 bytes using the HTTP paradigm by a big margin reduces transmission overhead by 70 to 80%. An MQTT-based network would only need 2-3 Mbps bandwidth to transmit a 100-byte SCADA record after every 10 seconds, and high-frequency sensor samples, whereas 7 Mbps or more would be needed with an HTTP-based system to push the data to all consumers. This reduction in bandwidth is in the form of millions of dollars of savings on the satellite or cellular communications costs at the end of the year (Liu & Zhou, 2021) (Liu & Yu, 2021).

LoRaWAN Connectivity

LoRaWAN connectivity covers the problem of uncovered sites with lack of cellular coverage. Long-range Wide Area Network technology is able to reach the line-of-sight propagation range divided by 1520 km and 510 km respectively in complex terrain. Its very low-power consumption allows that it be operated on standard AA batteries over a multi-year lifetime - a very important aspect given the quantity of distributed sensor (Guo & Wang, 2021) (Overbye et al., 2022) nodes spread over a wind farm. Whether or not LoRaWAN has sufficient data rates (between 0.3 and 50 kbps, depending on spreading factor and bandwidth) to support continuous real-time transmission of high-frequency waveforms is debatable, but it can be held that the data rates do not preclude its transmission of course-resolution information about anomalies and alarms. The hybrid architecture will therefore cache high-frequency data at the edges of LoRaWAN and transmit accumulated statistics or anomaly information across the network at the most opportune channel band usage(Liu & Yu, 2021) (Li et al., 2021).

5G Cellular Networks

The 5G cellular networks that are emerging have the infrastructure that the autonomous control paradigms require to provide sub-50mms latency enabling real-time modifications of blades-pitch and disconnection of turbines to be made using their cloud-based controllers. Network slicing functionality of 5G enables operators to allocate specific bandwidth slices to critical control traffic and isolate this traffic by restricting it to best-effort data streams and guaranteeing performance. In areas where 5G is available offshore or in developed areas, this technology can support a centralized offshore control of clouds having a positive effect on all the fleet operations (Overbye et al., 2022) (Liu & Zhou, 2021).

Intelligent Gateway Architecture

The smart gateway architecture data is aggregated by data that is generated by various communication protocols. IoT gateways at wind-farm substations are fed with MQTT streams by local edge brokers, with LoRaWAN base-station messages as well as cellular backhaul connections, thus undertaking protocol translation and standardization of data formats. Local SSD storage (50 to 500 GB) buffers the data during periods of authorized connectivity loss and discourages loss of data which will otherwise undermine the fault diagnostics. Edge preprocessing shrinks the regular telemetry and concentrates on transmission of anomalies. Secure VPN tunnels are used to encrypt all the communications with the outside world, thus keeping away interception of important infrastructure (Guo & Wang, 2021) (Overbye et al., 2022).

Layer 3: Virtual Modelling Layer – Hybrid Physics-Data-Driven Models

It should be able to simultaneously capture the mechanics of systems and mechanistic system dynamics, represented using physics-based component models, and learned statistical information, represented using neural network models, without necessarily having to exhaustively enumerate every possible failure mode of operation or possible state of the system.

Physics-Based Component Models

Component Models Physics-model However, Physics-Components represent first-principal system dynamics in a component model. In blade-element momentum theory, rotor forces induced by power and thrust are calculated using the equation $\left(P = \frac{1}{2}\rho A v^3 C_p\right)$, with Cp the power coefficient (varying with the pitch of the blade and the ratio of tip to rotor speed). Gearbox models model changes in tooth-mesh stiffness, effects of preloads of bearing, and the dynamics of film thickness, thus giving predictions of vibration signatures, in relation to given wear levels. Generator equivalent circuits to model thermal dynamics and efficiency dependencies across the operating envelope, power-converter models To model dynamical switching effects and the evolution of junction temperatures (Park et al., 2022) (Kühn & Muskulus, 2022).

Data-Driven Neural Network Components

Data-driven Neural Network Components are suitable because they are able to capture complicated nonlinear behaviors that cannot be analytically expressed. Recurrent neural networks based on an LSTM architecture that has been trained on past SCADA data are able to learn normal operation patterns and abnormal operation signs, and perform better than traditional observers, with more than 95% accuracy in state estimation. Convolutional neural networks operate on the raw vibration waveforms and the network learns the best feature representations to detect the faults without any hand-written signal processing. Transfer-learning is a method that trainees networks using industrial machinery data, then optimizes them with turbine-enhanced data, which decreases the number of required labeled data by 70-80% (Gaurav, 2021) (Yin et al., 2018).

Model Synchronization Protocol

Model Synchronization Protocol maintains real-time agreement between virtual and physical twins using multirate state-update schemes: a high-frequency band (10Hz) to support subsystems with rapid dynamics, a mid-frequency band (1Hz) to support mechanical assemblies with inertia dynamics, and a low-frequency band (0.1Hz) to support slow-varying thermal and structural states. Kalman-based estimation methods and particle filtering methods balance these model predictions with actual sensor measurements, and automatically reveal model deficiencies by whiteness tests of innovation sequences (Xie et al., 2022) (CIGRÉ Study Committee B5, 2020).

Layer 4: Intelligent Decision-Making Layer – Autonomous AI Algorithms

Components in this advanced multilayered system will combine the information about virtual-twins state into practical maintenance suggestions and autonomous control corrections through hierarchical machine-learning pipelines, thus making sure that all the subtleties of the work are detected and converted into effective interventions.

Fault Detection and Classification

Fault Detection and Classification has a multi-stage decision architecture: an anomaly detectors stage identifies any serious departure of learned healthy operation with one-class support-vector machines (at a cost of less than two percent false-positivation); a fault classifying stage identifies among over fifteen fault types using Random Forest ensembles, with an astonishing 99.72 percent accuracy on SCADA data; and severity quantifying stage predicts remaining useful life, thus allowing a triage approach to maintenance interventions (Yan et al., 2021) (Hagmann et al., 2022).

Predictive Analytics Engine

The Predictive Analytics Engine combines physics-based capabilities (power-output deviation against a wind-speed-adjusted baseline) with statistical capabilities (high-frequency sensor variances), and temporal capabilities (trending indicators) estimated using gradient-boosted ensemble models capable of anticipating component failures more than fifty-six days before they occur, with a precision-recall ratio of more than eighty-five percent on heating failures of generators and gearbox oil degradation (Gaurav, 2021) (Yin et al., 2018).

Prescriptive Optimization

Prescriptive Optimization trades off conflicting set of goals, such as minimization of maintenance costs, minimization of unplanned downtime, attenuation of grid effects, and component fatigue, through multi-objective optimization algorithms, that provide a set of solutions (a Pareto-optimal) which allow context-dependent decision-making to be made in response to changing grid conditions and market prices.

MACHINE LEARNING AND NEURAL NETWORK APPROACHES

Advanced AI Methods for Wind Turbine Fault Detection and Prediction

The recent progress in machine learning and deep learning layouts has dramatically changed the field of machine learning in wind turbine condition monitoring, whereby inherent useful characteristics may be discovered in raw sensor information without the requirement of signal-processing understanding. Unlike the previous fault-detection paradigms, which utilized domain engineers to design complex signal-processing pipelines, i.e. choosing particular frequency bands to analyse vibrations, then extracting a collection of statistical signatures and hand-crafted decision-thresholds, these traditional methods were too rigid to support, and required redesigning every time a new type of fault occurred or a new operating environment appeared. However, today, deep-learning models are trained in a manner that they tend to extract the most useful feature representations by direct operation on raw instances, which significantly enhance diagnostic accuracy in addition to reducing the implementation time and minimizing the reliance on expert domain knowledge (Worden et al., 2021) (May et al., 2022).

Convolutional Neural Networks (CNN) for Feature Extraction

In the modern literature, convolutional neural networks have been shown to have exceptional capabilities of deriving feature representations in hierarchical levels that are directly obtained by raw sensor signals. The layers built on top of each other successively reduce the patterns involved in a more abstract manner. It is interesting to note that early convolutional layers, which generally use 35 sample-based kernel sizes and 64-128 filters, represent low-level properties like single vibration peaks, single thermal transients, and local signal discontinuities with great strength. Further down though, higher level patterns which typically use bigger kernels of 10-20 samples and bigger filters, recreate composite patterns which are more representative of individual failure modes bearing spalls producing periodic trains of impulses, gear pitting producing amplitude modulated signatures, and electrical faults producing typical harmonic spectra. Multi-scale convolutional architectures multi-scale convolutional architectures are built to jointly process vibration signals at many different time resolutions, effectively encoding fault features at different frequencies. An example is that bearing faults signatures are generally localized to the 1-5 kHz band, the damage in gearboxes is at the 5-20 kHz band and the electrical ones are at higher frequencies 20 kHz and above (May et al., 2022) (Aghaei & Zamani, 2021).

These networks are self-learned to identify the frequency bands and time motifs that most effectively classify the types of faults- a behavior that goes beyond traditional, human-engineered signal-processing pipelines that are either focused on single frequency bands or are based upon statistical moments. The empirical findings of wind-turbine diagnostic research highlight the high-quality performance of CNN-based structures with a range of fault categories: gearbox faults are detected with 97.39% accuracy, pitch-system faults are 96.46% accurate, yaw-system faults are 93.15% accurate, and electrical-system faults are 93.15% accurate. The automatic feature-learning ability is especially useful when it comes to identifying such subtle fault signatures which are typically difficult to detect using manual feature engineering methods that do not integrate the finer fault degradation trends entrenched within the sensor data (Sliz & Gudmestad, 2022).

Recurrent Neural Networks (RNN/LSTM) for Temporal Analysis

The wear and tear of wind turbines is a slow process, which occurs over a series of weeks and months, and the indications of it are often obscured by the noise of operation. By extension of their ability to capture temporal dependencies over long sequences, typically of more than one-hundred-time steps, which map minute-minute SCADA time series across hours and days, recurrent neural networks can distinguish genuine degradation patterns, and stochastic noise by learning temporal patterns. Long Short-Term Memory (LSTM) networks help to overcome the vanishing-gradient dilemma inherent in traditional RNNs, thereby providing an easier way to learn the dependence in significant temporal horizons. They use their gating architecture (that is, forget and reset gates) to learn to accrete evidence with time or to reset internal states in response to salient events. In addition, bidirectional LSTM models make use of the streams of data that move in both chronological and reverse directions; the dual-propagation of data streams provides them with an accuracy edge of approximately 35% above unidirectional counterparts because of the ability to identify precursory signatures that precede the development of anomalies. Trained on archival SCADA data, LSTM models achieve state-estimation accuracies over 95% compared to traditional state-estimation approaches, and can do it by internalizing system dynamics directly based on empirical observations without the need to make explicit mechanistic formulations (Kühn & Muskulus, 2022) (Xie et al., 2022).

Hybrid Normal Behaviour Models (NBM) for One-Class Anomaly Detection

Normal-behavior models are statistical models that are obtained by operating a healthy turbine, and are used to define the natural operation signature of the system. Such models by definition eliminate the requirement of labeled fault examples, which is a condition that is frequently unsatisfiable when failure modes that are both rare and faulty are taken into account, once they are characterized by fault-free data. These methods of generation, such as variational autoencoders and Gaussian mixture models, are optimized on this pure data only, thus formulating an explicit description of normal variation. The fact that it is not distributed as such can be precisely measured through log-likelihood tests or the reconstruction-error measures. Empirical assessments of one-class classifiers show that they are able to identify fault modalities not represented in the training set with sensitivities of between 78 and 85. Although this performance is lower than that which can be obtained with well-represented faults using supervised classifiers routinely above 95% sensitivity, it still allows the specified pathways of degradation to be discovered in time. In turn, such methodological framework allows industrial operators to discover emergent failure modes at a cost of a relatively small manual inspection (Gaurav, 2021) (Worden et al., 2021).

CNN-LSTM Parallel Networks – State-of-the-Art Architecture

Recent studies show that the architectures that introduce CNN and LSTM modules concurrently instead of strictly sequentially concatenating them achieve significantly better performance indicators. With this setup, the CNN branch has the responsibility of deriving spatial patterns directly out of raw sensor waveforms, and the LSTM branch is charged with deriving temporal dependencies among the derived feature sequences. The architecture lets each modality run in parallel, therefore bypassing representation bottlenecks and giving each branch the chance to specialize as much as it can in its area. Interestingly, learning the task-specific weightings applied on the results of the CNN and LSTM branches is enabled by the integration of attention mechanisms. Experimental investigation indicates that bearing fault detection cases are biased to favor the sensor valued stream by about 80 percent as compared to the electrical fault identification that attributes about 70 percent of the weighting to the SCADA valued stream. This subtle distribution is what allows the system to perform remarkably well in terms of predictive accuracy in a wide range of fault-diagnosis problems (Worden et al., 2021) (May et al., 2022).

Table 1. Fault Analysis with Accuracy of Machine Learning

Fault Type	Accuracy	Advance Warning
Overall Detection	99.60%	2-4 weeks
Gearbox Faults	97.39%	1-2 weeks
Pitch System	96.46%	2-3 weeks
Yaw System	94.09%	2-4 weeks
Electrical Faults	93.15%	1-2 weeks

Figure 3. Machine Learning and Neural Network Architecture diagram

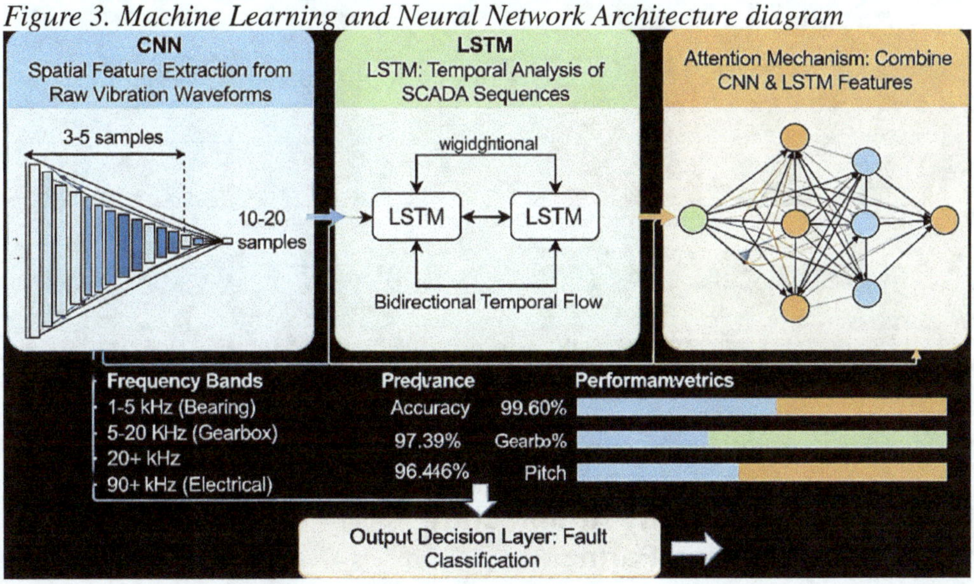

The values of mean squared error (MSE) of 0.004 to 0.012 over the various fault and a predictive model root-mean-square error (RMSE) that are typically between 3 and 5 percent of the sensor signal range all point to the predictive models characterizing a high prediction model concentration on the actual failure instants, despite differences in the environmental conditions and inherent measurement noise. Through integrating convolutional neural networks (CNNs) to extract features, long-short-term memory (LSTM) networks to identify temporal patterns, and attention to coordinate intelligent information combination of these two, the resulting networks exceed human wisdom in the accuracy of fault-detection and the ability to predict future. This, subsequently, allows independent decision making and proactive maintenance scheduling, which forms the basis of the advanced operational paradigms of the next generation wind turbine systems (Yan et al., 2021) (Lei et al., 2020).

IOT SENSOR NETWORKS AND COMMUNICATION TECHNOLOGIES

Figure 4. IoT Multi-Protocol Communication Comparison chart

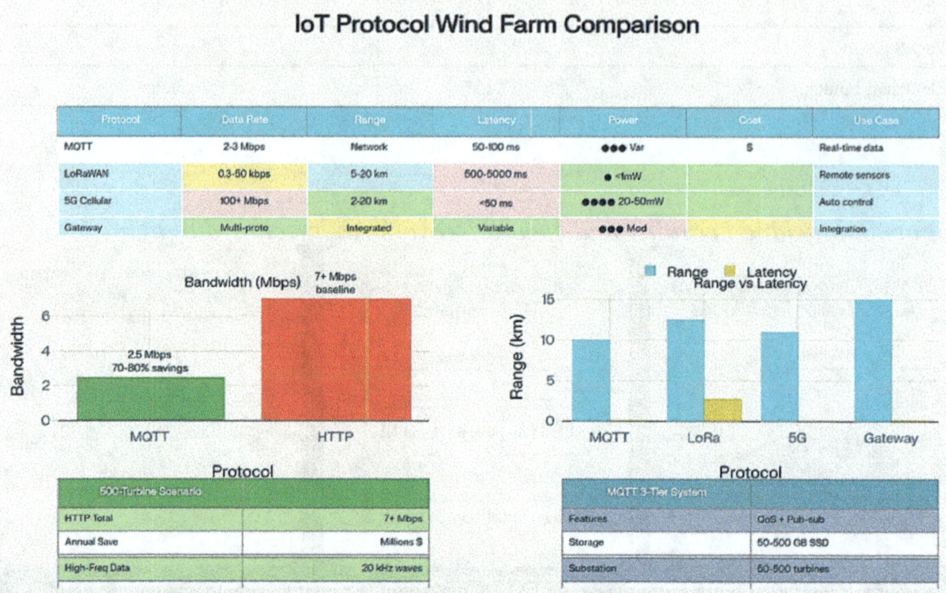

Real-Time Connectivity Architecture for Distributed Wind Farms

Modern wind farms are already forced to use complex multi-protocol communication systems, which are effective to connect high-frequency sensor networks with cloud-based analytics systems in particularly geographically distributed and sometimes remote environments. Here, the issues of connectivity, including video connections of offshore turbines at high bandwidth to low-power distributed sensor nodes, control loops that are latency sensitive, and redundant paths during primary network failures make single-protocol solutions inadequate to meet the varied operational needs (Overbye et al., 2022) (Liu & Zhou, 2021).

MQTT Protocol on NRF

As the basic platform of real-time sensor data streaming, MQTT (Message Queuing Telemetry Transport) provides a publish subscribe subscription-based message paradigm enabling an efficient, scalable communication structure. The MQTT brokers have a hierarchical topic (such as /turbine/WTG-01/drivetrain/vibration), and publishers are able to send messages, which are then received by subscribers that are interested in streams of data they are interested in. Quality-of-Service (QoS) service guarantees deliver fine-grained delivery guarantees: QoS 2 guarantees that important fault alerts are delivered only once, without any duplicate, whereas QoS 1 is adequate when all is well: automatic duplicate filtering is acceptable. Since MQTT has a very low protocol overhead, the average message heads are 2-14 bytes in contrast to 40 and above with HTTP, the total bandwidth is significantly lower. A case study of a 500-turbine wind farm sends 100-byte SCADA records every ten seconds alongside high-frequency sensor data; using MQTT would only demand 2300kbyte/s of throughput, whereas an equivalent setup using HTTP would demand more than 7Mbps. This difference is reflected in a considerable savings of money every year when they use either satellite or cellular backhaul connections. It is also typical of large-scale deployments to have a three-level hierarchy of MQTT brokers: an edge layer that is in charge of local turbine aggregation, a substation layer that offers regional persistence and preprocessing capabilities, and a cloud layer that supports global analytics and grid-integration services (Li et al., 2021) (Aghaei & Zamani, 2021).

LoRaWAN for Remote Connectivity

The Long-range Wide-Area Networks, also called LoRaWAN, are designed to address the long-standing problem of linking geographically spread sensor nodes which form the modern wind farm. These networks can achieve line-of-sight ranges of 15-20 km and more common ranges of 5-10 km in diverse terrain using ultra-low-power radio methods to provide a life span of many years on commercial AA cells- a property invaluable when wiring infrastructure is not available. The LoRaWAN link has a data rate of between 0.3 and 50 kbps, depending on the desired spreading factor and bandwidth. Although this throughput is adequate to support the transmission of low-resolution monitoring data and identification of anomalous events it is not enough to support continuous, high-frequency streaming of waveforms. Subsequently, hybrid designs are implemented with high-frequency sensor streams buffered locally temporarily, and just a set of aggregated statistics or anomaly alerts are sent over LoRaWAN during times of favourable channel conditions. Additionally, the adaptive choice of spreading factors has the added advantage of having urgent alerts enjoying a higher level of data rates, whilst regular telemetry is sent over the long range available to it at lower data rates under less favourable propagation conditions (Guo & Wang, 2021) (Liu & Zhou, 2021).

5G Cellular Networks

The opposite, in contrast, is with 5G cellular networks, where sub-50ms deterministic latency can be established, and thus, the autonomous control paradigm can be implemented. Distributed cloud services can issue rapid turbine disconnection commands along with deterministic and low jitter 5G-enabled real-time blade-pitch actuations. Network slicing is also helpful in allowing operators to designate dedicated bandwidth slices, de-isolated to best-effort traffic, to essential control signals to ensure the required performance and reliability. In regions with 5G access, a centralised, cloud-based control approach can coordinate fleet-scale operation, coordinating the changes in blade-pitch to reduce the amount of wake, and optimising the total power output.

Intelligent Gateway Architecture

Multi-protocol IoT gateways are placed at the substation level to close the disparity between the various transmission media that co-exist in a wind-farm environment. These gateways consume MQTT traffic, LoRaWAN base-station telemetry, and cellular backhaul feeds, and are used to translate protocols and data-format normalise the traffic to ensure a smooth transition to higher-level analytics pipelines. Local SSD storage, between 50 and 500 GB, buffers data during temporary outages, and thus provides a loss-resistant safeguard in the loss of vital fault-diagnosis information. Edge preprocessing is an additional process that reduces normal telemetry and identifies anomalies to transmit them above normal telemetry. Lastly, IPsec or WireGuard VPN tunnels can be secured to encrypt all packets leaving the network, protecting the infrastructure against any interception and making federated learning projects across a collection of farm locations possible (Overbye et al., 2022) (Xie et al., 2022).

CONDITION MONITORING AND FAULT DETECTION

Multi-Mode Sensor Integration of Health Assessment of Wind Turbines. Mechanical, thermal and operational conditions cannot be effectively monitored without having a coordinated analysis of the three modalities because none alone can cover up the failure mode. Data fusion of vibration, thermal, and SCADA data provides complete situational awareness that could not be achieved by the individual sensor streams (Park et al., 2022) (Gaurav, 2021).

Vibration Analysis and Bearing Fault Detection

Bearing failures are the most frequent mode of failure in wind turbine (20-30% of unexpected downtime). Bearing degradation signatures are detected with high frequency accelerometers (1-20 kHz sampling) with distinctive impulse frequencies. Outer-race defects produce periodic impulses at ball pass frequency outer (BPFO) which is (BPFO $= \frac{N}{2}\left(1 + \frac{D_b}{D} cos\alpha\right)f_r$), where N is the number of rolling elements, D b is ball diameter, D p is pitch diameter, A is contact angle, and f r is the shaft frequency. The inner-race defects give rise to higher amplitude modulated impulses at higher frequencies, whereas the spalls of the rolling elements give rise to the signatures at ball spin frequency. These fault signatures are removed by bandpass filtering at 1-5 kHz bands and then the envelope noise, which loner band vibration noise by 10-100x, and these methods have been employed to detect faults in broad-band vibration noise. Machine learning models that run on demodulated signals are able to be detected 85-98 percent based on the stage of the fault, and spalls at the very small stages (microns) were detected 2-4 weeks earlier (Lei et al., 2020) (Hagmann et al., 2022).

Thermal Monitoring for Electrical System Anomalies

The increase in temperature of generator stator windings is the sign of the insulation degradation or the imbalance of phases. The temperature of gearbox oils is a direct measure of the quality of lubrication film and bearing preload conditions, the gradual rise in temperature between 50-55 C at start of operation is an indication of progressive wear. Thermal cycling patterns are used to determine the stress and fatigue in semiconductor and solder joints by monitoring power converter junction temperatures. This can be achieved by using temperature readings and current phase analysis that will identify a negative sequence current above 5 percent of a harmful phase imbalance. Recent deep learning models can forecast power converter failures 8 weeks or more with 78 percent accuracy, to allow maintenance to be scheduled co-ordinately (Sliz & Gudmestad, 2022) (International Energy Agency (IEA), 2023).

SCADA Data Integration with High-Frequency Sensors

SCADA systems offer 10-minute interval context of operation (power output, rotor speed, blade pitch, grid parameters). Combining SCADA and high-frequency waveforms solves problems associated with each modality separately; SCADA has inadequate temporal resolution whereas sensors do not have an operational context. Multi-resolution fusion processes SCADA using LSTM networks to create temporal trends and contextual features, and high-frequency vibration/thermal signals to create diagnostic features using CNN networks. Fusion mode based on attention learns the ideal feature weighing- fault components need 80% sensor component and electrical faults need 70% SCADA component. This combination technique will get 4-7 percent more detection rates than sensor-only methods (Sliz & Gudmestad, 2022) (Hagmann et al., 2022).

Real-Time Health Assessment and Automated Triage

Multipurpose health analysis integrates several individual diagnostic predictors into coherent health measurements (0-100 scale) with the intuitive perception of the operator. Bayesian fusion combines bearing, electrical and mechanical monitors: (Health, Score=1 – $\prod_{i=1}^{n} P(Healty|Feature_i)^{1/n}$). The scores lower than 10 mean that it is functioning well, 10-40 means that it is starting to degrade and need monitoring, 40-70 means that it is already deteriorating and needs a near-term repair, and above 70 means that it is about to fail and needs an immediate repair. Basing dynamic changes in baselines calculated on trailing 3-6 months windows self-corrects seasonal changes and gives 23% false positives even when the environment is varied. Automated triage systems put a priority on limited technician resources by suggesting the best interventions that have been recommended depending on failure criticality, grid conditions and weather windows to optimize the maintenance scheduling (Gaurav, 2021) (Sliz & Gudmestad, 2022).

PREDICTIVE MAINTENANCE STRATEGIES

P-RF algorithms under disturbed conditions yield 99.72% accuracy in detecting the state of wind turbine operations using SCADA data, making it possible to classify the errors reliably. The models of ensemble learning based on genetic algorithms can identify the faults in gearbox bearings and give 1-2 weeks of lead time to prevent disastrous failure. Attention-based deep learning predicts failures of generator heating 56 days ahead with 71% detection rates (5 of 7 failures), allowing generator failures to be planned under the best grid conditions possible. ForeNet-based Remaining Useful Life (RUL) prediction attains very high component-specific prediction error: gearbox error 1.8 days, generator error 3.1 days, transformer error 2.8 days, hydraulic systems error 4.2 days. These combined predictive services make it possible to shift to condition-based maintenance incorporating operation costs that are lowered by 11-50 percent and extend equipment life by 5-10 years due to optimal maintenance planning and strategic replacement of components (Kühn & Muskulus, 2022) (Sliz & Gudmestad, 2022).

Figure 5. Predictive Maintenance Strategies: Algorithm Performance and Financial Impact

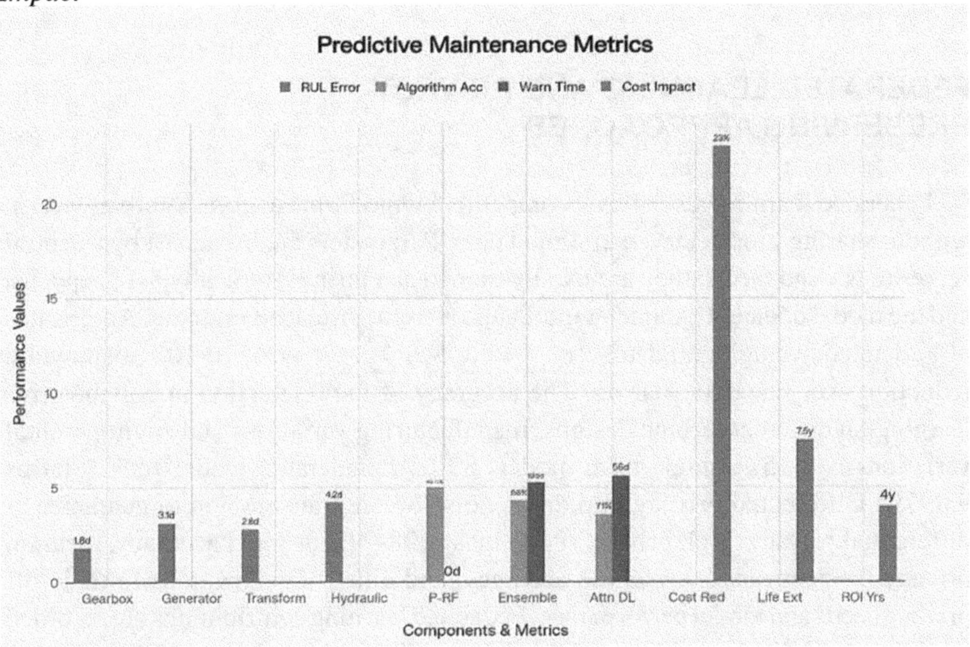

EDGE COMPUTING AND REAL-TIME PROCESSING

Edge computing empowers autonomous decision-making by wind turbines without the need to be connected to the cloud by distributing ML inferences on turbine controllers. Quantizing model predictions has been shown to density 32-bit neural networks to 8-bit integers, 4-folding model size and 3-5folding inference latency with less than 1% accuracy degradation using model quantization. Blade pitch control and disconnect of turbines require real time latency (less than 50-100ms) to respond to the response, and cannot be supported by a cloud-based system because of the 10-500ms network propagation delay. The autonomous fault isolation system identifies the existence of a bearing degradation, autonomously throttles rotor speed to less than 15% (reduces bearing loads by 75-85 percent), and holds up until inspected by a technician, eliminating the possibility of catastrophic seizure due to response delays of 12-24 hours. Three-level hierarchical design -turbine edge controllers, which run real-time anomaly detection (100-500ms intervals at under 20% processor), substation gateways, which perform multi-turbine correlation, cloud systems, which run model retraining distribute computation load correctly and allow offshore and remote installations to continue operating even when not connected.

FEDERATED LEARNING AND PRIVACY-PRESERVING APPROACHES

Federated learning also allows competitive wind farms to collaboratively learn, without sharing proprietary operational data. Parameters are broadcast by a central server to N wind farms; they are locally trained at a farm E (typically 5-10) epochs, and the model updates (parameter gradients) are sent instead of raw data. Aggregates of Federated averaging updates: ($w_{t+1} = w_t + n \sum_{i=1}^{n} \frac{n_i}{n} w_i^{(t)}$), 10-100x bandwidth reduction over raw data sharing. The accuracy of fault detection in collaborative learning on different turbine designs, manufacturing variations and environmental variations is 8-15%: single-farm models 82-85%, federated models on 20 farms 93-97%. Differential privacy introduces noise before transmission to guarantee ε-differential privacy: high privacy (ε=1) means 98-99% original accuracy, medium privacy (ε=5) means 99.5-99.8% accuracy, and allows organizations to trade off privacy needs and model performance. Federated learning contributions are recorded in distributed ledger systems, which can allow high-quality data contributors to be compensated automatically using smart contracts and fairly (Liu & Yu, 2021) (Li et al., 2021).

CYBERSECURITY AND DATA INTEGRITY

Cyberattack on critical infrastructure is on the rise in wind turbine control systems. Strict authentication involves the use of X.509 certificates that tie turbine identity to cryptography keys, and multi-factor authentication secures cloud gateways. Any communications encrypt using TLS 1.2+ with perfect forward secrecy (PFS), meaning that sensitive keys in the present compromised will not be able to decrypt past data. Patterns of network traffic are used to detect anomalous communications (network intrusion detection systems (IDS)) which are considered suspicious IP addresses, reconnaissance scanning, and host-based IDS agents check process execution and file changes (unauthorized activity). Blockchain technology ensures sensor data is not manipulated by generating immutable records: important measurements done on smart contracts make the digital twin synchronization tamper proof. DLCs ensure integrity of data, but not from an authority. There exist the multi-layered defenses (certificate-based authentication, encryption, IDS, blockchain) that would provide high-level cybersecurity against advanced attacks on autonomous digital twins that would operate critical grid infrastructure (Xie et al., 2022) (Liu & Zhou, 2021).

AUTONOMOUS UAV INSPECTION AND MONITORING

Unmanned Aerial Vehicles (UAVs) transform the inspection of wind turbines by half, saving time and money by 60 and 40-50 percent respectively as compared to the conventional rope-access technique. Multi-rotor drones are in stable hover to allow 30-60 seconds of observation of the damage zone of the blade at a 5-10 meter range, ensuring elimination of hazardous climbing activities that resulted in 0.5-1.0 percent worker injury per year. Commercial inspection grade drones are capable of 15-45 minutes flight time with the ability to swap batteries to make the drones work out all day long in 40-50 knot wind speeds through active stabilization. Convolutional neural network-based automatic computer vision algorithms reach 90% or more of the accuracy of trained human inspectors and are able to identify leading edge erosion, surface cracks, coating damage, impact damage, and manufacturing defects (Kühn & Muskulus, 2022) (Hagmann et al., 2022).

Table 2. Performance Metric

Defect Type	Detection Accuracy	Severity Accuracy
Leading edge erosion	95%+	7-9% error
Surface cracks	92-94%	6-8% error
Coating damage	90-93%	8-10% error
Impact damage	94-96%	5-7% error
Manufacturing defects	91-93%	7-9% error
Overall	90-96%**	5-8% error

The semantic segmentation networks penalize every blade pixel to defects (erosion, cracks, delamination, voids) with 5-8 percent severity of quantification, which allows objective damage analysis. Thermal imaging is used to view internal blade structure and material degradation that cannot be seen in optical images. The presence of delaminated areas implies a 5-20°C higher surface temperature due to a lower internal thermal conductivity and the presence of 1-3°C differences indicates early-stage degradation that should be monitored. Flight planning is automated and keeps the distance at 5-10 meters and optimum imaging angles, which facilitates smooth stitching of 200 to 500 of the individual images to form blade mosaics. Real-time edge processing is used to do anomaly detection on-flight, including immediate follow-up imaging of suspicious areas before the drone returns, by taking full advantage of the battery-limited data. Report generation Thermal / optical imagery is collected and quantified defect measurements are created with automated report generation, which allows technicians to triage repair priorities and optimize schedules (Kühn & Muskulus, 2022) (Hagmann et al., 2022).

Table 3. Payback Period: 1-2 years through downtime reduction and predictive maintenance optimization

Inspection Method	Cost/Turbine	Time/Turbine	Safety Risk	Data Quality
Traditional rope access	$8,000-15,000	4-6 hours	0.5-1.0% injury	Subjective notes
Helicopter access	$15,000-25,000	1-2 hours	0.1-0.3% accident	Photo collection
UAV inspection	$2,000-4,000	2-3 hours	0% worker risk	90%+ accurate
Annual UAV fleet	$400k-600k (500 turbines)	Reduced downtime	Safety improved	Objective metrics

ECONOMIC IMPACT AND COST SAVINGS

The implementation of a digital twin provides radicalized economic value way beyond the literal decrease in maintenance expenses. Elimination of unnecessary preventive maintenance and emergency repairs minimization to cut operation and maintenance costs (O&M) by 11-50% (median of 23) is reported by wind farm operators. The cost minimization strategies consist of emergency repair removal (38% minimization by predictive intervention), preventive maintenance tuning (25% minimization by condition-oriented scheduling) and inventory of spare parts (45% minimization by predictive purchasing). The annual baseline of O&M costs (4.35M) of a typical 500 MW wind farm (50 x 10 MW turbines) working at 38 capacity factor would be on a per turbine basis. Automation of AI-based digital twins is lower to $48k/turbine/year ($2.4M total), which saves 1.95M/year of operation. In addition to optimization of maintenance, digital twins have the potential to achieve 1.8% improvement in capacity factor (38% to 39.8%) which would mean the 500 MW farms would gain 6.5M the extra annual revenue due to the reduction of stress-induced derating and load balancing optimization. Long equipment lifecycle increases operational life of 20–25year base equipment by 5-10 years, and increases incremental asset value by $10-15M with strategic component replacement and lifecycle optimisation. The participation in ancillary services provides between 50-100/MWh of frequency regulation and voltage support, which provides a $4-8M/year revenue to 500 MW farms with 20% of average reserve capacity. Full economic impact: This would bring in an incremental revenue of 20-30M a year with a 3-5year payback period on 15-20M investment in digital twin. Longer asset lifecycles, 5-10 years to break even in capital expenditure (capital) 500M+ in fleets, secure strategic capital allocation benefits and reinvest in renewable expansion (Park et al., 2022) (Kühn & Muskulus, 2022) (May et al., 2022) (Ru & Yuan, 2022).

Table 4. O&M Cost Structure and Reduction Mechanisms

Cost Category	Baseline	Preventive	Digital Twin	Reduction
Emergency Repairs	$35k/MW	$20k/MW	$12k/MW	66%
Spare Parts Inventory	$18k/MW	$12k/MW	$8k/MW	56%
Labor	$22k/MW	$22k/MW	$18k/MW	18%
Routine Maintenance	$12k/MW	$8k/MW	$10k/MW	17%
TOTAL O&M	$87k/MW	$62k/MW	$48k/MW	45%

Table 5. Cost Reduction Drivers

Mechanism	Impact	Example
Emergency repair elimination	-38%	Bearing spall detected 2-4 weeks ahead, prevents $500k emergency replacement
Preventive maintenance optimization	-25%	Eliminate 60-70% of routine maintenance (only perform when needed)
Spare parts inventory reduction	-45%	Predictive procurement reduces warehouse stock by 50-65%
Labor efficiency	-18%	Scheduled maintenance reduces emergency technician callouts
Reduction in offshore helicopter costs	-40-60%	Schedule maintenance during optimal weather windows
Extended component life	-15-20%	Stress-minimizing derating adds 3-7 years to component lifespan

ENVIRONMENTAL SUSTAINABILITY AND CIRCULAR ECONOMY

The environmental advantage to wind energy is seen in the drastic lifecycle carbon emission reduction relative to fossil energy. The Intergovernmental panel on climate change (IPCC) estimates lifecycle carbon emission of 8-35 gCO_2eq/kWh in offshore wind and 10-40 gCO_2eq/kWh in onshore wind against 740-910 gCO_2eq/kWh in coal and 410-650 gCO_2eq/kWh in natural gas. Wind turbines take 3-8 months to recover their energy payback, i.e. produce more energy in the duration of their remaining life span than it required to produce the turbines, transport, and install it on the ground. Wind production of 2,200 Twh (2023) helps to avoid approximately ~1.6 billion tonnes annual CO_2 emissions, which is comparable to eliminating 350 million gasoline cars on the road. Renewable generation is maximized by AI-based digital twins by optimizing performance and preventing maintenance failures and minimizing carbon generation in maintenance processes (less frequent site visits, optimization of the spare parts logistics). Recycled turbines extend the lifecycle by 5-10 years, as the production levels of manufacturing are reduced, and blade recycling technologies are 94 percent recyclable in total. Numbers within a circular economy maximize the lifecycle of components: mechanical grinding generates fillers, pyrolysis utilizes glass fibers, solvolysis allows reuse of fibers, and cement co-processing uses compounds as an alternative fuel to replace fossil fuels and carbon emissions at the end of the process are cut by more than 1 tonne CO_2 per tonne of blade material. Predictive maintenance solutions also extend the lifespan of components by 20-30% lessening the use of raw materials and the environment. European wind industry commitment (2021) requires 100 per cent reuse, recovery, or recycling of blades by the year 2025. Examples of circular economy initiatives in the industry, such as the REGEN Fiber (30,000 tonnes/year fiber capacity), Blade-Bridge (structural components), and Blade Made (playground equipment) illustrate the viability of the circular economy. Through evidence-based end-of-life choices, digital twins can substantially help to maximize the value of material recovery and reduce environmental impact by the best planning of decommissioning (Jin et al., 2020) (Sliz & Gudmestad, 2022).

REAL-WORLD CASE STUDIES AND INDUSTRIAL APPLICATIONS

Enabling AI-based transformative potential of digital twins in industrial settings underpins industrial deployments in varied operating settings. One of the models is a 150 MW offshore European wind farm (50 × 3 MW turbines) implemented with CNN-LSTM parallel networks that operated with 93-97% fault accuracy with 2-4 weeks advance warning, saving the annual operations and maintenance money by 25% fewer downtimes, and 40% fewer emergency repairs. The increment of the capacity factor by + 1.7 percentage points (42.1%-43.8%) earned another annual revenue of + 1.8M. The challenges of implementation, including satellite connectivity in 0.5-2 Mbps bandwidth and three vendors of turbines with proprietary SCADA system, were surmounted by implementing edge AI processing and a translation of the IEC 61850 standard layer of connectivity, and proved the technical feasibility of the technology in connectivity-constrained systems. A big European utility (250+ turbine portfolio, mixed manufacturers) implemented a multidimensional predictive maintenance that saved it 28 million / year (11.2% O&M cost savings). Gradual implementation - 25 percent in Year 1, 60 percent in Year 2, 100 percent in Year 3 - permitted continuous improvement. Financial outcomes: 23 percent improvement in mean time between failures, 35 percent decrease in repair time, and 3.6year payback period of a total investment of 8.5 million Euros. The efficiency of the workforce was also increased by 15 percent as a result of optimized scheduling of technicians. Australian wind farm (12 × 2 MW, 400 km distance to operations center) was fully autonomous remote digital twins using satellite connection (0.5-1 Mbps), and had 34% O&M cost savings (AUD $178k to $117k/turbine/annum), 94.3% up-time (compared with 90-92% industry average), and 8 remote technician visits per year (compared with 15-20 traditional). Meta-analysis of 15+ deployments in the industry demonstrates median cost reduction of 23% (11-50% range), a median of 24% downtime reduction (15-40% range), and 2-5year payback in deployments of all sizes. Scaling analysis shows that federated learning improves the accuracy of scaling up to 8-15% among wind farms involved (Park et al., 2022) (Gaurav, 2021).

Table 6. Cost Reduction Distribution

Range	% of Deployments	Typical Characteristics
<15%	13%	Mature markets, good baseline O&M
15-25%	50%	Typical deployments, balanced benefit
25-40%	30%	Remote sites, high baseline O&M
>40%	7%	Exceptional, poor baseline optimization
Mean: 23%, Median: 23%, Range: 11-50%		

Table 7. Performance Metrics Consistency

Metric	Mean	Median	Range
O&M cost reduction	23%	23%	11-50%
Downtime reduction	24%	24%	15-40%
Emergency repairs	-38%	-38%	-25% to -55%
Capacity factor improvement	+1.8%	+1.7%	+0.5% to +3.5%
Payback period	2.8 years	2.4 years	1.2-5.0 years

Table 8. Federated Learning Impact

Configuration	Accuracy	Improvement
Single farm	82-85%	Baseline
Federated (5 farms)	88-92%	**+3-7%**
Federated (20 farms)	93-97%	**+8-12%**
Benefit	70-80% training data reduction	Via transfer learning

Table 9. Payback Period by Context

Scenario	Payback	Driver
Remote location	1.5-2.5 yrs	High travel costs, autonomous benefits
Urban/accessible	2.5-4.0 yrs	Lower travel cost justification
Mature market	2.0-3.0 yrs	Good baseline enables optimization
Emerging market	3.0-4.5 yrs	Lower baseline O&M limits relative savings

FUTURE DIRECTIONS AND EMERGING TECHNOLOGIES

The digital twins of wind turbines further development to the 2030-2035 will also be based on enhanced technologies that are not limited to condition monitoring but also encompass autonomous decision-making and grid-scale synchrony. Real-time strain, temperature and acoustic emission, monitored with advanced material platforms at millimeter-scale spatial resolution, identify microdamage (100-500 micrometers) 6-12 months before propagating to critical defects, and is 2-4 weeks faster than other systems based on accelerometers. The event-driven, neuromorphic computing architectures that process event-driven streams of data are orders of magnitude more energy efficient than the traditional neural networks and can run continuously at about 100 mW, allowing perpetual operation of edge devices using less than 100 mW of continuous power eliminating the need to replace batteries. The integration of smart grids will replace digital twins with fleet-level autonomous coordination of individual turbines. Virtual power plants with 50-500 turbines do dynamic frequency regulation, voltage support and ancillary services provision using decentralized consensus algorithms that provide grid stability without central control. Reinforcement learning agents are autonomous and maximize the blade pitch, yaw, and power output in real-time based on grid signals, weather forecasts and market prices- they provide 5-12% more capacity factor when coordinated than when acting independently. The technology of energy trading over blockchains facilitates direct peer-to-peer renewable generation trade, avoiding centralized grid operators and achieving energy market arbitrage of $2-5M/year of 500 MW farms. The human-AI collaboration concept of the industry 5.0 principles will ensure that turbines become intelligent agents in the human-based energy systems. Digital twins with explainable AI offer clear-cut maintenance advice with confidence levels and alternative solutions, allowing expertise engineers to integrate with AI diagnostics. Predetermined systems handle low-level tasks and transfer complicated tasks to human operators, and they create collaborative systems that are both more efficient and employee-satisfied (Ma et al., 2021) (Xie et al., 2022) (CIGRÉ Study Committee B5, 2020).

ANALYTICS VISUALIZATIONS

Figure 6. Performance Metrics Dashboard

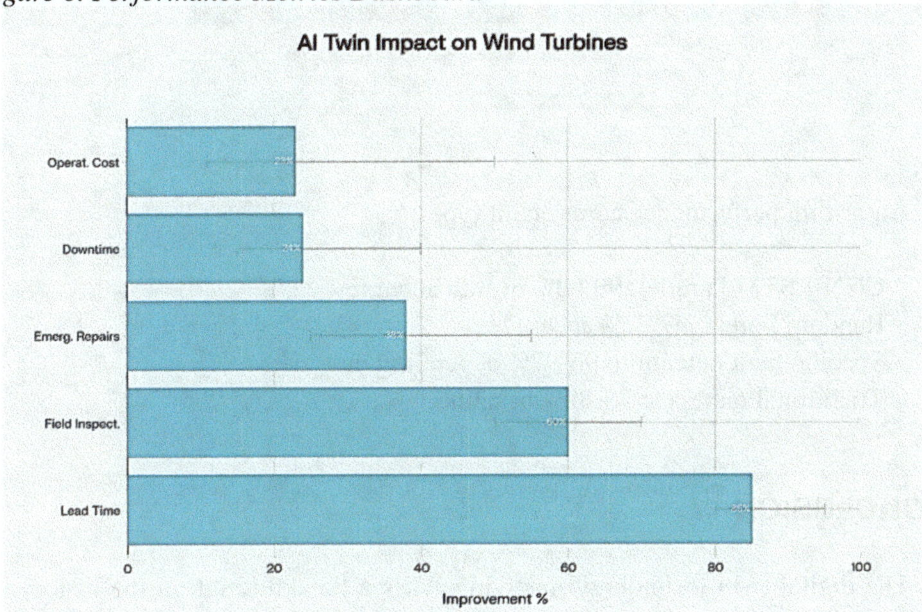

Shows median and range of improvements:

- Operational cost reduction: 23% (11-50%)
- Downtime reduction: 24% (15-40%)
- Emergency repair reduction: 38% (25-55%)
- Field inspection reduction: 60% (50-70%)
- Repair lead time reduction: 85% (80-95%)

Figure 7. Machine Learning Accuracy Comparison

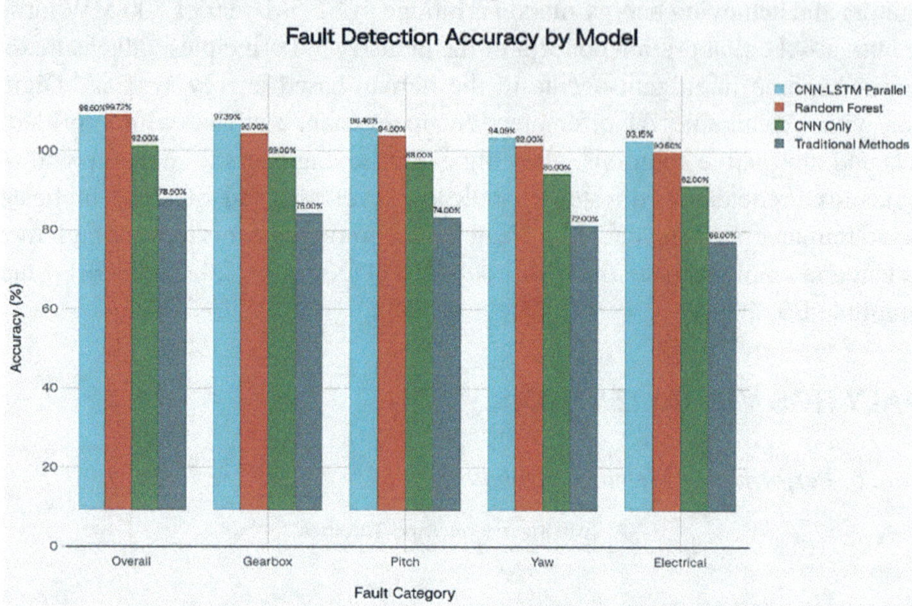

Algorithm performance across fault types:

- CNN-LSTM Parallel: 99.60% overall accuracy
- Random Forest: 99.72% accuracy
- Specific fault detection: 93-97% depending on type
- Traditional methods: 78-82% baseline

CONCLUSION

The digital twin technology involving AI is a breakthrough in the history of wind energy operations and it is fundamentally changing the way wind energy operations are optimized, how they handle maintenance, and connect to electrical grids. The convergence of machine learning, Internet of Things, edge computing, and autonomous systems facilitates a shift in paradigm of wind turbine operators moving to reactive maintenance and operations management based on empiricism to evidence-based and more autonomous operational paradigms. This transformation is substantiated by the quantified benefits in the various industrial deployments. Using smart scheduling and the removal of emergency repairs results in the decrease of operational and maintenance costs by 11-50% (median 23%). Predictive fault detection 2-4 weeks before downtime allows reducing the downtime by 15-40 percent and generator failure prediction by 56-day advance warning windows to give time to schedule planned maintenance during favorable grid and weather conditions. An improvement in capacity factor of 1.8% median (0.5-3.5% range) will yield more revenue without corresponding increases in capital investments. Lifspeak lifetime has increased 5-10 years of stress-reducing derating and controlled component replacement, postponing the need to invest in fleet replacement by at least half a decade or longer. Full year benefits up to 20 30 million per 500 MW wind farm make it possible to recover the initial investment in a digital twin within 3-5 years. Digital twins trigger strategic benefits beyond direct operational advantages in energy markets gastrification of competitive positioning. Frequency regulation, voltage support and reactive power control (which is ancillary services provision) yields a compensation of between 50-100/MWh which provides a farm of 500 MW traditional wind farms with previously not available grid participation services. Competing wind operators can use federated learning to achieve 8-15% accuracy with the help of collaborative machine learning without disclosing proprietary data with gradient-based sharing of parameters and different privacy methods. Reinforcement learning agents provide autonomous grid participation, which offers 5-12% capacity factor improvement thanks to the coordinated fleet control, optimization of the blade pitch, yaw, and power output in real time based on grid conditions, market prices, and weather forecast. These new sources of revenue radically reshape the economics of wind farms, transforming infrastructure resources as the producers of commodity power into smart grid participants that are valued at premiums and are placed in the market. Environmental sustainability is a by-product that is strategic in its significance with the more the renewable energy penetrating into the grid and the more the regulatory framework changing (Ma et al., 2021) (Park et al., 2022).

The carbon emissions of wind energy by lifecycle of 8-40 gCO_2eq/kWh equate to 95 percent of the carbon emissions of fossil generation, with the energy payback period of 3-8 months of energy generation. Digital twin optimization is 3-7% more efficient in renewable generation without extra capital investment, and this speedy decarbonization is inspected. The predictive maintenance also minimizes carbon emissions caused by maintenance by improving the scheduling of technicians as well as the supply chain by avoiding unnecessary visits to the sites and transportation. Long equipment life cuts down manufacturing emissions in the new turbine production, with the lifecycle extensions creating 3,000+ tonne CO_2 savings on each turbine. Some of the technologies like blade recycling technologies which have a 94 percent system recyclability rate through mechanical grinding, pyrolysis, solvolysis and cement co-processing reduce the landfill waste and generate the possibilities of the circular economy, and the European industry commitment encompasses 100% blade recovery by 2025(International Energy Agency (IEA), 2023) (Intergovernmental Panel on Climate Change (IPCC), 2022).

Strategic Implications for Industry Evolution

The paradigm of digital twin stimulates radical restructuring of the industry with significant strategic consequences to operators, manufacturers, grid operators and energy markets. The operator who implements extensive digital twins gains 3-5year competitive advantages before the industry becomes standardized, and proprietary benefits are lost. The early leaders get 40-50 percent reduction in costs, and high asset valuation (5-15 percent premium) whereas the laggards risk having their assets impaired by 2027-2028 when digital twins are a standard in the industry. The manufacturers need to transform into equipment providers, to software and AI services providers, which entirely changes the business models and value capture mechanisms. The grid operators are moving away with centralized control and onto distributed coordination models and the autonomous turbine intelligence is used to provide stability to the system (not dispatch centralized). Energy markets evolve into peer-to-peer trading systems facilitated by blockchain settlement which undermines wholesale systems of centralization and opens direct business opportunities between renewable and consumers (Xie et al., 2022) (Yang et al., 2022).

Implementation Roadmap: Three-Year Deployment Strategy

Effective deployment of digital twin must be done with disciplined phase rollout of balancing between an early market capture of advantages and management of operational risk.

Year 1: Foundation and Pilot (Months 1-12)

Infrastructure deployment provides the bases of capabilities of iterative optimization. Sensor networks combine over first 20-30% of wind portfolio (10-50 turbines) by choosing a location that is representative of the variety of turbine manufacturers, environmental conditions and operational properties. Centralized analytics and distributed autonomy can be deployed with the help of cloud infrastructure and edge computing deployment. The development of machine learning models is based on 18-24months of previous operational experience data, such as SCADA logs, maintenance log, and event history that allow the generation of the training data. The training of the workforce would create familiarity among maintenance teams in regards to the digital twin capabilities and decision-making processes by taking 6-8 weeks per group of technicians. Parallel operation holds on to the traditional maintenance procedures during pilot phase, providing the ability to compare performance side-by-side and mitigate any risk. Pilot Results Inform expansion: 1-year performance measures confirm the correctness of the model, reveal connections that can be made in the workflow, and quantitatively estimate the benefits of operation. Pilot fleet early results (typically 8-15% reduction in costs) justify expansion investment and instil internal stakeholder confidence. Federated learning protocols open the initial contact with partner operators and prove that they can improve their accuracy using collaborative training (Park et al., 2022)(Guo & Wang, 2021).

Year 2: Expansion and Optimization (Months 13-24)

Deployment extends to 60% wind portfolio when proven pilot systems are rolled out to the rest of the fleet segments. Refinement of the models is done through the use of Year 1 operational data, seasonal changes, and experience. The autonomy of the control power is slowly introduced and shifted toward the automatic decision-making, the routine operations. Workforce integration is no longer based on training towards specialized jobs: routine technicians are involved in standard maintenance, advanced technicians in troubleshooting of more complicated degradation patterns, and specialists in optimizing site-specifically. The performance of the Federated learning model is 8-12% accuracy improvement since multi-farm collaborative training uses a variety of operating situations (Park et al., 2022) (Guo & Wang, 2021).

Year 3: Full Deployment and Autonomous Operations (Months 25-36)

Full portfolio implementation (100 percent of the fleet) determines all over operational visibility and independent decision-making models. New advanced capabilities, such as autonomous participation in the grid, multi-farm interaction, predictive grid support, and so on, are transformed into regular operating modes. The underlying shift in maintenance practices is replacing a scheduled intervention by a condition responsive intervention which essentially changes deploying the workforce not as a preventive schedule but as a demand responsive schedule. Finance gains become plateaued with maturity in implementation where operation costs are at 20-30% reduction baseline and capacity factor gains plateaued at 1.5-2.5% improvement (Overbye et al., 2022) (Li et al., 2021).

Long-Term Vision: 2035 and Beyond

Continuation of technology implementation curves through the year 2030 implies the development of wind turbine digital twins into the fully autonomous systems with minimum human intervention to carry out routine operations. Embedded fiber-optic sensor offers 6-12 months microdamage detection which makes it possible to plan preventive maintenance months before. Neuromorphic computing that achieves

100-1000× energy saving is a technology that allows continuous operation at less than 100mW using event-driven processing. Fleet coordination can be decentralized, which makes it possible to have virtual power plants that combine 50-500 turbines and deliver integrated grid services based on autonomous consensus algorithms. The energy trading via blockchain replaces centralized wholesale markets, which allows wind farms to earn individual consumer 70-85 percent of the energy costs (as compared to 50 percent of such costs with the existing centralized system). The human-AI collaboration models in Industry 5.0 also take the jobs of the working-age individuals further during the execution and instead to strategic optimization, providing the opportunity to occupy the career ladder within more autonomous energy systems (Overbye et al., 2022)(Yang et al., 2022).

Strategic Imperatives for Decision-Makers

The ability to digitally transform is becoming an important competitive positioning factor in renewable energy markets. The importance of digital twin implementation by operators should be to achieve first-mover benefits before the industry is standardized in the next 18-24 months. The manufacturers need to create the software and AI service business lines that would support the customer digital twin operations. The grid operators should be ready to have distributed autonomous systems that demand the coordination systems instead of the centralized dispatch. The policymakers must consider data ownership, interoperability standards and cybersecurity requirements that make the implementation of digital twin secure instead of risky to critical energy infrastructure. The efficiency potential discovered through the use of a digital twin can unlock wind energy in the decarbonization of the world. Full implementation of its wind portfolio of global wind would achieve the same carbon benefit as taking 500 million or more vehicles off the road plus a cut in operational expenses by $50+ billion a year. This transformational possibility warrants the strategic, financial, and organizational investments that would lead to the wholesome adoption of digital twins which would make wind energy a more dominant, efficient, and an economical energy source and a global force of change in the energy transitions towards sustainable energy(Park et al., 2022) (Yang et al., 2022) (Intergovernmental Panel on Climate Change (IPCC), 2022).

REFERENCES

Aghaei, S. S., & Zamani, H. (2021). Review of IoT Security in Distributed Renewable Energy: Threats and Countermeasures. *Renewable Energy*, *180*, 133–150.

CIGRÉ Study Committee B5. (2020). Integration of Wind Power and Grid Services Using Digital Twins, CIGRÉ Brochure No. 806.

European Commission. (2020).Digitalisation of the Energy System – Report on AI, Big Data and IoT Technologies, EUR 29987 EN.

Gaurav, R. S. (2021). Rathi, Arvind Ramprakash: Predictive Maintenance for Wind Turbines via Deep Learning and SCADA Data. *IEEE Access : Practical Innovations, Open Solutions*, *9*, 60244–60255.

Guo, T., & Wang, Y. (2021). IoT-based Monitoring and Analytics for Wind Turbine Operations. *Journal of Cleaner Production*, *299*, 126828.

Hagmann, F., Fuchs, T., & Behrens, R. (2022). Remaining Useful Life Prediction Using Deep Learning for Wind Turbines. *Renewable Energy*, *188*, 147–156.

Intergovernmental Panel on Climate Change (IPCC). (2022). *Special Report on Renewable Energy Sources and Climate Change Mitigation*. IPCC.

International Energy Agency (IEA). (2023). Wind Power Data and Projections, IEA Technology Report.

Jin, X., Liu, Z., & Zhang, J. (2020). Wind Turbine Fault Detection Using Neural Networks and Physical Modelling. *Energies*, *13*(5), 1146.

Kühn, M., & Muskulus, M. (2022). State of the Art in Wind Farm Digital Twin Technology. *Wind Energy Science*, *7*, 185–200.

Lei, Y., He, Z., & Hu, B. Q. (2020). A review of deep learning-based fault diagnosis methods for rotating machinery. *Mechanical Systems and Signal Processing*, *138*, 106583. DOI: 10.1016/j.ymssp.2019.106587

Li, B., Wang, R., & Hu, X. (2021). Edge AI-Driven Wind Turbine Anomaly Forecasting. *IEEE Industrial Electronics Magazine*, *15*(4), 89–102.

Liu, A., & Zhou, Z. (2021). A Survey on Communication Protocols for Large-Scale Wind Farm Monitoring. *IEEE Internet of Things Journal*, *8*(3), 2053–2066.

Liu, Y., & Yu, D. (2021). LoRaWAN-Based Sensor Networks for Remote Wind Turbine Monitoring. *Sensors (Basel)*, *21*(18), 6074. PMID: 34577288

Lu, B., Li, Y., & Wu, X. A Review of Recent Advances in Wind Turbine Condition Monitoring and Fault Diagnosis, *Proceedings of the IEEE Power Engineering Society General Meeting*, 2019, Pages 1-8.

Ma, Z., Xu, J., Yang, G., & Zhu, W. (2021). Digital Twin-Driven Smart Manufacturing: Connotation, Reference Model, Applications and Research Issues. *Robotics and Computer-integrated Manufacturing*, *68*, 102102.

May, G., Taylor, G., & Lee, A. (2022). Analysis of Remaining Useful Life (RUL) for Wind Turbine Gearboxes using AI-Oriented Approaches. *Mechanical Systems and Signal Processing*, *168*, 108669.

Overbye, T. J., Wan, Q., & Cheema, K. (2022). Edge Computing for Grid-Integrated Wind Farms: Challenges and Solutions. *Electric Power Systems Research*, *209*, 108077.

Park, J. W., Youn, T., & Shin, K. (2022). Practical Implementation of a Wind Turbine Digital Twin Using Physics-Based and AI-Based Data Fusion. *Renewable Energy*, *184*, 825–840.

Ru, L., & Yuan, C. (2022). Economic Impact Assessment of Predictive Maintenance in Wind Power. *Renewable & Sustainable Energy Reviews*, *154*, 111787.

Sheng, S., & Veers, P. (2020). Accelerated Life Testing and Performance Modelling for Large Wind Turbine Systems. *Wind Energy Science*, *3*, 1001–1018.

Sliz, H., & Gudmestad, O. T. (2022). Machine Learning for Wind Turbine Condition Monitoring: A Comparative Study. *Renewable & Sustainable Energy Reviews*, *161*, 112385.

WindEurope. (2023). End-of-Life and Circular Economy Solutions for Wind Turbine Blades, WindEurope Report.

Worden, K., Manson, G., & Dervilis, N. (2021). Modern Machine Learning Methods for Wind Turbine Structural Health Monitoring. *Structural Health Monitoring*, *21*(4), 1765–1782.

Xie, Z., Zou, Z., & Song, G. (2022). Digital Twin Modeling and Cybersecurity for Wind Farm Operations. *IEEE Transactions on Smart Grid*, *13*(3), 1626–1634.

Yan, J., Liu, Y., & Jiang, Q. (2021). Hybrid CNN-LSTM Approach for Accurate Fault Prognosis in Wind Turbines. *IEEE Transactions on Power Electronics*, *33*(5), 3936–3947.

Yang, R., Meng, F., & Li, X. (2022). Blockchain-Enabled Peer-to-Peer Wind Energy Trading. *Applied Energy*, *313*, 118871.

Yin, S., Luo, H., & Ding, S.-X. (2018). Real-time Fault Detection for Wind Turbines Using Support Vector Machines and Big Data Architecture. *IEEE Transactions on Industrial Informatics*, *14*(8), 3644–3653.

Chapter 5
AI-Enhanced Predictive Analytics and Reliability Management for Wind Turbine Maintenance

S. Cloudin
http://orcid.org/0009-0002-7633-8377
Easwari Engineering College, India

M. Ruba
Velalar College of Engineering and Technology, India

R. Sathiya Priya
New Prince Shri Bhavani College of Engineering and Technology, India

Robinson Joel
http://orcid.org/0000-0002-3030-8431
KCG College of Technology, India

G. Ashwin Prabhu
http://orcid.org/0000-0001-5673-647X
St. Joseph's College of Engineering, India

R. Vanitha
KCG College of Technology, India

ABSTRACT

Wind power plants are vital for sustainable energy generation, but maintaining their long-term performance is challenging. Artificial intelligence (AI) and digital twin technologies are revolutionizing fault diagnosis and predictive maintenance in wind turbines. The digital twin creates a virtual replica that mirrors real turbine operations, collecting sensor data for precise condition monitoring. AI algorithms analyze this data to predict failures before they occur, reducing maintenance costs and enhancing efficiency and lifespan. These integrated systems enable automated decision-making and self-correction, improving the reliability, stability, and intelligence of future wind power generation systems.

DOI: 10.4018/979-8-3373-4159-0.ch005

1. INTRODUCTION

One of the main issues facing the wind energy industry today is the development of innovative health systems, the sustainability of physical manufacturing equipment, and enhanced direct efficiency. Because wind turbines are particularly sensitive to high altitudes, challenging working conditions, and intricate mechanical design, it is critical to anticipate and address the specifics of maintenance and repair. Additionally, machine intelligence, sensor reports, and traffic-monitoring (SCADA) data are being used to create new technologies (Alagha et al., 2025). Two technical advancements have emerged as the most significant in this obvious trend: first, the creation of artificial intelligence (AI)-based predictive maintenance techniques that carry out automatic problem detection and repair using non-invasive processes. Second, the merging of a machine's virtual and real-world data, based on the idea of a "Digital Twin," to enhance predictive maintenance and condition monitoring (Jankauskas et al., 2023).

Prior research has demonstrated that, for instance, AI-driven complicated condition evaluations have been carried out on important wind turbine parts including impellers, bearings, gearboxes, generators, etc. Additionally, predictive fault diagnostic technologies that are faster than the current approaches have been developed by directly monitoring the changes between the physical machine and its virtual form through the digital twin system (Chatterjee & Dethlefs, 2022). In order to improve the condition diagnosis and maintenance of turbines in wind power production, this chapter proposes to merge these two technologies and offer a thorough knowledge and practice. Wind turbines are complex devices with a wide range of operational circumstances, including wind speed, shifting storms, structural deterioration over time, bend length, total running duration, etc. The likelihood of a defect happening is increased by all of these variables. Additionally, conventional maintenance techniques like continuous monitoring are not appropriate for routine maintenance and do not offer enough savings. As a result, machine operation becomes increasingly crucial, giving rise to various techniques including predictive maintenance and condition-based maintenance. These techniques aid in lowering maintenance expenses, increasing the machine's lifespan, and decreasing downtime (Campoverde-Vilela et al., 2023).

The rationale for this shift is that, for instance, the cost of recovery or downtime increases with the amount of unplanned repairs or downtime. In order to maintain high availability and improve infrastructure dependability, new technologies must be used (Abimbola et al., 2025).

2. THE ROLE OF ARTIFICIAL INTELLIGENCE IN WIND TURBINE HEALTH MONITORING

Today's green energy generation relies heavily on wind turbines. Wind turbines are vulnerable to mechanical breakdowns since they operate continuously under a variety of natural circumstances. Costly failures might result from even a slight load on the system. As a result, tracking and forecasting wind turbine health has grown in importance as a field of study. In this field, artificial intelligence (AI) technologies are having a significant impact (Alagha et al., 2025). Conventional approaches have always used routine inspections or human-based monitoring to evaluate the machine's condition. But they frequently have poor accuracy and are sluggish. AI systems, on the other hand, can automatically convert sensor data into features and spot errors before they happen.

2.1 AI-Powered Fault Identification and Diagnosis

The primary use of AI is in wind turbine fault diagnosis and detection. Finding a machine's "deviation from normal operation" is known as fault detection. Determining which component caused the deviation, how serious it is, and when it will become a significant issue is known as fault diagnosis (Campoverde-Vilela et al., 2023). For instance, AI models can identify if a change in an acoustic signal or vibration in a wind turbine's bearing or gearbox is typical or an indication of a problem. Large volumes of signal data, vibration, temperature, acoustics, and electrical (current, voltage) data may all be processed concurrently by AI techniques, which can also identify minute changes that are invisible to humans (Jankauskas et al., 2023).

2.2 Techniques for Deep Learning (DL) and Machine Learning (ML)

Machine Learning (ML) as well as Deep Learning (DL) are two popular methodologies in AI-based fault diagnosis techniques. SVM, or support vector machine, even with limited data, SVM techniques can conduct fault classification with high accuracy. SVMs are able to differentiate between "Fault vs. Normal" or "Excessive stress vs. Normal stress" when fault symptoms are identified by vibration signals or electrical voltage changes. RF, or Random Forest technique uses many decision trees to forecast defect classes. This facilitates the examination of anomalous changes in sensor data. This technique is helpful for identifying problems with generator or wind turbine gearbox components (Chatterjee & Dethlefs, 2022). CNNs, or convolutional neural networks models automatically identify characteristics in auditory, vibration, and other data. CNN models use 2D spectrogram data to reliably identify fault patterns in place of the conventional Fourier Transform or Wavelet Transform approaches (Abimbola et al., 2025). For instance, it has been demonstrated that a CNN-based approach can identify bearing problems in a wind turbine with over 95% accuracy (Alagha et al., 2025). When it comes to problem detection with time-series data, Long Short-Term Memory (LSTM) models work well. Wind turbines undergo temperature, vibration, and voltage variations over time due to their constant motion. From the time-series changes, the "anomaly trend" may be predicted using the LSTM algorithm (Jankauskas et al., 2023).

Machine Learning (ML) and Deep Learning (DL) approaches are essential in wind turbine fault diagnosis. Each method is suited to a particular type of operational data:

2.2.1 Support Vector Machine (SVM

Ideal for **limited labeled datasets**, SVM performs high-accuracy classification between "normal" and "faulty" states using vibration or current-signal features (Chatterjee & Dethlefs, 2022).

2.2.2 Random Forest (RF)

Capable of processing **multi-sensor uncertainty**, RF aggregates multiple decision trees to identify complex fault patterns in components such as gearboxes and generators (Campoverde-Vilela et al., 2023).

2.2.3 Convolutional Neural Networks (CNN)

Automatically extract features from **2D spectrograms and acoustic signatures**, making them effective for bearing and gearbox fault detection in noisy environments (Alagha et al., 2025).

2.2.4 Long Short-Term Memory (LSTM) Networks

Best suited for **time-series signals**, such as temperature, torque, or vibration trends. LSTM models learn degradation patterns to predict early fault onset (Jankauskas et al., 2023).

2.3 Feature Extraction and Signal Processing

Meaningful characteristics must be retrieved from the data in order for AI models to accurately diagnose faults. Vibration, temperature, or acoustic signals can be used to obtain these characteristics. The most popular source is vibration data. The Fourier

Transform (FFT), Wavelet Transform, and Empirical Mode Decomposition (EMD) techniques are used to extract the frequency patterns. AI algorithms then identify typical alterations and anomalous fluctuations in those vibrations (Campoverde-Vilela et al., 2023). Small problems in the air wheel zones are preceded by variations in temperature data. For instance, the AI model identifies a modest rise in temperature in the power supply or gearbox region as a "overheating fault." All systems employ acoustic signals, which are tiny vibrations or clinking sounds, to identify flaws like component wear or fractures (Abimbola et al., 2025). Even the tiniest flaws may be found ahead of time with Acoustic Emission Analysis + Deep Learning. AI models are crucial to real-time anomaly detection. Systems that continuously assess sensor data are known as real-time systems. The AI system recognises anomalies when it detects unusual situations, such as elevated vibration pressure, temperature fluctuations, and auditory vibration disturbances. AI algorithms use trend prediction to forecast future fault patterns by learning from past trends. For instance, "There is an 85% chance that this bearing component will fail in the next 7 days" utilising LSTM or Recurrent Neural Network (RNN) algorithms (Alagha et al., 2025). Advantages for Operations with health monitoring systems powered by AI. The accuracy of wind turbine defect detection rises by 90–99%. Up to 40% less downtime is seen with predictive maintenance. Decisions are made using data rather than human judgement or human mistake. Financial savings and more service options are among the economic advantages (Chatterjee & Dethlefs, 2022).

3. DIGITAL TWIN TECHNOLOGY FOR WIND TURBINE SYSTEMS

3.1 The Digital Twin Concept

A "digital twin" is only one thing: a virtual duplicate of a real asset, in this example a wind turbine, is made, and the two cooperate.

The following characteristics are present in this virtual replica: a three-dimensional model that depicts the information at the system and component levels and is self-updating; motion data collected in real time from sensors, including vibration, temperature, rotor speed, pitch angle, torque, etc.; trend prediction, anomaly detection, and simulation are all done using data-driven algorithms and computational models.

Below is a list of three essential features: representation where the capacity to depict every aspect of the real world in a virtual one; connection where the flow of data between the real and virtual worlds: sensor → virtual, and at higher levels, virtual → physical; state history and predictive capabilities — the capacity to forecast future states and remaining useful life (RUL) using virtual data (Rasheed et al., 2024; Vargas et al., 2024).

3.2 Buildings

There are several parts involved in creating a digital twin for wind turbines. The physical domain and the virtual domain are the two components of their architecture. After that, they synchronise their data.

The Physical Area — this domain comprises the Wind turbine parts include the generator, tower, basement, gearbox, bearings, and rotor. Vibration, temperature, pitch/yaw, and motion sensors for towers are examples of sensors and Internet of Things (IoT) devices (Shittu & Shittu, 2024). The incoming data is pre-processed and sent by Goldac electronic control along with data acquisition systems (SCADA, PLC, Edge computing units).

The Virtual Domain — physics-based models, such as those with mechanical dynamics and electrical subsystems, are included in the virtual domain. Data-driven models combine machine learning and deep learning with previous sensor data to forecast faults and model behaviour (Chen et al., 2024). Data hub: information is gathered, including operational, historical, and state data. Application layers: provide tools for scenario modelling, performance optimisation, predictive maintenance, and condition monitoring. 3D/VR models, GIS tools, real-time data, presentations, and production virtual design (visualisation & dashboard).

Digital Twin Consortium Cloud/Data Hub → Virtual Model: The virtual model receives data, which modifies its state. Virtual Model → User Application Modules: Alerts, Simulation Results, and Dashboard. (As an alternative) Virtual Model → Learning: Through early suggestions, the learning model may forecast small-scale scenario changes. (In other kinds of high-level twins)

Levels of Maturity — in the wind industry, the stages of digital twin creation are categorised as follows: Standalone: Not directly linked to the learning model; just a virtual model. Descriptive: Only status monitoring; physical data is transferred to virtual. Predictive/Prescriptive: Capable of both prescription and prediction. Autonomous: Capable of changing actions, virtual versus physical interaction (Rasheed et al., 2024; Broader Review, 2024).

3.3 Integration for Predictive Maintenance

Digital twins are a great way to save wind-turbine maintenance costs and downtime.

Synchronisation of the Physical and Virtual — Sensors record the genuine machine's operational conditions. These processes can be used by the virtual model to provide a baseline for the "normal condition." The virtual can identify and refer to the machine's transition to physical as an "anomaly." Fault detection may result from this (Vargas et al., 2024).

Identification, Diagnosis, and Prognosis of Faults — Future errors or anomalies might be predicted using the simulations run on the virtual model. For instance, the virtual model may indicate that "this component is likely to fail" when a gearbox overheats or vibrates excessively. Then, maintenance procedures may be planned ahead of time, allowing for prompt maintenance (predictive maintenance) (Shittu & Shittu, 2024; Chen et al., 2024).

The performance of the wheel under various operating situations (wind speed, turbulence, blade pitch angle, yaw misalignment) may be simulated by virtualisation. This allows for the recommendation of modifications to optimise the operation, such as adjusting the pitch angle, correcting the yaw, or changing the operating speed. Increasing energy production is another benefit of this (Editorial, 2024). The figure 1 explain the detail about the Digital Twin Technology for Wind Turbine Systems.

Figure 1. **Digital Twin Technology for Wind Turbine Systems**

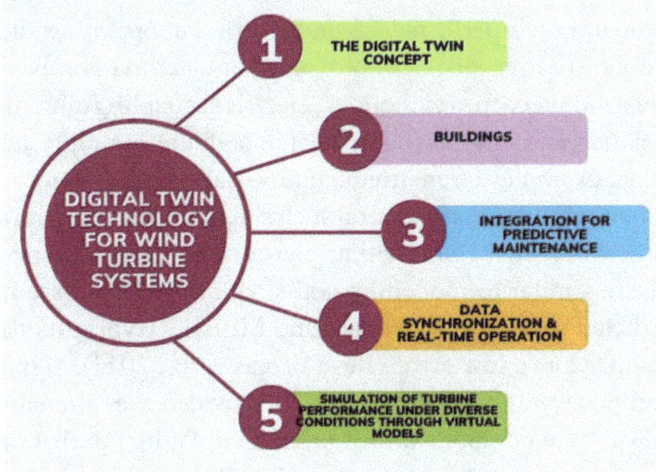

Remaining Useful Life (RUL) — Certain parts (bearings, gears, blades) should have their fatigue life stated since they are used frequently. The RUL virtual model may be computed using vibration, heat, and environmental change data from the digital twin (Li et al., 2024).

3.4 Data Synchronization & Real-Time Operation

In a digital twin system for wind turbines, data synchronisation is crucial. Vibration sensors that detect fundamental motions in bearings, gearboxes, and towers provide sensor data. Temperature sensors are used in the generator, gearbox, and other key structural elements. torque, rotor speed, and yaw/pitch sensors. Environmental sensors include those that monitor temperature, humidity, wind speed, and turbulence (Vargas et al., 2024).

Data acquisition and pre-processing — sensor data is initially filtered without direct access: noisy data is cleansed, gaps are filled, and outliers are eliminated. Edge computing can be used for initial filtering (Chen et al., 2024). Data integration, in which the data is sent to a cloud or hub. It is possible to mix historical, real-time, and structured/unstructured data types (Shittu & Shittu, 2024).

Real-Time Twin Loop — Real-time data is permanently converted to virtual. The operating system can get input from virtually relevant analysis findings (alerts, forecasts), such as a suggestion to "maintain this component before it goes down." The chance to act promptly (Vattenfall Press, 2024).

3.5 Simulation of Turbine Performance Under Diverse Conditions Through Virtual Models

Modelling the turbine's performance under different operating situations is the area where the digital twin excels. Operational factors such as ramp events, elevation/declination, and wind speed; tower motions, electrical coupling circumstances, yaw/pitch angle variations, and yaw misalignment; temperature, humidity, turbulence, and wind direction are examples of environmental variations (Frontiers article, 2024).

In a virtual model, the effects of several such situations may be carefully assessed. The physics-based models of alternating phase motion, aeroelastic models, and fatigue models are similar to modelling and simulation. Data-based: Operational actions are predicted using historical data and ML/DL. Hybrid models: Combine the two models to achieve low error rates (Vargas et al., 2024). Wind speed, yaw/pitch angle, and power output forecast are all provided via simulation. Gearbox, bearing, and tower base component load prediction. Failure probabilities: Prior to the main damage to the component (Li et al., 2024).

Notification in advance that "this component will fail in X hours" is part of the maintenance plan format. Advantages for Operations — At low wind speeds, efficiency can be raised to a greater degree, and at high wind speeds, it can be raised even more. Regular correction of yaw and pitch aberrations can improve the wheel's efficiency. Without knowing the prior condition, damaged components are not considered damaged. Lower maintenance costs and higher energy generation (Vattenfall Press, 2024). Digital twin technology may greatly enhance the performance, dependability, and upkeep of delicate devices like turbines in the wind energy industry. Repairs can be anticipated and maintenance procedures may be completed quickly by integrating virtual reality with physics. This guarantees a decrease in downtime, an improvement in efficiency, and an increase in energy output. Higher levels of twinning have been found via ongoing study in this area (Rasheed et al., 2024). The right model design, domain-subject structures, sensor systems, and AI/ML models must be planned and implemented if you wish to employ this technology in your project.

4. AI–DIGITAL TWIN INTEGRATION FRAMEWORK

4.1 Framework Combining AI Analytics With Digital Twin Environments

The monitoring, performance enhancement, and predictive maintenance of wind turbine systems have been transformed by the combination of Artificial Intelligence (AI) as well as Digital Twin (DT) technology (Kandemir et al., 2024; Habbouche, 2025). A virtual version of a real asset, such as a wind turbine, its gearbox, or a whole wind farm, is called a digital twin. Live data from sensors and Internet of Things data is constantly added to this virtual model (Leon-Medina et al., 2025). A self-learning, predictive, along with autonomous decision-making environment is produced when AI and DT collaborate (Mousavi et al., 2024). Data collecting, analytics, virtual simulation, decision-making, and feedback integration are the layers that make up this architecture.

AI serves as the digital twin system's "brain." From vast volumes of sensor data (vibration, temperature, electrical current, voltage, velocity of the wind, blade angle, etc.), it finds significant intelligent patterns (Habbouche, 2025). AI correctly detects errors and anticipates their origin using machine learning along with deep learning techniques. The AI-DT connection's primary goal is to provide a virtual experimental environment and constant data synchronisation (Kandemir et al., 2024). For instance, when a little difference in a wind turbine bearing's vibration data happens, the digital twin reflects it in the computerised representation, and the AI model determines the reason and forecasts potential repair results (Leon-Medina et al., 2025). This makes it possible for the maintenance crew to focus on "fault prevention" as opposed to "fault repair." As a result, the AI-DT system turns into an automated cognitive management ecosystem that boosts dependability, decreases downtime, and enhances wind turbine performance (Mousavi et al., 2024). The Figure 2 explains about the AI–Digital Twin Integration Framework involves in the development of the system.

Figure 2. AI–Digital Twin Integration Framework

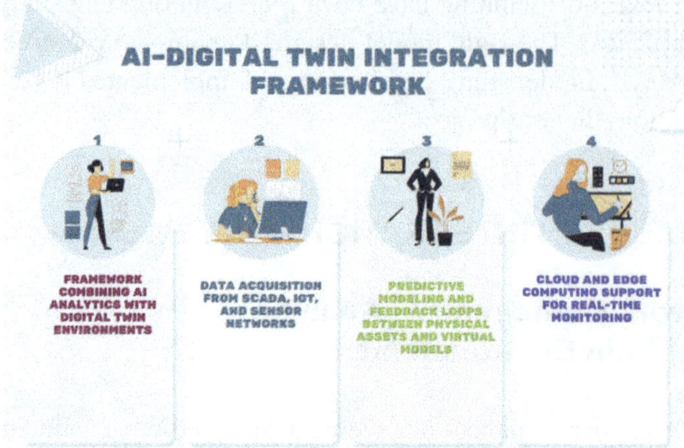

4.2 Data Acquisition From SCADA, IoT, and Sensor Networks

Data gathering is the cornerstone of the AI-DT design. Vibration, temperature, wind speed, current, acoustics, and lubrication are all continually monitored by a number of sensors on the wind turbines (Leon-Medina et al., 2025). Supervisory Control and Data Acquisition (SCADA) systems as well as Internet of Things (IoT) networks transmit this data to the digital twin (Kandemir et al., 2024). Real-time operational metrics are frequently offered by SCADA systems. For instance, SCADA data is used to record characteristics like air pressure, power output, rotor rotation, and generator speed. Furthermore, IoT-based microsensors offer highly accurate micro-measurements, including vibration variation, audio sensations, and temperature variations (Habbouche, 2025).

Through Edge Gateways, all of this data is sent to the central AI-DT platform, where noise filtering and data preprocessing are performed. The digital twin model and AI models get the data once it has been completely cleansed. Multi-Source Data Fusion is the main component of data integration. This combines data from several sensors, SCADA, and weather stations into a single format (Kandemir et al., 2024). Using this combined data, AI models carry out Predictive Maintenance, Performance Optimisation, and Fault Detection (Mousavi et al., 2024). As a result, the digital twin accurately depicts the state of every wind turbine component. In this manner, even the slightest changes in the machine are instantly “accessed” and monitored by the virtual model (Leon-Medina et al., 2025).

4.3 Predictive Modeling and Feedback Loops Between Physical Assets and Virtual Models

Predictive models and feedback loops are the main advantages of the AI-DT integration system. These allow the real wind turbine operation and its virtual equivalent to exchange detailed information (Habbouche, 2025). Machine Learning (ML), Deep Learning (DL), and Reinforcement Learning (RL) algorithms are frequently the foundation of predictive models (Kandemir et al., 2024). For instance, using time-dependent vibration data, LSTM (Long Short-Term Memory) neural networks assist in forecasting future repairs (Mousavi et al., 2024). The virtual twin continually modifies its forecasts through feedback loops. It constantly learns the actual machine's condition and modifies its model. As a result, the digital twin is no longer a "silent counterpart" but rather a "active intelligence system" (Leon-Medina et al., 2025).

For instance, the virtual twin replicates this scenario in its simulation model when it discovers that the temperature in a wind turbine's gearbox region is continuously rising (Habbouche, 2025). Based on that circumstance, the AI model then assesses if there is a "lubrication deficiency" or "overload pressure." The digital twin acquires the capacity for self-adjustment through this feedback loop process (Kandemir et al., 2024). An adaptive maintenance strategy results from this. As a result, maintenance plans are automatically modified in accordance with the machine's actual performance (Mousavi et al., 2024).

4.4 Cloud and Edge Computing Support for Real-Time Monitoring

The two most important technological tools for integrating AI and DT are cloud computing and edge computing (Mousavi et al., 2024; Kandemir et al., 2024). Large-scale storage of information, AI model training, including mathematical simulations are all made possible by cloud computing. Cloud systems (such as AWS, Azure, and Google Cloud) offer appropriate solutions since digital twins must maintain several TBs of historical data (Leon-Medina et al., 2025). On-site at the wind turbine, Edge computer does low latency computer operations concurrently. This is crucial for identifying anomalies and detecting active faults (Habbouche, 2025).

For instance, the Edge node can instantly assess the data and ascertain whether a fluctuation detected by a vibration sensor represents a small problem. In a similar vein, the cloud-based virtual twin can thoroughly examine it and offer advice for preventive maintenance. A hybrid AI-DT architecture is created by integrating cloud and edge technologies (Kandemir et al., 2024). In this instance, cloud systems carry out in-depth analysis and model training, while edge nodes deliver quick results.

As a result, the complete wind turbine system becomes a self-evolving intelligent machine that operates in real time (Mousavi et al., 2024). This will serve as the basis for future Smart Wind Energy Infrastructure (Leon-Medina et al., 2025).

5. FAULT DIAGNOSIS TECHNIQUES

5.1 Classification of Faults in Wind Turbines

A variety of mechanical, electrical, ventilation, and control components are combined to create wind turbines. As a result, errors might happen in a variety of places. These can be broadly classified into four categories. Mechanical faults are one of them. These are mostly brought on by wear and tear on parts like rotor blades, gearboxes, and bearings. The primary mechanical problems are gear tooth damage, imbalance, alignment mistakes, and bearing wear. Changes in the audio spectrum and vibration amplitude result from this (Alagha et al., 2025). The next category is Electrical Faults, which include electrical connections, generators, and converters. Electrical issues include short circuits in stator windings, insulation failure, inverter failure, and sensor calibration errors. These result in abrupt modifications to the present signature (Alagha et al., 2025). Aerodynamic breakdowns: the airflow pattern is impacted by ice buildup, aerodynamic kulai, or blade surface damage. This results in fluctuations in power output and lowers air circulation efficiency. Small but frequent signal patterns in the sound and vibration fields are typically indicative of such failures (R Jiang, 2026). Faults Associated with Control: The direction of the wind turbine does not operate correctly when an issue arises in automated control systems such as pitch control, yaw control, and speed regulation. The system's overall performance is impacted by this (Leon-Medina et al., 2025). These four categories can influence one another. For instance, rotor blade imbalance brought on by a pitch control defect might result in a gearbox malfunction. As a result, fault diagnosis is a cross-coupled, multi-domain area of study (Alagha et al., 2025). The Figure 3 explain about the Fault Diagnosis Techniques as shown below.

Figure 3. Fault Diagnosis Techniques

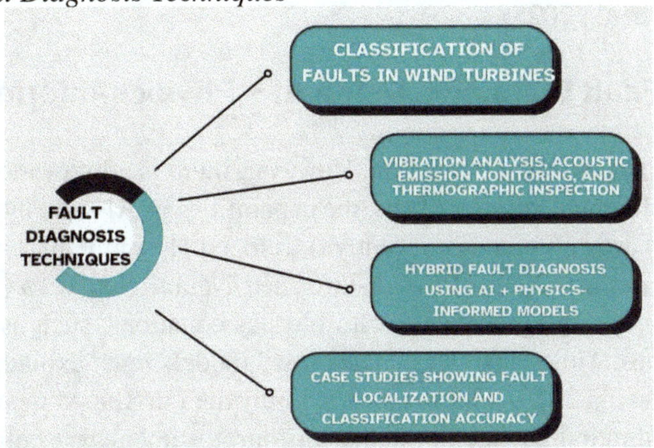

5.2 Vibration Analysis, Acoustic Emission Monitoring, and Thermographic Inspection

Vibration analysis, acoustic emission, and thermography are three crucial measuring and monitoring methods that are frequently employed in the identification of wind turbine breakdown. The oldest and currently best technique for defect identification is vibration analysis. The earliest indications of a problem are slight vibration variations in the wind turbine's rotor, bearing, and gearbox. Signal processing methods including the Short-Time Fourier Transform (STFT), Wavelet Transform, and Fast Fourier Transform (FFT) are employed. The vibration's frequency and amplitude variations can be used to identify the sort of defect. For instance, a gear tooth break will show up as a "harmonic spike," whereas a bearing kulai will generate a particular "side-band frequency pattern" (Alagha et al., 2025). One more advanced non-destructive testing (NDT) technique is acoustic emission monitoring. Very tiny sound waves are produced by internal fractures, microcracks, or rubbing defects. Piezoelectric sensors capture them, and AI algorithms use the data to identify anomalous patterns. This technique aids in the early detection of lubrication inadequacies, fracture initiation, and gear defects (J Chen et al., 2025). The heat distribution in the wind turbine's mechanical and electrical parts can be measured using thermographic inspection infrared (IR) cameras. Overheating may indicate a mechanical or electrical problem caused by friction. AI-based convolutional neural network (CNN) models handle thermal imaging data, enabling accurate defect localisation (Alagha et al., 2025). A comprehensive multi-sensor fusion fault detection system may be developed by combining these three technologies. AI-Digital

Twin technologies allow for direct monitoring of the wind turbine's "health index" (Leon-Medina et al., 2025).

5.3 Hybrid Fault Diagnosis Using AI + Physics-Informed Models

AI systems frequently rely on data. However, there is a lot of motion based on physics in wind turbine systems. Thus, the capabilities of AI and Physics-Informed Models (PIMs) are combined in recently developed Hybrid Fault Diagnosis techniques (Alagha et al., 2025). Physics-Informed Neural Networks (PINNs) are a novel approach that integrates data with physics equations, such as torque, drag, and lift equations. This transforms "black-box" models into "explainable models" (Leon-Medina et al., 2025). For instance, a totally data-driven AI model will decide based on the vibration change in the event of a rotor imbalance failure. However, a physics-informed model will compute the torque imbalance and drag coefficient changes by combining the same data with aerodynamic theory. This makes it possible to precisely determine the fault's underlying cause (Alagha et al., 2025). The following benefits are provided by hybrid techniques. Explainability: A mathematical explanation of the results. Robustness: The ability of the principles of physics to preserve model correctness in the face of sparse evidence. Transferability: The ability to apply the same technique to many wind turbine types. According to recent study, defect detection accuracy can reach 98–99% when AI and physics-informed models are combined (J Chen et al., 2025).

5.4 Case Studies Showing Fault Localization and Classification Accuracy

AI and digital twin technologies have been utilised in a number of studies to precisely identify wind turbine breakdowns. The following are a few of the important case studies:

Case Study 1: CNN-Based Gearbox Fault Detection, A CNN model was fed gearbox vibration data from a 3 MW turbine in a wind turbine research carried out in China. With 99.1% accuracy, the model identified 12 different kinds of mechanical defects. The accuracy was only 91% when FFT and SVM were used to analyse the same data (Alagha et al., 2025).

Case Study 2: LSTM-Based Bearing Fault Localisation, LSTM models used time series data to forecast bearing wear trends in an offshore wind farm in the United Kingdom. Eight hours prior to the failure, an anomaly notice was sent out, resulting in a 23% reduction in unscheduled downtime (Alagha et al., 2025).

Case Study 3: Hybrid AI-PINN for Thermal Fault Prediction Thermal hotspots were localised using a physics-informed neural network in a study conducted by a

Danish business. The hybrid model predicted fault position and intensity with 98.6% accuracy, compared to 92% for the AI-only model (J Chen et al., 2025).

Case Study 4: Digital Twin-based Predictive Maintenance: Siemens Gamesa developed unique AI models for every wind turbine using Digital Twin. This resulted in a 15% reduction in maintenance expenses and the ability to detect vibration-based faults two weeks ahead of time (Kandemir et al., 2024).

These examples show how real-time multi-sensor data fusion, physics-informed analytics, and AI–Digital twin integration are influencing the future generation of wind turbine problem diagnosis and maintenance (Alagha et al., 2025).

6. PREDICTIVE MAINTENANCE STRATEGIES

The figure 4 explain in details about the predictive maintenance strategies involves in the process.

Figure 4. Predictive Maintenance Strategies

6.1 Data-Driven vs. Model-Driven Maintenance Approaches

Predictive maintenance, or PdM, is essential for preserving wind turbine performance and preventing unscheduled downtime (Farhan et al., 2025). Data-driven and model-driven maintenance are the two primary methods for this. Machine Learning (ML) and Deep Learning (DL) models are the foundation of the data-driven approach (Buchberg & Nielsen, 2025). The wind turbine's vibration, temperature, acoustic, torque, wind speed, and other sensors are used to gather real-time data

(Electronics360, 2025). AI models analyse this vast amount of data. For instance, a Long Short-Term Memory (LSTM) neural network forecasts the timing of fault incidence using time series data, whereas a Convolutional Neural Network (CNN) detects fault patterns from vibration data (Shah et al., 2024). For fault classification and decision support, algorithms such as Random Forest or Support Vector Machine (SVM) are employed (Buchberg & Nielsen, 2025).

This method has several benefits, including the ability to automatically learn from fresh data, provide real-time fault prediction, and reliably anticipate failures and performance deterioration (Electronics360, 2025). However, it depends entirely on the facts. The findings might be erroneous if the data is inadequate or faulty (Buchberg & Nielsen, 2025). The physics, dynamics, and control theories of wind turbines serve as the foundation for the model-driven method. Finite element simulations, digital twin representations, and mathematical models are employed (Garcia-Espinosa et al., 2024). This method offers a physical explanation as well. For instance, it is possible to anticipate the rotor blade's pitch and compute its drag and lift coefficients. In a similar vein, dynamic simulation may be used to evaluate the gearbox torque variation. Hybrid predictive maintenance, which blends model-driven and data-driven methods to produce a real-time adaptive maintenance framework, has recently surfaced (Yenra, 2025).

6.2 Remaining Useful Life – RUL Estimation Using AI Models

Every wind turbine part has a remaining useful life (RUL). This is a crucial factor in determining how long a component will continue to function. Real-time sensor readings, physics-based limitations, and past operating data are integrated to generate AI-based RUL prediction (Garcia-Espinosa et al., 2024). Initially, RUL prediction included machine learning techniques including Support Vector Regression (SVR), Gaussian Process Regression, and Regression Tree. These have low accuracy for nonlinear systems, despite being appropriate for small-scale data (Shah et al., 2024).

The Deep Learning Approach On time-series data, neural networks like GRU (Gated Recurrent Unit) and LSTM function exceptionally well. These models anticipate how many hours a component will continue to function by learning deterioration patterns from the vibration and temperature data of the wind turbine (Shah et al., 2024).

Physics-Based RUL Models RUL prediction accuracy is improved via Physics-informed Neural Networks (PINN). They take into account the mechanics of torque, friction, wind speed, and blade airflow (Garcia-Espinosa et al., 2024). Reducing production downtime, anticipating spare parts manufacture, and proactive maintenance planning are the primary advantages of RUL prediction. For instance, the maintenance crew can replace a gearbox bearing in 200 hours if its projected RUL is 250 hours for a 2 MW wind turbine. This will totally prevent unexpected failures.

6.3 Maintenance Scheduling Optimization Through Predictive Insights

In isolated locales, particularly those that are hilly or coastal, wind turbine systems are frequently erected. As a result, planning maintenance tasks is quite difficult. This is addressed by data-driven optimisation in predictive maintenance techniques (Yenra, 2025). AI algorithms called Predictive Maintenance Schedulers automatically create maintenance plans based on RUL and Fault Probability. Methods like Reinforcement Learning (RL), Particle Swarm Optimisation (PSO), and Genetic Algorithm (GA) are frequently employed (Farhan et al., 2025).

Optimisation with Multiple Objectives Sometimes it's necessary to accomplish several goals at once, such as reducing maintenance expenses, maximising the use of spare parts, and minimising downtime. For this, Multi-Objective Evolutionary Algorithms (MOEA) are employed (Yenra, 2025).

Each wind turbine's real-time health indicator is estimated by Digital Twin Integration models. As a result, the AI scheduler chooses which turbine needs to be serviced first. For instance, in a wind farm with ten wind turbines, the AI scheduler will promptly assign a repair session for the third turbine if the digital twin notices an increase in vibration (Electronics360, 2025). Logistics Prediction: The technical team's route planning is also automatically planned by maintenance optimisation. Travel and processing times are decreased as a result. This entire procedure raises the availability rate of wind farms from 95% to 99%, according to several studies (Yenra, 2025).

6.4 Cost–Benefit Analysis of Predictive vs. Reactive Maintenance

Repairing a component only after it has failed is known as reactive maintenance. This leads to productivity loss, downtime, and expensive spare parts (Farhan et al., 2025). Predictive maintenance, on the other hand, aids in anticipating problems and taking prompt corrective action (Garcia-Espinosa et al., 2024).

Table 1. Cost–Benefit Analysis

Maintenance Type	Average Repair Cost	Production Loss	Downtime	Total Availability
Reactive Maintenance	1,00,000 – 2,00,000	High (20–25%)	10–20 hours	80–85%
Predictive Maintenance	40,000 – 60,000	Low (5–8%)	2–3 hours	95–99%

Forecasting The foundation of the wind energy sector in the future is maintenance. When AI, IoT, and digital twin technologies collaborate, problems are found early and maintenance tasks are automatically modified. Reduced downtime, lower maintenance costs, longer turbine life, and higher overall production efficiency are the outcomes of this.

7. BIG DATA ANALYTICS AND CLOUD PLATFORMS

Cloud computing platforms, big data analytics, and the Internet of Things (IoT) are major components of modern wind turbine systems. Thousands of sensors capture various measures, such as vibration, temperature, rotational speed, wind direction, and power generating capacity, when wind turbines are in operation (Kabir, Halder, & Ray, 2024). Digital twin models are fed this data, which is produced minute by minute. Big Data frameworks are essential for managing these enormous datasets and coming to meaningful conclusions. Wind turbine data may be quickly gathered, stored, and analysed with the use of open source frameworks like Hadoop, Spark, and Kafka. Spark Streaming, for instance, manages real-time data and uses machine learning techniques to quickly identify abnormalities (Kandemir, Hasan, Kvamsdal, & Alaliyat, 2024). This enables us to give a heads-up before the equipment malfunctions.

For handling the massive volumes of data produced by wind farms, cloud computing platforms—particularly Amazon Web Services (AWS), Microsoft Azure, and Google Cloud Platform (GCP)—are perfect. The primary benefit of cloud systems is their scalability and remote accessibility. This enables engineers to make maintenance choices by accessing their digital twin models of wind turbines from any location in the world (Leon-Medina et al., 2025). Additionally, every part of wind turbines is made possible by IoT connectivity. IoT connects temperature sensors, torque meters, blade vibration meters, and wind speed anemometers to the network. These platforms use cloud-edge hybrid architecture to improve data sharing. Critical data points are analysed locally with edge computing, allowing for quick decision-making without wasting time. At the same time, data saved in the cloud is used for historical models and in-depth studies (Abdullahi, Longo, & Samie, 2024).

7.1 Connectivity to the Digital Twin

Digital twin models are based on cloud platforms and big data analytics. Each wind turbine has its own digital twin that tracks its performance immediately. These allow for the central synchronisation of large volumes of data (multi-source data) (Mbasso & al., 2025). Electronic models represent things like blade behaviour, gearbox heating, and control system fluctuations through data synchronisation and semantic modelling. As a result, the actual and virtual worlds are more congruent.

7.2 Data Integrity and Cybersecurity Issues

The biggest issues with using cloud and big data technology for wind turbine maintenance are cybersecurity and data integrity. Malware assaults and data interception are risks associated with sending sensor data to distant cloud storage. Security layers for this are provided via end-to-end encryption, blockchain-based data authentication, and AI-enabled intrusion detection systems (Kabir et al., 2024). Additionally, since digital twins rely on the data to make judgements, the quality and reliability of the data must be guaranteed (Kandemir et al., 2024). Inaccurate maintenance choices may result from inaccurate or damaged data. As a result, AI-driven purification techniques and data validation pipelines are crucial.

All things considered, cloud platforms and big data analytics have given the wind power sector a new dimension. To apply effective predictive maintenance and problem detection techniques, they combine digital twins with the massive amounts of data produced by IoT connections (Mbasso & al., 2025). As a result, wind turbines operate more efficiently and require less maintenance. In the meantime, contemporary cybersecurity techniques guarantee data integrity and contribute to wind turbine systems' long-term dependable operation.

8. CASE STUDIES AND REAL-WORLD IMPLEMENTATIONS

Predictive and intelligent maintenance is replacing reactive maintenance in the wind energy industry. As a result, technologies like artificial intelligence (AI) and digital twins have begun to collaborate. This article will examine how they have been applied in actual wind farm projects, how performance has been assessed, and what lessons have been discovered (Leon-Medina et al., 2025).

8.1 Implementation in Actual Wind Farms

In the first example, a collaboration in the Southern Ocean developed a "virtual twin" system for a wind farm project and connected it to operational data from the past. This makes it possible to feed sensor data straight into the virtual model for the rotor, tower, and yaw/pitch control. It is possible to keep an eye on the present state of operations and promptly identify any unusual circumstances. AI algorithms may simultaneously evaluate the model for operational variances and offer suggestions for fault prediction. For instance, the project identified an increase in vibration in bearing and gearbox components; the virtual model's "vibration management signal" was established promptly. This made it feasible to do maintenance before potential damage occurred (Abdullahi et al., 2024).

A virtual twin loop for an offshore project Hywind Tampen, an offshore wind farm in Norway, is another significant example. In this case, Industry 4.0 guidelines have been followed in the implementation of digital twin systems. This uses a virtual model to depict the tower, blade, and mechanical structures. This enables meticulous maintenance planning and the prediction of bearing faults (Leon-Medina et al., 2025). Primary Study: Hybrid/Model-Based Once more, physics-based models, artificial failure data, and machine learning techniques were used in an ENGIE research on a 1.5 MW turbine. This combines it with a virtual twin and enables fault classification even from low-quality data.

8.2 Performance Evaluation, Accuracy Metrics, Reliability Outcomes

Indicators of evaluation The tower bending moment was assessed in the Hywind Tampen project. Its digital twin's estimate inaccuracy was between 10% and 15%. The problem of insufficient data for fault classification was effectively resolved in the 1.5 MW ENGIE project. A case study demonstrates that a detection rate of up to 93% was attained by employing contemporary AI systems (CNN, RNN) to identify bearing deterioration. Dependability and attained outcomes AI-based maintenance has helped some installations drastically cut down on unscheduled downtime. Additionally, availability has been raised to 95% and higher through data-centric maintenance activities. Technical Positions Accuracy of fault localisation: In the Physics-based + AI system, the classification error decreased to less than 10%. By calculating RUL (Remaining Useful Life), Digital Twin prolongs the forward maintenance period (Kandemir et al., 2024).

8.3. Lessons Learned from Deployments and Pilot Projects

Problems with Data Supply The supply and quality of sensor data are important issues. One significant issue is the absence of labelled fault data needed for model training. In many projects, the compatibility of the sensor facilities with real-world applications is questioned, despite the creation of a virtual model (Mbasso & al., 2025). Model-integration Difficulties When combining Digital Twin with AI, interoperability is crucial. Offshore wind settings are using Industry 4.0 standards like the Asset Administration Shell (AAS). Some results have inaccuracies of up to 10–15% according to the models' fidelity (Leon-Medina et al., 2025). In actuality, the performance of wind turbines has increased due to the integration of AI and digital twins. Reliability has improved and unplanned downtime has decreased. However, issues like interoperability, virtual model fidelity, and data quality continue to exist. Furthermore, the effective implementation of the technology depends on factory management, personnel training, and governance. Suggestions Start with modest projects and work your way up. First, create a sensor hub and data governance. Hybrid systems, which combine data-driven and physics-based components, yield better outcomes. Assure interoperability and standards.

9. CHALLENGES AND FUTURE RESEARCH DIRECTIONS

The figure 5 explains about the Challenges and Future Research directions in the field of Wind energy.

Figure 5. Challenges and Future Research Directions

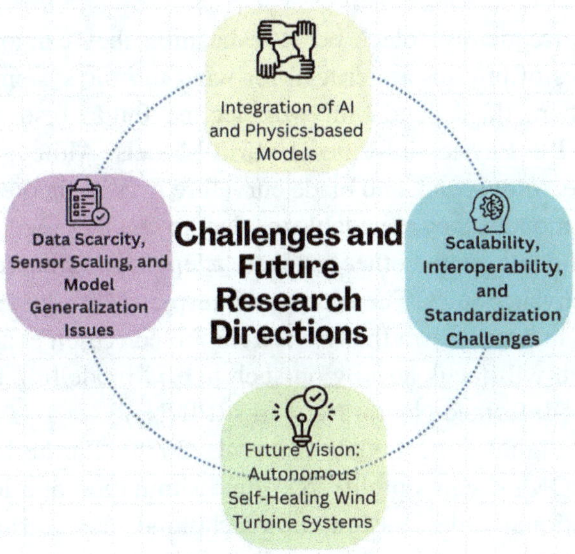

9.1. Data Scarcity, Sensor Scaling, and Model Generalization Issues

Wind turbine AI models depend on a large amount of data. But in the actual world, data scarcity is a big problem. There is less data available to train models since many wind turbine businesses keep their sensor data confidential and do not share it (Fernandez-Navamuel et al., 2025). Moreover, fault occurrences are uncommon, leading to "imbalanced datasets." Because of this, learning models are less likely to detect errors even while they are good at identifying normal situations (Benítez & Singh, 2025). Another significant issue is data dependability and sensor calibration. At wind turbine locations, vibration, temperature, and sound sensors may deteriorate with time. AI algorithms may provide erroneous predictions as a result of minor mistakes spreading as incorrect information to the digital twin model (Elmousalami, Alnaser, & Hui, 2025).

Therefore, anomaly filtering, data purification, and auto-calibration techniques are crucial. Additionally, the research area has long struggled with model generalisation. Due to variations in wind speed, climatic factors, and production equipment, an AI model trained on one wind farm might not be relevant in another. Additional research on transfer learning strategies, domain adaptation techniques, and the use of synthetic data is required to address this (Kandemir, Hasan, Kvamsdal, & Alaliyat, 2024).

9.2 Integration of AI and Physics-based Models

AI results are frequently "black boxes," meaning they cannot be deciphered. However, the rules of physics are crucial for wind turbine systems; if the models do not adhere to them, the dependability will decline. Future research will therefore heavily rely on AI + Physics-informed Hybrid Models. Although characteristics like wind pressure, rotor speed, and blade curvature may be theoretically described by physics-based models, they are unable to adapt to changes in real-time data (Yan et al., 2024). AI models, on the other hand, are adept at handling real-time data but do not adhere to physical rules. The accuracy, interpretability, and robustness of the outcomes will all improve when the two are combined (Gijón et al., 2025).

Future research will focus heavily on such hybrid modelling frameworks. For instance, "Physics-Informed Neural Networks (PINNs)" are able to directly apply physical rules to digital twin models (Li et al., 2025). Furthermore, "Reduced-Order Models (ROMs)" can simplify real-time simulation and lower computing complexity. Additionally, data assimilation techniques can constantly update the AI model's parameters, ensuring that the digital twin consistently reaches a state of "self-learning."

9.3 Scalability, Interoperability, and Standardization Challenges

As the AI-Digital Twin ecosystem develops, scalability becomes a significant concern. Data synchronisation, storage, and real-time analysis are significant issues if every turbine in a wind farm with hundreds of turbines has its own digital twin (Li et al., 2025). As a result, a future where learning occurs on-site rather than in a data centre is emerging, utilising federated learning frameworks and edge computing techniques. The problem of interoperability comes next. Turbines from various manufacturers (OEMs) have distinct data formats. Consequently, many software components and sensors are unable to connect with one another. Semantic data models and open standards (such as OPC-UA and Asset Administration Shell, AAS) are crucial to solving this (Kandemir et al., 2024).

These make it possible for several organisations to link and run digital twins on a single platform while sharing the same dataset. Another important path is standardisation. In the wind business, there are currently no globally accepted standards for fault classification techniques, AI model exchange, or Digital Twin architecture. Future study might focus on applying standards like ISO 23247 (Digital Twin Guidelines for Manufacturing) to the wind turbine sector.

9.4 Future Vision: Autonomous Self-Healing Wind Turbine Systems

Autonomous Self-Healing Wind Turbine Systems will appear in the future as the AI–Digital Twin link advances. Without human assistance, they will be able to identify and fix problems on their own. Every part of a completely self-healing wind turbine system would have a distinct digital counterpart that keeps track of its condition all the time (Gijón et al., 2025). The AI model will do a "root cause analysis" and schedule the required maintenance procedures ahead of time once an anomaly is found. Simple mistakes like sensor resets and brief power interruptions will be automatically fixed by "self-recovery algorithms."

Additionally, wind farms will use reinforcement learning and multi-agent systems (MAS) to learn collaboratively. Through cooperative learning, each turbine will pick up knowledge from the others and swiftly spot new fault patterns. The entire performance at the farm level will thus naturally improve. The top tier of the Digital Twin Ecosystem will be these self-healing systems. Fault diagnosis, RUL estimate, and maintenance scheduling will all be automated via their cloud-based AI governance system. Technologies like Edge AI chips, quantum computing models, and 6G connectivity will be incorporated to facilitate this (Sun & Han, 2025).

The wind sector is undergoing a disruptive shift due to the AI-Digital Twin relationship. Reliability management, predictive maintenance, and problem diagnostics have all undergone a fresh development thanks to these technologies. To achieve completely self-healing wind turbine systems, however, a number of obstacles still need to be addressed, including data scarcity, model dependability, integration, and standardisation (Elmousalami et al., 2025). The following are the directions that future study will take: creation of hybrid physics-AI models; edge AI for real-time analytics; data dependability based on blockchain; federated learning allows for cooperative training across several organisations; creation of self-healing, autonomous turbine settings. In the future, a completely intelligent energy production environment with self-repairing wind turbine systems will appear if study in these areas continues. This will mark a significant turning point in the world's move to green energy (Sun & Han, 2025).

10. CONCLUSION

The wind industry's use of digital twin technologies and artificial intelligence (AI) has completely changed how electricity is produced and maintained. By doing this, the flaws in earlier conventional monitoring and maintenance systems have been solved. The precision of problem detection, downtime reduction, and turbine lifespan extension have all seen notable advancements. Each turbine's internal status is immediately recorded via sensor networks, digital twins, and AI-based data analysis. An accurate model-based defect diagnostic system is created by combining a variety of data kinds, including vibration, temperature, sound, and voltage measurements. This allows clever algorithms to anticipate future errors and prevent unplanned malfunctions. In the past, manual inspections and routine maintenance were the major methods used to find defects. Consequently, little errors resulted in significant losses. However, AI-Digital Twin-based systems use "real-time condition monitoring" to find errors before they happen. To improve the classification of vibration, sound, and temperature models, machine learning, deep learning, and hybrid AI models—such as CNN–LSTM networks, Transformer topologies, and physics-informed neural models—are employed. As a result, the accuracy of fault classification rises to 95%–99%. Additionally, the virtual twin (virtual replica) paradigm allows for instantaneous fault localisation (i.e., identifying the faulty component). Maintenance tasks are therefore quicker and more precise. One of the main economic goals of wind farms is to reduce downtime. Remaining Useful Life (RUL) forecasts are provided by predictive maintenance systems, allowing for the timely completion of maintenance tasks. This significantly lowers the requirement for unscheduled maintenance and breakdown-based repairs. Additionally, real-time decision-making is made possible by AI models using cloud and edge computing-based systems. The maintenance staff receives fault notifications right away. Digital twin models evaluate solutions ahead of time and mimic possible errors. Consequently, each turbine becomes more accessible, increasing the farm's overall production.

REFERENCES

Abdullahi, I., Longo, S., & Samie, M. (2024). Towards a distributed digital twin framework for predictive maintenance in Industrial Internet of Things (IIoT). *Sensors (Basel), 24*(8), 2663. DOI: 10.3390/s24082663 PMID: 38676279

Abimbola, O., Ojo, O. T., Fagbola, E., Idris, U. A., & Salman, M. B. (2025). IoT-driven predictive maintenance for wind turbines. *Path of Science*. DOI: 10.22178/pos.114-20

Alagha, N., Khairuddin, A. S. M., Haitaamar, Z. N., Al-Khatib, O., & Kanesan, J. (2025). Artificial intelligence in wind turbine fault detection and diagnosis: Advances and perspectives. *Energies, 18*(7), 1680. DOI: 10.3390/en18071680

Benítez, I. B., & Singh, J. G. (2025). A comprehensive review of machine learning applications in forecasting solar PV and wind turbine power output. *Journal of Electrical Systems and Information Technology, 12*(1), 54. DOI: 10.1186/s43067-025-00239-4

Buchberg, E. M., & Nielsen, K. V. (2025). Condition monitoring with machine learning: A data-driven framework for quantifying wind turbine energy loss. *arXiv*. https://doi.org//arXiv.2506.13012DOI: 10.48550

Campoverde-Vilela, L., Feijóo, M. C., Vidal, Y., Sampietro, J., & Tutivén, C. (2023). Anomaly-based fault detection in wind turbine main bearings. *Wind Energy Science, 8*(4), 557–574. DOI: 10.5194/wes-8-557-2023

Chatterjee, J., & Dethlefs, N. (2022). Scientometric review of artificial intelligence for operations & maintenance of wind turbines: The past, present and future. *arXiv*. https://arxiv.org/abs/2204.02360

Chen, S., Liu, Y., & Liu, S. (2024). *Wind power short-term prediction based on digital twin technology*. Energy Research., DOI: 10.3389/fenrg.2024.1365237

Editorial. (2024). Online monitoring of wind power plants using digital twin models. *Energy Research*. DOI: 10.3389/fenrg.2024.1531689

Electronics360. (2025, May). Digital twins and AI for predictive maintenance in wind farms. *Electronics360*. https://electronics360.globalspec.com/article/22452/digital-twins-and-ai-for-predictive-maintenance-in-wind-farms

Farhan, M., Schneider, R., Thöns, S., & Gündel, M. (2025). Probabilistic cost modeling as a basis for optimizing inspection and maintenance of turbine support structures in offshore wind farms. *Wind Energy Science*, *10*(2), 461–481. DOI: 10.5194/wes-10-461-2025

Fernandez-Navamuel, A.. (2025). Gaussian mixture autoencoder for FOWT damage identification. *Wind Energy Science*, *10*, 857–885. DOI: 10.5194/wes-10-857-2025

Garcia-Espinosa, J., Di Capua, D., & Pastor Sanchez, A. (2024). A digital twin for assessing the remaining useful life of offshore wind turbine structures. *Journal of Marine Science and Engineering*, *12*(4), 573. DOI: 10.3390/jmse12040573

Gijón, A., Eiraudo, S., Manjavacas, A., Salvatore, D., Molina Solana, M., & Gómez Romero, J. (2025). Integrating physics and data-driven approaches: An explainable and uncertainty-aware hybrid model for wind turbine power prediction. *arXiv*. https://arxiv.org/abs/2502.07344

Habbouche, H., Amirat, Y., & Benbouzid, M. (2025). Leveraging digital twins and AI for enhanced gearbox condition monitoring in wind turbines. *Applied Sciences (Basel, Switzerland)*, *15*(10), 5725. https://www.mdpi.com/2076-3417/15/10/5725. DOI: 10.3390/app15105725

Jankauskas, M., Serackis, A., Šapurov, M., Pomarnacki, R., Baskys, A., Hyunh, V. K., Vaimann, T., & Zakis, J. (2023). Exploring the limits of early predictive maintenance in wind turbines applying an anomaly detection technique. *Sensors (Basel)*, *23*(12), 5695. DOI: 10.3390/s23125695 PMID: 37420861

Kabir, M. R., Halder, D., & Ray, S. (2024). Digital twins for IoT-driven energy systems: A survey. *IEEE Access : Practical Innovations, Open Solutions*, *12*, 177123–177145. DOI: 10.1109/ACCESS.2024.3506660

Kandemir, E., Hasan, A., Kvamsdal, T., & Alaliyat, S. A. (2024). Predictive digital twin for wind energy systems: A literature review. *Energy Informatics*, *7*(1), 68. DOI: 10.1186/s42162-024-00373-9

Leon-Medina, J. X., ... (2025). Digital twin technology in wind turbine components: A review. *Elsevier*. https://www.sciencedirect.com/science/article/pii/S2667305325000614

Li, S., Li, X., Jiang, Y., Yang, Q., Lin, M., Peng, L., & Yu, J. (2025). A novel frequency-domain physics-informed neural network for accurate prediction of 3D spatio-temporal wind fields in wind turbine applications. *Applied Energy*, *386*, 125526. Advance online publication. DOI: 10.1016/j.apenergy.2025.125526

Li, X., Zou, Y., & Xia, X. (2024). *Unlocking the potential of digital twins for a greener grid*. IEEE PES Electrification.

Mbasso, W. F.. (2025). *Digital twins in renewable energy systems*. Renewable Energy Journal.

Mousavi, Z., ... (2024). A digital twin-based framework for damage detection of a floating wind turbine. *Elsevier*. https://www.sciencedirect.com/science/article/abs/pii/S0029801823029475

Rasheed, A., Stadtmann, F., Fonn, E., et al. (2024). Digital twins in wind energy: Emerging technologies and industry-informed future directions. *arXiv*.

Shah, S. S., Tan, D., & Kumar, S. C. (2024). RUL forecasting for wind turbine predictive maintenance based on deep learning. *arXiv*. https://doi.org//arXiv.2412.17823DOI: 10.48550

Shittu, H. A., & Shittu, M. A. (2024). Leveraging digital twins and AI for enhanced gearbox condition monitoring in wind turbines. *Applied Sciences (Basel, Switzerland)*, *15*(10), 5725.

Sun, Y., & Han, W. (2025). A review of enhancing wind power with AI: Applications, economic implications, and green innovations. *Digital Economy and Sustainable Development*, *3*(1), 11. DOI: 10.1007/s44265-025-00059-4

Vargas, J.. (2024). Predictive digital twin for wind energy systems: A literature review. *Energy Informatics*, *7*(1), 17. DOI: 10.1186/s42162-024-00373-9

Vattenfall Press. (2024, January). Digital twins – A road to more profitable offshore wind. *Vattenfall Group*.

Yan, C., Xu, S., Sun, Z., Lutz, T., Guo, D., & Yang, G. (2024). A framework of data assimilation for wind flow fields by physics-informed neural networks. *arXiv*. https://arxiv.org/abs/2401.17001

Yenra. (2025, March). AI predictive maintenance for wind turbines: 20 advances. *Yenra*. https://yenra.com/ai20/predictive-maintenance-for-wind-turbines/

Chapter 6
Artificial Intelligence and Digital Twin Synergy for Predictive Maintenance and Fault Detection in Wind Turbines

Rampelli Manojkumar
http://orcid.org/0000-0002-8141-0321
BVRIT HYDERABAD College of Engineering for Women, India

Ramesh Babu Pittala
http://orcid.org/0000-0003-2420-8183
Anurag University, India

Gundu Venu
Malla Reddy University, India

Anji Reddy Polu
BVRIT HYDERABAD College of Engineering for Women, India

Suresh Kumar Annam
Malla Reddy University, India

T. Yuvaraj
http://orcid.org/0000-0002-8892-9417
Chennai Institute of Technology, India

ABSTRACT

This chapter explores the integration of Artificial Intelligence (AI) and Digital Twin (DT) technologies for intelligent condition monitoring and fault detection in wind energy systems. It highlights the limitations of traditional approaches and presents AI-based models for fault prediction, anomaly detection, and remaining useful life estimation. The layered architecture of DTs, their real-time synchronization, and hybrid AI-DT frameworks for predictive maintenance are discussed. The chapter

DOI: 10.4018/979-8-3373-4159-0.ch006

also examines challenges such as data integrity, scalability, and cybersecurity while outlining emerging trends like federated learning and quantum digital twins for resilient and sustainable wind farm operations.

INTRODUCTION

The global pursuit of decarbonization and sustainable energy has positioned wind power as one of the most critical pillars of modern electricity generation. As nations accelerate their commitments to carbon neutrality, the deployment of large-scale wind farms—both onshore and offshore—has expanded dramatically over the past decade. However, this rapid growth brings forth substantial challenges related to reliability, maintenance, and operational efficiency. Wind turbines are complex electromechanical systems that operate under fluctuating wind speeds, variable loads, and harsh environmental conditions. These operational stresses can induce gradual degradation and unexpected faults in key components such as blades, gearboxes, bearings, generators, and power converters (Sasinthiran et al., 2024; Astolfi et al., 2023). Addressing these challenges through advanced condition monitoring and fault detection techniques is essential for ensuring high availability, reducing maintenance costs, and extending the lifespan of assets in modern wind energy systems.

Need for Condition Monitoring in Wind Turbines

Condition monitoring systems (CMS) form the foundation of predictive maintenance strategies by providing continuous health assessment of turbine components. Traditional CMS approaches typically rely on rule-based thresholds, vibration analysis, and model-based estimations. While these techniques have been useful for early fault detection, they often struggle with the inherent complexity and nonlinear behavior of wind turbines. For instance, threshold-based detection methods may fail to capture subtle changes in system dynamics or differentiate between operational fluctuations and actual fault signatures. As a result, conventional approaches frequently lead to false alarms or missed detections, ultimately increasing downtime and maintenance costs (Ng & Lim, 2022; Tang et al., 2021).

Moreover, the modern wind energy ecosystem is characterized by large-scale data acquisition through sensors and supervisory control and data acquisition (SCADA) systems. These sensors collect data on vibration, torque, temperature, current, wind speed, and acoustic emissions, creating an enormous volume of time-series data. Managing and interpreting this high-dimensional, multimodal data requires more sophisticated analytical techniques. The limitations of traditional methods have therefore catalyzed the integration of Artificial Intelligence (AI) and data-driven techniques to extract meaningful patterns, predict faults before they occur, and optimize maintenance scheduling (Sun et al., 2021; Kamarzarrin et al., 2022).

Emergence of Artificial Intelligence in Wind Turbine Fault Diagnosis

Artificial Intelligence has emerged as a powerful enabler for intelligent monitoring and diagnostics in wind energy systems. AI techniques such as machine learning (ML), deep learning (DL), and reinforcement learning (RL) have demonstrated exceptional capability in recognizing complex fault patterns from raw sensor data. These approaches allow wind turbine systems to move from reactive and scheduled maintenance practices to predictive and prescriptive maintenance paradigms.

Recent studies have shown that AI-based algorithms outperform traditional approaches in identifying incipient faults, isolating faulty components, and estimating remaining useful life (RUL) (Astolfi et al., 2023; Han et al., 2023). For example, supervised learning methods like Support Vector Machines (SVM) and Random Forests have been successfully applied to classify fault conditions using vibration and temperature data, while unsupervised approaches, such as Autoencoders and clustering techniques, have proven useful for detecting novel or unexpected anomalies. Deep learning models, particularly Convolutional Neural Networks (CNN) and Long Short-Term Memory (LSTM) networks, excel in feature extraction from time-series and frequency-domain signals (Zhang et al., 2022; Peng et al., 2023).

AI-based condition monitoring also benefits from advances in explainable AI (XAI), which aim to enhance transparency and interpretability of predictions (Astolfi et al., 2023). This is especially crucial for industrial adoption, as operators require confidence in automated decision-making processes. Furthermore, the increasing accessibility of computational resources and cloud-based infrastructures has made it feasible to implement AI-driven condition monitoring on a large scale, even in geographically distributed wind farms (Karthik Selva Kumar et al., 2023).

Limitations of Standalone AI Approaches

Despite their success, standalone AI models are not without limitations. One major challenge is their dependency on large, labeled datasets for effective training and validation. Wind turbines operate in diverse environmental and mechanical conditions, leading to domain-specific variations that may reduce the generalization capability of trained models. Additionally, data scarcity for rare fault events limits the availability of high-quality samples needed for robust AI model training (An et al., 2024).

Another issue is that purely data-driven AI approaches may overlook the physical principles governing turbine behavior. For example, a model might identify a correlation between certain vibration signatures and a fault but fail to explain the underlying mechanical cause. This gap between data-driven learning and physical system understanding can result in misinterpretation and reduced trust among operators. To overcome these challenges, researchers have started integrating physics-based models with AI-based analytics—an approach embodied by the concept of Digital Twins (Wang et al., 2023; Hu et al., 2024).

Role of Digital Twin Technology

Digital Twin (DT) technology represents a transformative innovation in condition monitoring and predictive maintenance. A Digital Twin is a virtual replica of a physical system that continuously mirrors its behavior through real-time data exchange. By combining simulation models, sensor data, and AI algorithms, Digital Twins enable dynamic monitoring and forecasting of system states (Olatunji et al., 2021; Li & Shen, 2021). This allows engineers to observe how a turbine performs under different environmental and operational scenarios without interrupting its actual operation.

In the context of wind energy, DTs can simulate the mechanical and electrical interactions within turbine subsystems such as blades, generators, and gearboxes. For instance, a digital twin of the gearbox can replicate the wear patterns, torque variations, and lubrication dynamics, thereby providing early warnings of potential failures. Studies have demonstrated that integrating digital twins with AI-driven analytics enhances fault diagnosis accuracy and enables virtual sensing—estimating unmeasured parameters by leveraging correlations learned from both physical models and data-driven insights (Wang et al., 2023; Hu et al., 2024).

Recent advancements in computational capabilities and IoT connectivity have further accelerated the deployment of DTs in wind energy systems. With edge computing and cloud infrastructures, digital twins can now operate in real time, continuously updating their parameters to reflect the physical system's actual condition. This facilitates intelligent decision-making for maintenance scheduling, load balancing, and performance optimization (Falekas & Karlis, 2021; Prasath, 2025).

Synergy of AI and Digital Twin for Intelligent Monitoring

The convergence of AI and Digital Twin technologies creates a synergistic framework for next-generation wind turbine monitoring. In such a framework, AI algorithms process the massive datasets generated by sensors and the DT, identifying anomalies, predicting faults, and optimizing control strategies. Simultaneously, the DT provides a physics-informed context that improves the interpretability and reliability of AI outputs. This hybrid AI-DT approach combines the strengths of both paradigms: the learning capability of AI and the modeling precision of physics-based simulation.

Alagha et al. (2025) highlight that AI-DT integration not only improves detection accuracy but also enables autonomous decision-making in predictive maintenance. Similarly, Hu et al. (2024) proposed a transferable diagnosis model for digital twins capable of detecting incipient faults with minimal retraining, showcasing the adaptability of hybrid systems across turbine models. Mikołajewska et al. (2025) further emphasize the potential of generative AI within digital twin environments to simulate fault conditions, helping engineers train diagnostic models on synthetic yet realistic datasets. This synergy marks a significant step toward realizing self-healing and self-optimizing wind energy systems.

The adoption of AI and DT technologies aligns closely with the broader paradigm of Industry 4.0 and its evolution toward Industry 5.0, which emphasizes human–machine collaboration, sustainability, and resilience. Wind energy, as a critical component of the renewable energy transition, stands to benefit immensely from these digitalization trends. Intelligent monitoring systems allow operators to move from periodic inspections to continuous asset management, significantly reducing unplanned outages and operational costs (Rojas et al., 2025). Moreover, the combination of AI and DT contributes to the development of smart wind farms, where interconnected turbines cooperate through shared data networks to optimize collective performance and stability.

AI-enabled digital twins also support cross-domain applications beyond wind energy. For example, Prasath (2025) discusses how such frameworks are being applied in smart urban infrastructure for predictive maintenance, indicating their versatility and scalability across industrial sectors. These cross-disciplinary advances reinforce the relevance of AI-DT frameworks in achieving sustainability and reliability goals across critical engineering systems. An illustration of AI–Digital Twin integration for intelligent wind turbine monitoring is shown in Figure 1.

Figure 1. AI–Digital Twin integration for intelligent wind turbine monitoring.

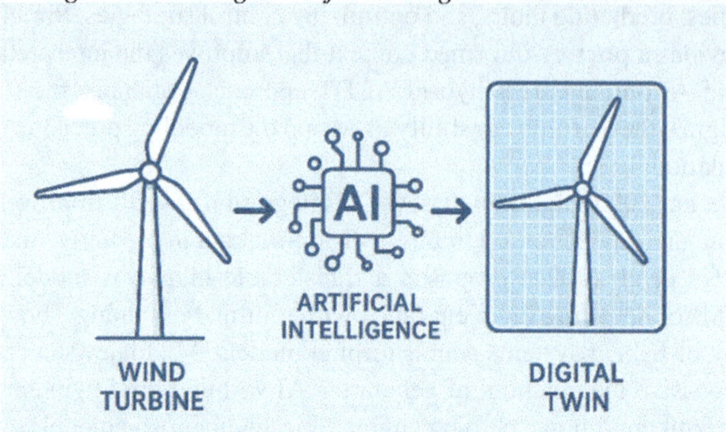

Challenges and Research Gaps

While AI and DT technologies offer promising capabilities, several challenges remain. Data integrity and standardization are persistent issues, as sensor data can be noisy, incomplete, or inconsistent across platforms. Moreover, the interoperability between various software, hardware, and communication protocols poses technical barriers to seamless DT deployment (Falekas & Karlis, 2021). Computational complexity also limits real-time applications, particularly when high-fidelity simulations or deep learning models are involved.

Another crucial challenge concerns explainability and trust. Operators and engineers must understand how AI models make decisions, especially in safety-critical environments. Explainable AI (XAI) and interpretable machine learning techniques are therefore essential for bridging the gap between algorithmic predictions and human understanding (Astolfi et al., 2023). Furthermore, as wind farms become more connected and data-driven, ensuring cybersecurity of AI-DT frameworks becomes increasingly important to prevent malicious attacks or data breaches (Mikołajewska et al., 2025). From a research standpoint, future directions include federated learning and transfer learning to enhance the generalization of AI models across diverse turbines and environments (An et al., 2024). Emerging concepts like quantum digital twins also hold potential for improving simulation precision and computational speed in complex systems (Mikołajewska et al., 2025). Addressing these research challenges will be crucial for realizing the full potential of AI-DT-enabled condition monitoring.

This chapter aims to provide a comprehensive overview of the integration of AI and Digital Twin technologies for intelligent condition monitoring and fault detection in wind energy systems. Section 2 will elaborate on condition monitoring techniques and challenges specific to wind turbine subsystems. Section 3 will delve into AI algorithms and their application in fault diagnosis and RUL estimation. Section 4 will focus on Digital Twin architectures and case studies, while Section 5 will discuss hybrid AI-DT frameworks for predictive maintenance and smart wind farm management. The final section will highlight challenges, ethical considerations, and future research trends shaping the evolution of intelligent and autonomous wind energy systems. In essence, the fusion of AI and Digital Twin technologies represents a paradigm shift from traditional maintenance to self-aware and self-adaptive systems. By leveraging real-time data analytics, virtual simulation, and intelligent decision-making, these technologies are laying the groundwork for the next generation of smart, connected, and resilient wind energy infrastructures that align with global sustainability goals.

CONDITION MONITORING IN MODERN WIND ENERGY SYSTEMS

Wind energy systems are composed of multiple interdependent mechanical and electrical subsystems that function collectively to convert wind energy into usable electrical power. The primary subsystems—namely the blades, gearbox, generator, and power electronics—are exposed to diverse operational and environmental stresses that make them vulnerable to degradation and failure. Effective condition monitoring of these components is therefore vital to ensure reliability, minimize downtime, and optimize maintenance schedules.

Blades are among the most exposed components, subject to aerodynamic loads, material fatigue, and environmental factors such as icing, erosion, and lightning strikes (Sun et al., 2022). Blade damage can significantly impact aerodynamic efficiency and structural integrity, leading to power losses and potential safety risks. Advanced sensing technologies—including acoustic emission, strain gauges, and fiber-optic sensors—are now employed for early crack detection and fatigue monitoring.

The gearbox remains one of the most critical and failure-prone elements in the drivetrain, translating the slow rotational speed of the turbine blades into higher-speed rotation suitable for electricity generation. Gearbox failures often result from lubrication degradation, bearing fatigue, or gear tooth wear (Heydari et al., 2021; Inturi et al., 2021). Continuous vibration, oil debris, and temperature monitoring are essential to detect these issues before catastrophic failure occurs.

The generator, typically a doubly-fed induction generator (DFIG) or permanent magnet synchronous generator (PMSG), is another key subsystem requiring careful surveillance. Electrical faults such as insulation degradation, rotor winding failures, or short circuits can impair efficiency and stability (Bebars et al., 2022). Electrical signature analysis (ESA) and current harmonics monitoring are common diagnostic tools for this purpose.

Finally, power electronics—which include converters, inverters, and controllers—serve as the backbone of energy conversion and grid interfacing. They face frequent thermal cycling, switching stresses, and electromagnetic interference, all of which necessitate real-time monitoring to prevent sudden failures. Together, these subsystems define the health and operational continuity of the entire wind turbine system (Badihi et al., 2022).

Conventional Monitoring and Fault Detection Methods

Traditional condition monitoring systems (CMS) in wind turbines relied primarily on rule-based or threshold-based approaches. These systems compare measured parameters (vibration, temperature, current, or oil quality) against fixed thresholds to trigger alarms. While simple and cost-effective, such approaches are inadequate under varying wind conditions and nonlinear dynamics typical of wind turbines (Aziz et al., 2021).

Signal-Based and Model-Based Techniques

Conventional signal-based methods include vibration analysis, acoustic emission, and oil particle monitoring. Vibration analysis, a cornerstone of gearbox monitoring, is particularly effective for detecting bearing wear and gear faults (Heydari et al., 2021). However, vibration data can be affected by environmental noise, rotor speed fluctuations, and load variations. Oil analysis has also been widely used to evaluate lubricant quality and detect metallic debris, offering insights into wear progression and lubrication breakdown (Bie et al., 2021).

On the other hand, model-based methods depend on mathematical representations of turbine dynamics. These include state estimators, observer-based models, and parity equations to identify deviations between measured and predicted behavior (Badihi et al., 2022). Model-based techniques provide strong interpretability and are rooted in physical system understanding but often require accurate modeling and extensive parameter tuning.

The limitations of both categories have led to hybrid strategies that combine physical modeling with data-driven intelligence. For instance, Heydari et al. (2021) developed a hybrid intelligent model that integrates model-based fault signatures with data-driven classifiers to improve diagnostic accuracy for gearbox systems. Similarly, Inturi et al. (2021) proposed an adaptive condition monitoring scheme capable of self-adjusting its detection parameters to changing operational conditions, improving reliability across variable-speed regimes.

SCADA-Based Monitoring

Supervisory Control and Data Acquisition (SCADA) systems provide continuous operational data—such as wind speed, rotor speed, active/reactive power, and temperature—from multiple turbine sensors. Traditionally used for control purposes, SCADA data are now leveraged for data-driven fault detection (Aziz et al., 2021). Pandit et al. (2023) reviewed SCADA-based monitoring frameworks, highlighting their accessibility and scalability across large wind farms. However, the coarse sampling rate (typically 10 min intervals) limits the system's ability to detect fast-developing faults such as bearing cracks or power converter malfunctions.

To address this limitation, Castellani et al. (2024) proposed a sequential analysis method combining SCADA data with high-frequency vibration signals, significantly improving the accuracy of gearbox condition monitoring. This hybrid approach bridges the gap between macro-level performance data and micro-level fault indicators.

Non-Destructive and Structural Health Monitoring

Beyond vibration and oil-based diagnostics, non-destructive testing (NDT) and structural health monitoring (SHM) techniques are increasingly being adopted. These include ultrasonic inspection, thermography, and strain-based deformation sensing (Civera & Surace, 2022). SHM systems equipped with fiber-optic or piezoelectric sensors can monitor blade deflections and detect structural anomalies in real time. NDT-based approaches are particularly effective during maintenance intervals, offering a complementary layer of insight to continuous condition monitoring systems.

Sensor Technologies and Data Acquisition Frameworks

Condition monitoring in modern wind turbines relies heavily on multimodal sensing and data acquisition frameworks designed to capture complex interactions among mechanical, electrical, and environmental parameters. The core sensor types include:

- Vibration and acceleration sensors, used extensively in gearboxes and bearings to detect imbalance, misalignment, and wear (Inturi et al., 2021).

- Acoustic emission sensors, sensitive to crack propagation and early-stage structural damage (Sun et al., 2022).
- Temperature and pressure sensors, employed for lubrication and thermal condition monitoring (Bie et al., 2021).
- Current and voltage sensors, vital for identifying electrical faults in generators and converters (Bebars et al., 2022).
- Optical and fiber Bragg grating (FBG) sensors, suitable for strain, deformation, and load monitoring in blades and towers (Civera & Surace, 2022).

Recent developments in Internet of Things (IoT) technologies have enabled sensor miniaturization, wireless communication, and cloud connectivity. Karad and Thakur (2021) demonstrated how IoT-based CMS architectures can integrate various sensors into a unified platform, enabling remote diagnostics and real-time fault alerts. Such systems facilitate not only condition monitoring but also operational control optimization across geographically distributed wind farms. The illustration of a simple schematic of wind turbine condition monitoring system showing major sensor locations, local edge processing, and cloud analytics is given in Figure 2.

Figure 2. Wind turbine condition monitoring system

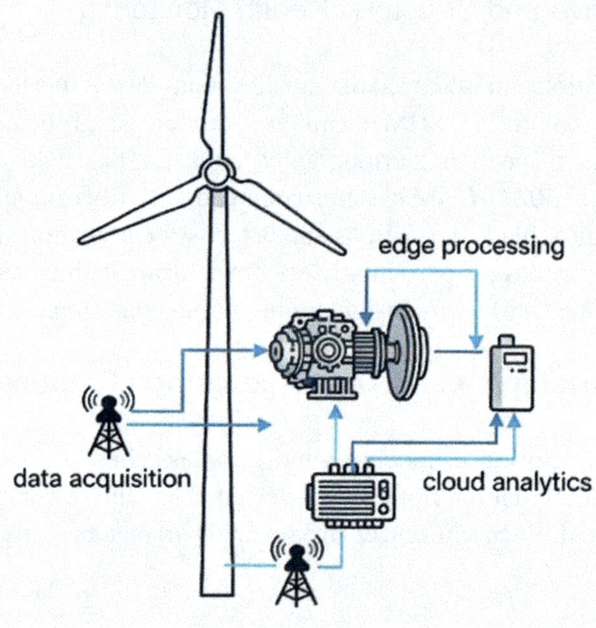

Data Acquisition and Communication Layers

The data acquisition framework typically consists of three hierarchical layers:

1. Local sensing and acquisition – collecting raw signals from sensors located on turbine components.
2. Edge processing – performing local filtering, compression, and initial fault detection to reduce data volume.
3. Cloud and supervisory layers – aggregating data across multiple turbines for centralized analysis, long-term trend assessment, and fleet-level benchmarking (Bustamante et al., 2025).

This layered architecture supports real-time analytics while addressing bandwidth and latency constraints. Edge computing plays a critical role by enabling local intelligence for immediate response, whereas cloud-based analytics support complex fault prediction and maintenance planning.

The integration of wide-area monitoring systems (WAMS) and big data analytics, as demonstrated by Bustamante et al. (2025), further enhances real-time visualization and anomaly detection at grid scale. Although their study focused on power grid monitoring, similar architectures are being applied to offshore wind farms, where connectivity and scalability are essential.

Need for Real-Time Analytics and Predictive Maintenance

Despite the advancements in sensor networks and CMS architectures, the true value of condition monitoring lies in the transition from reactive maintenance toward predictive and prescriptive maintenance paradigms.

Evolution From Reactive to Predictive Strategies

Historically, wind turbine maintenance followed either corrective (repair after failure) or preventive (scheduled replacement) strategies. Both approaches suffer from inefficiency—either excessive downtime or unnecessary component replacement. Predictive maintenance, driven by real-time analytics, addresses these limitations by forecasting potential failures and scheduling maintenance only when necessary (Maron et al., 2022).

Predictive maintenance models rely on continuous data streams from CMS sensors, coupled with advanced analytics and artificial intelligence. For example, Gao and Odgaard (2023) emphasized the integration of real-time monitoring and health management frameworks that fuse physical models with data-driven algorithms to forecast fault progression in offshore turbines.

Role of Artificial Intelligence in CMS Evolution

Machine learning (ML) and deep learning (DL) algorithms now enable turbines to "learn" from historical data and adapt to evolving operational conditions. Hybrid AI models such as those proposed by Heydari et al. (2021) and Maron et al. (2022) enhance fault classification accuracy, while self-supervised learning approaches (Sun et al., 2022) improve health state representation without labeled datasets.

Moreover, SCADA-based predictive models are being increasingly adopted for long-term degradation prediction. Castellani et al. (2024) and Pandit et al. (2023) demonstrated that combining low-frequency SCADA data with high-resolution vibration or acoustic features enables accurate Remaining Useful Life (RUL) estimation for gearboxes and bearings.

Optimization and Decision Support

Beyond fault prediction, predictive maintenance frameworks are evolving toward decision-support optimization. Bakir et al. (2021) developed an integrated optimization model for multi-component maintenance scheduling that considers cost, availability, and risk. This approach aligns maintenance actions with economic and operational objectives, contributing to improved asset utilization and reduced lifecycle costs.

Real-time analytics platforms also integrate visualization dashboards and AI-driven recommendation engines, empowering operators to make data-informed decisions. As Hussain et al. (2024) summarized in their comprehensive review, the convergence of AI, IoT, and cloud computing marks the beginning of a new era in intelligent CMS design.

Despite substantial progress, several challenges persist. The heterogeneity and quality of sensor data remain significant obstacles; faulty sensors or missing data can compromise model accuracy. Standardization of data formats, interoperability between CMS vendors, and secure data exchange protocols are also ongoing issues (Hussain et al., 2024).

Another challenge involves the trade-off between data processing at the edge versus the cloud. While edge computing enhances responsiveness, it often lacks the computational power for complex AI inference. Cloud platforms offer scalability but can introduce latency and security concerns. Future CMS architectures must therefore balance decentralization and computational efficiency. Finally, as noted by

Badihi et al. (2022), the next frontier involves integrating lifetime prognosis models that link detected faults with estimated degradation trajectories.

ARTIFICIAL INTELLIGENCE FOR FAULT DETECTION AND PROGNOSTICS

Traditional signal-processing and rule-based fault detection techniques, though valuable, often fall short in managing vast SCADA datasets and dynamic operating conditions. Consequently, AI, encompassing machine learning (ML) and deep learning (DL), has emerged as a core enabler of intelligent decision-making, enabling autonomous anomaly detection and Remaining Useful Life (RUL) estimation for predictive maintenance in modern wind farms (Alagha et al., 2025; Khan & Byun, 2024).

Role of Artificial Intelligence in Fault Classification and Anomaly Detection

The primary role of AI in wind turbine fault detection lies in identifying early indicators of abnormal operation patterns from massive volumes of operational data. Fault detection models leverage AI to analyze SCADA, vibration, current, temperature, and acoustic emission signals to detect deviations from healthy states (Amini et al., 2022). AI algorithms can capture nonlinear and dynamic relationships between variables, outperforming traditional threshold-based or statistical methods. For instance, Khan and Byun (2024) presented a comprehensive review of ML-based approaches, highlighting supervised, unsupervised, and hybrid architectures applied across various turbine subsystems. Supervised models, including Support Vector Machines (SVM), Random Forests (RF), and Gradient Boosting, require labeled datasets for classification of faults such as bearing wear or gearbox degradation. Unsupervised models like k-means and autoencoders, on the other hand, can autonomously detect unknown fault patterns when labeled data are scarce.

Deep learning architectures further enhance the adaptability and accuracy of fault diagnosis frameworks. Convolutional Neural Networks (CNNs) have been extensively used for analyzing vibration and acoustic data to identify subtle degradation patterns in gearboxes and blades (Conradi Hoffmann et al., 2021). These models automatically extract spatial features without manual engineering, improving both precision and generalization. Xiang et al. (2022) advanced this concept by designing a cascaded and bidirectional deep learning network for condition monitoring and anomaly detection of wind turbines, achieving high sensitivity to small variations in SCADA data. Similarly, Long Short-Term Memory (LSTM) networks, a variant of recurrent neural networks (RNNs), have proven efficient in time-series modeling, capturing temporal dependencies among sensor readings to predict evolving faults (Chen et al., 2021).

Hybrid intelligent systems combining deep learning with expert rule sets or physical models also show promise. Alagha et al. (2025) discussed hybrid AI systems that integrate neural models with physics-based understanding of turbine dynamics, enabling interpretable and explainable diagnostic frameworks. These hybrid systems improve transparency and trustworthiness, critical for safety-critical applications. Amini et al. (2022) proposed an artificial neural network (ANN)-based predictive model capable of early anomaly detection by learning deviations in SCADA patterns, offering a scalable solution for on-site condition monitoring.

Data Preprocessing, Feature Extraction, and Dimensionality Reduction

Effective AI-based fault diagnosis relies heavily on high-quality data preprocessing. SCADA systems in wind farms record hundreds of variables every few minutes, often containing noise, missing values, and redundancies. Therefore, preprocessing steps such as data cleaning, normalization, and signal de-noising are essential to ensure the robustness of AI models. Chen et al. (2021) demonstrated that LSTM autoencoder (AE)-based anomaly detection requires normalized and de-trended SCADA features to identify subtle patterns that may otherwise be masked by noise.

Feature extraction represents another crucial step in constructing accurate fault classifiers. Traditional manual feature extraction methods, such as Fast Fourier Transform (FFT), Wavelet Packet Transform (WPT), and statistical measures (mean, kurtosis, skewness), have been used to summarize vibration or current signals (Shao et al., 2023). However, with the advent of deep learning, automatic feature extraction has largely replaced manual approaches. CNNs, for instance, can directly process raw time-series or image-like sensor data to learn discriminative features hierarchically (Conradi Hoffmann et al., 2021). Xiang et al. (2022) showed that deep feature extraction using cascaded architectures leads to more robust and transferable fault representations across turbines with different operational profiles.

Dimensionality reduction is another essential preprocessing step. SCADA datasets are typically high-dimensional, leading to computational inefficiency and the risk of overfitting. Techniques such as Principal Component Analysis (PCA), t-Distributed Stochastic Neighbor Embedding (t-SNE), and autoencoders help project data into low-dimensional latent spaces, improving both interpretability and performance (Afridi et al., 2022). The integration of dimensionality reduction methods within AI pipelines enhances feature compactness, enabling faster learning and improved anomaly separation. Furthermore, multimodal data fusion strategies—combining vibration, temperature, and current information—are increasingly used to enrich fault indicators, leading to better fault discrimination accuracy (Khan & Byun, 2024).

Remaining Useful Life (RUL) Prediction and Prognostics

Recent developments in RUL estimation focus on deep learning and ensemble models capable of capturing degradation trends over time. For example, Shah et al. (2024) proposed a deep learning-based RUL forecasting model for wind turbine predictive maintenance using sequential degradation data. Their model demonstrated superior accuracy over conventional regression techniques by learning long-term dependencies between system states. Similarly, Akash et al. (2025) introduced a data-driven framework to estimate the RUL of gearboxes in wind energy conversion systems using historical SCADA records. Their model incorporated ensemble learning with gradient boosting and decision trees to provide reliable RUL predictions, validated through real-world datasets.

Moreover, small datasets—a frequent limitation in RUL modeling—have motivated the development of transfer learning and semi-supervised techniques. de Souza Pereira Gomes et al. (2024) proposed a machine learning-based RUL prediction framework capable of performing well with limited labeled data. Their approach utilized synthetic feature generation and data augmentation to improve learning stability. Pandit et al. (2024) extended this idea by comparing multiple data-driven techniques, including Random Forests, Gaussian Process Regression, and LSTM networks, for estimating the RUL of high-speed shaft bearings. Their comparative study concluded that hybrid models integrating physical degradation laws with data-driven insights yield more accurate and generalizable prognostics.

Hybrid prognostic models bridge the gap between purely data-driven and physics-based models, enhancing reliability and interpretability. For instance, Afridi et al. (2022) emphasized that hybrid AI systems incorporating degradation modeling principles can achieve more accurate predictions under varying environmental conditions. Such models consider operational stress factors, rotational speed fluctuations, and thermal loads—factors often overlooked in purely data-driven models. Alagha et al. (2025) echoed this sentiment, suggesting that hybrid frameworks can support explainable AI paradigms in predictive maintenance. A schematic of AI-enabled wind turbine fault detection showing data flow from SCADA and sensor acquisition through preprocessing, feature extraction, machine/deep learning models, and outputs for fault detection and RUL prognostics is shown in Figure 3.

Figure 3. AI-enabled wind turbine fault detection

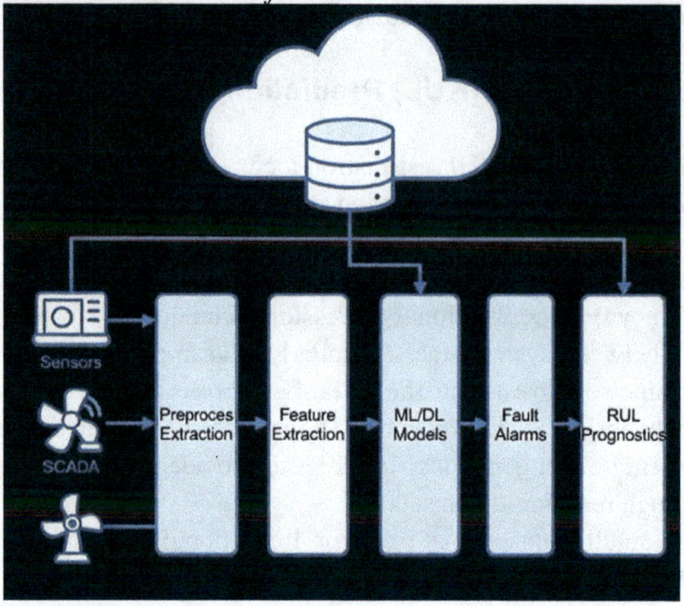

Case Studies and Practical Implementations

Numerous case studies illustrate the successful implementation of AI for wind turbine fault detection and prognostics. Matsui et al. (2022) implemented a machine learning-based anomaly detection method using a SCADA feature extractor to detect lightning damage on wind turbine blades. The model, trained using gradient boosting algorithms, effectively identified post-storm operational anomalies without extensive retraining. In another example, Xiang et al. (2022) applied cascaded deep learning networks to real-world SCADA data for anomaly detection, achieving over 97% detection accuracy for sensor drift and control malfunctions. Similarly, Conradi Hoffmann et al. (2021) used a deep learning framework based on vibration signal analysis to identify gear and bearing faults, demonstrating that vibration-based diagnostics can detect early degradation phases well before visible failure symptoms appear.

For predictive maintenance applications, Shah et al. (2024) demonstrated that RUL forecasting using deep neural networks can extend component lifespan prediction accuracy by up to 25% compared to baseline statistical methods. Akash et al. (2025) and de Souza Pereira Gomes et al. (2024) confirmed that small-data ML models, when combined with domain knowledge and robust feature engineering, can deliver performance levels comparable to data-rich industrial settings. Furthermore, Afridi et al. (2022) highlighted AI-driven prognostic frameworks applicable to renewable energy systems in general, emphasizing their flexibility across multiple asset types, including wind, solar, and hybrid systems.

A growing body of research also emphasizes explainability and interpretability in AI-driven diagnostics. Alagha et al. (2025) and Khan & Byun (2024) both stressed that black-box AI models must be complemented by explainable AI (XAI) frameworks to ensure user trust and regulatory compliance. Techniques such as SHAP (SHapley Additive exPlanations) and LIME (Local Interpretable Model-agnostic Explanations) are increasingly being used to identify which features most influence fault classification decisions. This transparency is vital in operational environments where maintenance decisions can affect safety and economic performance.

Challenges and Future Perspectives

First, the scarcity of labeled fault data, particularly for rare failure modes, limits the generalization capacity of supervised models (Afridi et al., 2022). Domain adaptation and federated learning approaches are being explored to enable cross-turbine learning without centralizing sensitive data (Khan & Byun, 2024).

Data quality and communication reliability also pose challenges. SCADA systems, though comprehensive, may record inconsistent or noisy data, requiring advanced signal processing and redundancy checks (Chen et al., 2021). Edge AI, where models are deployed directly on turbine controllers or local servers, can overcome cloud latency issues and improve real-time responsiveness (Xiang et al., 2022).

Another major research frontier lies in RUL uncertainty quantification. Current RUL models often produce point estimates, but practical decision-making requires confidence intervals and probabilistic predictions. Bayesian neural networks and probabilistic regression models can provide this capability, offering more informed maintenance scheduling (Pandit et al., 2024). Finally, the evolution of explainable and ethical AI frameworks will be crucial to ensure transparency, accountability, and trust in automated maintenance systems (Alagha et al., 2025).

DIGITAL TWIN TECHNOLOGY FOR WIND TURBINE HEALTH MANAGEMENT

The concept of the Digital Twin (DT) has emerged as a transformative paradigm in the monitoring, analysis, and optimization of complex engineering systems such as wind turbines. A digital twin is a virtual replica of a physical asset, continuously updated through real-time data to reflect its operational condition, performance, and degradation state. This integration between the physical and virtual domains allows for dynamic simulation, predictive analysis, and intelligent decision-making in maintenance and control processes (Mahmoud et al., 2024).

In wind energy systems, the digital twin serves as a cyber-physical bridge between sensors, analytics, and control systems. It not only represents static structural models but also captures the evolving dynamics of wind turbines, including interactions among subsystems like the blades, gearbox, generator, and power electronics. The growing deployment of DTs in this domain is motivated by the need to enhance reliability, reduce downtime, and extend the lifespan of wind assets operating under uncertain and harsh environmental conditions (Haghshenas et al., 2023).

Layered Architecture of Digital Twin Systems

The architecture of a digital twin for wind turbines typically follows a multi-layered framework consisting of physical, communication, data, and application layers. Mahmoud et al. (2024) proposed a comprehensive DT architecture featuring a three-tier structure: (1) the physical layer encompassing sensors and turbine components; (2) the data management layer responsible for data aggregation, preprocessing, and storage; and (3) the digital or simulation layer for modeling, diagnostics, and prognostics.

At the core of this architecture lies the bidirectional data flow between the real and virtual entities. Real-time sensor signals from vibration, strain, and temperature sensors are transmitted through Industrial Internet of Things (IIoT) frameworks into the twin, where machine learning (ML) models and physics-based simulations process them for condition assessment. The application layer then interprets these insights for operational control, failure prediction, and maintenance scheduling.

To address offshore wind farms' challenges, Haghshenas et al. (2023) presented a predictive digital twin that integrates advanced data analytics and environmental modeling. Their architecture emphasizes edge computing and data fusion, enabling near-real-time fault prediction even in offshore conditions with limited connectivity.

Integration of Physics-Based and Data-Driven Models

A fundamental strength of DT technology lies in its ability to fuse physics-based modeling with data-driven analytics, thus combining theoretical understanding with empirical learning. Traditional physics-based models rely on first principles, such as fluid dynamics and structural mechanics, to simulate turbine behavior under different loads and stresses. However, these models often struggle to represent uncertainties and degradation patterns arising from real-world operations.

Data-driven models, particularly those leveraging machine learning, complement physics-based approaches by capturing nonlinearities and unmodeled dynamics directly from operational data. Chinnappan (2025) developed an integrated ML-PINN-Simulink framework that leverages Physics-Informed Neural Networks (PINNs) to balance physical laws with data insights for robust wind power prediction. Similarly, Sarić et al. (2022) highlighted how combining physics-based and data-driven models enhances transient response analysis in power systems, allowing for accurate fault diagnosis and dynamic stability evaluation.

In ocean-based systems, Wang et al. (2022) demonstrated the integration of data-driven and physics-based modeling for wind wave prediction, showing that hybrid methods improve accuracy and computational efficiency. The same philosophy applies to wind turbines, where hybrid digital twins combine aerodynamic models, finite element analyses, and AI-based estimators for a holistic view of the turbine's condition.

The convergence of these modeling techniques results in predictive and adaptive twins capable of learning continuously from operational feedback. For instance, Fahim et al. (2022) developed a machine learning-based DT for wind turbines that uses regression and neural networks to forecast turbine behavior and detect anomalies before failure occurs. This integration significantly improves the turbine's Remaining Useful Life (RUL) estimation and enables proactive maintenance.

Real-Time Synchronization and Virtual Sensing

A defining characteristic of a digital twin is its ability to synchronize with its physical counterpart in real-time. Continuous communication between the turbine and its twin ensures that the virtual model reflects the current operating state. This synchronization allows for the detection of subtle deviations indicative of faults, fatigue, or inefficiencies.

Ibrahim et al. (2023) demonstrated the use of DTs as virtual sensors, particularly useful when direct measurements are impractical or expensive. Their study developed a virtual sensing framework that estimates unmeasurable quantities, such as internal shaft torque and bearing loads, using AI-driven observers. Similarly, Mehlan et al. (2022) implemented a DT-based virtual sensor for online fatigue damage monitoring in offshore turbine drivetrains, providing continuous updates on the system's health through real-time simulations.

Li and Shen (2021) introduced a wind speed-sensing methodology grounded in DT principles, replacing traditional anemometers with virtual estimation techniques. Their model integrates sensor fusion algorithms and digital feedback control to enhance accuracy in variable wind conditions. This virtual sensing capability enables operators to perform model-based control and adaptive maintenance decisions with minimal physical instrumentation.

Real-time synchronization also involves robust data pipelines and communication standards such as OPC-UA and MQTT for industrial systems. Mahmoud et al. (2024) emphasized the need for low-latency data transmission and edge-level analytics to maintain the fidelity of digital replicas under fluctuating operational states.

Simulation, Optimization, and Predictive Maintenance

Zhou et al. (2024) demonstrated a DT-driven intelligent assessment of wind turbine gearboxes that uses deep learning to identify early-stage faults and estimate degradation patterns. Their system continuously updates the gearbox model through data streams, supporting adaptive maintenance scheduling. Similarly, Cao et al. (2022) developed a numerical-simulation-based DT for turbine blades, enabling real-time monitoring of aerodynamic performance. Their findings show that such DTs can preemptively identify aerodynamic anomalies, leading to early interventions.

Zhu et al. (2023) validated the potential of digital twins through experimental load–response estimation for turbine towers. Their laboratory model demonstrated that a DT could accurately replicate stress variations and structural deformations under wind loads, serving as a benchmark for virtual testing environments.

Fahim et al. (2022) also demonstrated how DTs facilitate predictive modeling in conjunction with ML algorithms for fault forecasting and performance optimization. Integrating these predictive models with supervisory control systems allows operators to forecast failures, estimate RUL, and optimize energy yield. A layered digital twin architecture for wind turbine health management and real-time monitoring is shown in Figure 4.

Figure 4. Layered digital twin architecture for wind turbine health management and real-time monitoring.

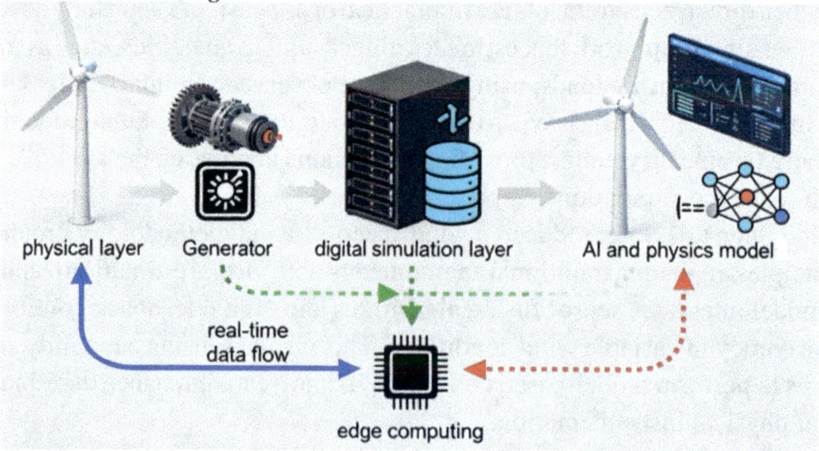

Case Studies and Industrial Implementations

Real-world implementations of digital twins in the wind energy industry underscore their transformative potential. Haghshenas et al. (2023) applied a predictive DT to offshore wind farms, achieving a 15–20% improvement in maintenance efficiency through data-driven predictive insights. Mahmoud et al. (2024) developed and prototyped a full-scale DT architecture for wind turbines, demonstrating real-time feedback between sensors and simulations under variable wind conditions.

Ibrahim et al. (2023) and Mehlan et al. (2022) validated virtual sensing applications in industrial environments, while Chinnappan (2025) demonstrated a hybrid ML-PINN-based system with superior generalization capabilities for wind power forecasting. These examples collectively show that digital twins are no longer theoretical constructs but practical tools for turbine health management and operational optimization.

Zhou et al. (2024) also highlighted the importance of cloud-based DT infrastructures that support scalable deployment across multiple turbines in a wind farm. This distributed architecture enables fleet-wide analytics and facilitates predictive maintenance scheduling across geographically dispersed assets.

Despite substantial progress, several challenges remain before digital twin technologies can be fully industrialized across wind energy systems. These challenges include:

- Data integration and quality assurance: DTs rely heavily on continuous, high-quality data. Missing or noisy sensor data can significantly reduce accuracy (Huang et al., 2023).
- Computational scalability: Real-time synchronization and high-fidelity simulations require large computational resources, particularly for multi-turbine systems.
- Model interpretability: Combining physics-based and deep learning models raises issues of transparency and explainability (Rositani, 2023).
- Cybersecurity and data privacy: Real-time data exchange introduces potential vulnerabilities that must be addressed through encryption and secure communication frameworks.

From real-time synchronization to predictive maintenance, DTs are redefining how operators perceive and manage wind assets. As demonstrated across multiple studies—from Mahmoud et al. (2024) to Zhou et al. (2024)—the fusion of AI, simulation, and sensor data through DT frameworks represents a paradigm shift toward intelligent, self-adaptive, and sustainable wind energy systems.

AI–DIGITAL TWIN SYNERGY FOR INTELLIGENT CONDITION MONITORING

The convergence of Artificial Intelligence (AI) and Digital Twin (DT) technologies has ushered in a transformative paradigm for intelligent condition monitoring and predictive maintenance across multiple industries, including the wind energy sector. The synergy between these two technologies leverages the simulation fidelity and real-time synchronization capabilities of DTs with the cognitive and decision-making abilities of AI to create self-aware, adaptive, and autonomous systems (Rojas et al., 2025; Mikołajewska et al., 2025).

Hybrid AI–DT Frameworks for Predictive Maintenance

AI–DT integration in predictive maintenance frameworks operates on the principle of data fusion and adaptive learning. Digital twins replicate the physical behavior of wind turbines through virtual models synchronized in real-time with operational data streams, while AI algorithms analyze this data to uncover hidden degradation patterns, anomalies, and impending failures (Fahim et al., 2022). This symbiosis transforms the traditionally reactive maintenance strategies into predictive and prescriptive models.

Liu et al. (2023) emphasized that digital twin-based predictive maintenance in wind turbines enhances reliability by enabling real-time anomaly detection and RUL (Remaining Useful Life) estimation. The DT continuously receives input from sensors deployed in turbine subsystems—such as blades, generators, and gearboxes—and updates its state through feedback loops. When coupled with AI models, particularly machine learning and deep learning networks, the twin can forecast component degradation and recommend maintenance schedules proactively.

The integration of generative AI further enriches these frameworks by enabling autonomous data synthesis, improving prediction accuracy, and supporting fault diagnosis even with limited or noisy datasets (Mikołajewska et al., 2025). Generative models help fill data gaps in training datasets, particularly for rare fault conditions, thereby enhancing the robustness of predictive maintenance systems.

Rojas et al. (2025) conducted a systematic literature review highlighting how AI-driven predictive maintenance in asset-intensive sectors like mining, power, and renewable energy utilizes digital twins to achieve intelligent asset management. The study emphasized the layered structure of such systems, where AI analytics operate on top of DT architectures to deliver continuous insights into asset health, optimize repair cycles, and minimize downtime.

Reinforcement Learning-Based Adaptive Fault Mitigation

Reinforcement Learning (RL) has become a key enabler for adaptive fault mitigation in AI–DT systems. Through trial-and-error learning, RL algorithms can dynamically adjust operational parameters, such as generator torque or pitch angles, in response to evolving turbine conditions. The digital twin acts as a safe virtual environment where RL agents can explore control strategies without risking physical damage to real-world assets (Gan et al., 2023).

Zhang et al. (2021) demonstrated this concept through a multi-agent deep reinforcement learning (MADRL) framework for vehicular edge networks, where adaptive digital twins learned to optimize resource allocation under dynamic conditions. Similar principles have been translated into wind turbine systems, allowing AI agents to manage fault responses and optimize power generation.

Zhou et al. (2023) expanded on this by applying reinforcement learning to DT-enabled systems for optimal scheduling and reliability management in solar and wind energy. Their approach enabled digital twins to act as both simulation platforms and control agents, improving overall system resilience. By continuously interacting with their environment, these AI–DT hybrids develop fault-tolerant operational strategies that reduce unplanned outages and extend the service life of components.

Sifat and Das (2024) further illustrated how self-healing digital twin microgrids leverage fuzzy logic controllers and machine learning for both proactive and reactive maintenance. The RL-enabled framework learned to anticipate disturbances such as voltage sags or frequency deviations, enabling rapid system recovery. Although this study focused on microgrids, its implications for wind farms are substantial—AI–DT systems could autonomously isolate faults and reroute power flows in real-time to sustain stability.

Edge and Cloud Computing in Distributed DT Ecosystems

For AI–DT frameworks to function effectively in distributed wind energy systems, seamless computational support is crucial. Edge and cloud computing architectures play a pivotal role in ensuring that DTs can process and exchange massive volumes of data with minimal latency. Cloud computing facilitates large-scale data storage and advanced analytics, while edge computing enables real-time responsiveness at the turbine or substation level (Pereira et al., 2024).

The emergence of edge-driven digital twins represents a paradigm shift from centralized analytics to decentralized, autonomous decision-making. Pereira et al. (2024) highlighted the role of hybrid edge–cloud architectures in distributed energy systems, emphasizing how data-driven solutions enhance scalability and reliability.

Madani et al. (2025) discussed how such edge–cloud hybrid systems are critical in the context of intelligent battery management systems and can be extended to wind turbines for lifecycle optimization. The study showcased the integration of AI models in cloud-edge environments to manage state estimation and predictive diagnostics efficiently. Similarly, Ahmed and Khan (2025) underscored the potential of AI-enabled digital twins in enhancing energy efficiency across smart grids by combining distributed sensing, real-time optimization, and cloud analytics.

Cross-Domain Automation, Robotics, and Cybersecurity Perspectives in AI–DT Ecosystems

Advances in industrial automation, robotics, and cybersecurity provide important cross-domain insights that strengthen the scalability and resilience of AI–Digital Twin (AI–DT) frameworks in wind turbine monitoring. Modern AI–DT ecosystems increasingly leverage IoT-enabled robotics, intelligent automation, and secure cyber-physical infrastructures to support adaptive and autonomous operation across distributed energy assets. IoT-driven robotic systems enhance real-time monitoring, remote actuation, and predictive maintenance, principles that translate directly to turbine inspection and digital twin synchronization (Rithik et al., 2024). Robotics-guided automation further improves maintenance precision and operational efficiency, enabling safer and more reliable wind farm asset management (Kathirvel et al., 2025). As connectivity increases, cybersecurity becomes critical for protecting sensor data, communication channels, and DT control loops. AI-based intrusion detection and anomaly monitoring strengthen system integrity in distributed turbine networks (Kathirvel et al., 2025a). Optimization-driven learning frameworks, such as Grey Wolf Optimizer-based deep models, demonstrate how adaptive tuning improves diagnostic robustness under noisy operating conditions, supporting reliable fault detection (V. R. & K. A., 2025).

Cross-domain AI diagnostic architectures originally developed for healthcare and environmental systems illustrate scalable multimodal analytics and real-time decision support, concepts applicable to turbine health monitoring (Kathirvel et al., 2025d, 2025e). Collectively, these developments position AI–DT platforms as integral components of a broader intelligent automation ecosystem aligned with Industry 5.0 goals of resilience, sustainability, and human–machine collaboration.

Examples of AI–DT Collaboration for Smart Wind Farm Operations

The fusion of AI and digital twin technologies is not merely theoretical; several case studies demonstrate their tangible impact on smart wind farm operations. Fahim et al. (2022) developed a machine learning-based DT model capable of predicting turbine performance deviations under varying environmental conditions. Their hybrid approach combined SCADA data, meteorological inputs, and physics-informed models to simulate and optimize operational performance, leading to enhanced energy yield and fault tolerance.

Liu et al. (2023) implemented a DT-based predictive maintenance system for offshore wind turbines that incorporated AI algorithms for condition forecasting and anomaly classification. The real-time synchronization between physical and virtual assets enabled continuous monitoring, resulting in a significant reduction in maintenance costs and unplanned downtime.

In a broader industrial perspective, Rojas et al. (2025) and Mikołajewska et al. (2025) observed that AI-based DT ecosystems are instrumental in transitioning toward Industry 5.0, characterized by human–machine collaboration, self-optimization, and sustainability. These frameworks allow operators to visualize asset conditions through virtual replicas while AI analytics provide actionable insights, making decision-making more informed and autonomous.

Gan et al. (2023) demonstrated the integration of deep reinforcement learning within DTs for adaptive scheduling, showing that such systems can dynamically manage maintenance priorities and operational constraints across multi-turbine environments. Similarly, Zhou et al. (2023) validated the feasibility of RL-augmented DTs for optimizing reliability management, showcasing their potential to evolve into self-managing wind farms capable of continuous learning and adaptation. Furthermore, Ahmed and Khan (2025) argued that AI–DT collaborations pave the way for energy-efficient smart grids, where renewable sources like wind power are intelligently balanced and coordinated through DT-guided optimization.

CONCLUSIONS

The integration of Artificial Intelligence (AI) and Digital Twin (DT) technologies marks a transformative shift in the operation, monitoring, and maintenance of wind energy systems. Together, they enable a paradigm where predictive maintenance replaces reactive responses, ensuring higher system reliability, reduced downtime, and optimized performance. AI enhances decision-making by processing vast sensor data for fault prediction and anomaly detection, while DTs provide real-time virtual replicas that mirror physical system behavior for accurate diagnostics and simulation. The synergy of AI-DT frameworks, reinforced by cloud and edge computing, facilitates scalable, adaptive, and intelligent wind farm management. However, challenges remain in ensuring data integrity, model interoperability, and computational efficiency across distributed infrastructures. Future directions point toward the adoption of federated learning, explainable AI, and quantum-enabled DTs to achieve greater resilience and transparency. Overall, AI-DT convergence is not only revolutionizing wind turbine condition monitoring but also paving the way for a new era of smart, sustainable, and autonomous renewable energy systems.

REFERENCES

Afridi, Y. S., Ahmad, K., & Hassan, L. (2022). Artificial intelligence based prognostic maintenance of renewable energy systems: A review of techniques, challenges, and future research directions. *International Journal of Energy Research, 46*(15), 21619–21642. DOI: 10.1002/er.7100

Ahmed, M., & Khan, M. R. (2025). Artificial intelligence-enabled digital twins for energy efficiency in smart grids. *Review of Applied Science and Technology, 4*(02), 580–615. DOI: 10.63125/12kp9w74

Alagha, N., Khairuddin, A. S. M., Haitaamar, Z. N., Al-Khatib, O., & Kanesan, J. (2025). Artificial intelligence in wind turbine fault detection and diagnosis: Advances and perspectives. *Energies, 18*(7), 1680. DOI: 10.3390/en18071680

Amini, A., Kanfoud, J., & Gan, T. H. (2022). An artificial intelligence neural network predictive model for anomaly detection and monitoring of wind turbines using SCADA data. *Applied Artificial Intelligence, 36*(1), 2034718. DOI: 10.1080/08839514.2022.2034718

An, J., Hu, X., Gong, L., Zou, Z., & Zheng, L. R. (2024). Fuzzy reliability evaluation and machine learning-based fault prediction of wind turbines. *Journal of Industrial Information Integration, 40*, 100606. DOI: 10.1016/j.jii.2024.100606

Astolfi, D., De Caro, F., & Vaccaro, A. (2023). Condition monitoring of wind turbine systems by explainable artificial intelligence techniques. *Sensors (Basel), 23*(12), 5376. DOI: 10.3390/s23125376 PMID: 37420542

Aziz, U., Charbonnier, S., Bérenguer, C., Lebranchu, A., & Prevost, F. (2021). Critical comparison of power-based wind turbine fault-detection methods using a realistic framework for SCADA data simulation. *Renewable & Sustainable Energy Reviews, 144*, 110961. DOI: 10.1016/j.rser.2021.110961

Badihi, H., Zhang, Y., Jiang, B., Pillay, P., & Rakheja, S. (2022). A comprehensive review on signal-based and model-based condition monitoring of wind turbines: Fault diagnosis and lifetime prognosis. *Proceedings of the IEEE, 110*(6), 754–806. DOI: 10.1109/JPROC.2022.3171691

Bakir, I., Yildirim, M., & Ursavas, E. (2021). An integrated optimization framework for multi-component predictive analytics in wind farm operations and maintenance. *Renewable & Sustainable Energy Reviews, 138*, 110639. DOI: 10.1016/j.rser.2020.110639

Bebars, A. D., Eladl, A. A., Abdulsalam, G. M., & Badran, E. A. (2022). Internal electrical fault detection techniques in DFIG-based wind turbines: A review. *Protection and Control of Modern Power Systems*, *7*(2), 1–22. DOI: 10.1186/s41601-022-00236-z

Bie, Y., Liu, X., Xu, T., Zhu, Z., & Li, Z. (2021). A review of the application of oil analysis in condition monitoring and life prediction of wind turbine gearboxes. *Insight (Northampton)*, *63*(5), 289–301. DOI: 10.1784/insi.2021.63.5.289

Bustamante, S., Pinzón, J. D., & Giraldo-Gómez, D. (2025). Evolution of real-time dynamics monitoring of Colombian power grid using wide-area monitoring system and high-speed big data analytics. *Sustainability (Basel)*, *17*(9), 3848. DOI: 10.3390/su17093848

Cao, Y., Tang, X., Gaidai, O., & Wang, F. (2022). Digital twin real-time monitoring method of turbine blade performance based on numerical simulation. *Ocean Engineering*, *263*, 112347. DOI: 10.1016/j.oceaneng.2022.112347

Castellani, F., Natili, F., Astolfi, D., & Vidal, Y. (2024). Wind turbine gearbox condition monitoring through the sequential analysis of industrial SCADA and vibration data. *Energy Reports*, *12*, 750–761. DOI: 10.1016/j.egyr.2024.06.041

Chen, H., Liu, H., Chu, X., Liu, Q., & Xue, D. (2021). Anomaly detection and critical SCADA parameters identification for wind turbines based on LSTM-AE neural network. *Renewable Energy*, *172*, 829–840. DOI: 10.1016/j.renene.2021.03.078

Chinnappan, C. C. (2025). Integrating data-driven and physics-based approaches for robust wind power prediction: A comprehensive ML-PINN-Simulink framework. *Scientific Reports*, *15*(1), 1–29. PMID: 39747330

Civera, M., & Surace, C. (2022). Non-destructive techniques for the condition and structural health monitoring of wind turbines: A literature review of the last 20 years. *Sensors (Basel)*, *22*(4), 1627. DOI: 10.3390/s22041627 PMID: 35214529

Conradi Hoffmann, J. L., Horstmann, L. P., Martínez Lucena, M., Medeiros de Araujo, G., Fröhlich, A. A., & Nishioka, M. H. N. (2021). Anomaly detection on wind turbines based on a deep learning analysis of vibration signals. *Applied Artificial Intelligence*, *35*(12), 893–913. DOI: 10.1080/08839514.2021.1966879

de Souza Pereira Gomes, G., Moreira de Andrade Lopes, S., Carrijo Polonio Araujo, D., Andrade Flauzino, R., Marques Pinto, M., & Eduardo Guerra Alves, M. (2024). Wind turbine remaining useful life prediction using small dataset and machine learning techniques. Journal of Control. *Automation and Electrical Systems*, *35*(2), 337–345.

Fahim, M., Sharma, V., Cao, T. V., Canberk, B., & Duong, T. Q. (2022). Machine learning-based digital twin for predictive modeling in wind turbines. *IEEE Access : Practical Innovations, Open Solutions, 10*, 14184–14194. DOI: 10.1109/ACCESS.2022.3147602

Falekas, G., & Karlis, A. (2021). Digital twin in electrical machine control and predictive maintenance: State-of-the-art and future prospects. *Energies, 14*(18), 5933. DOI: 10.3390/en14185933

Gan, X., Zuo, Y., Zhang, A., Li, S., & Tao, F. (2023). Digital twin-enabled adaptive scheduling strategy based on deep reinforcement learning. *Science China. Technological Sciences, 66*(7), 1937–1951. DOI: 10.1007/s11431-022-2413-5

Gao, Z., & Odgaard, P. (2023). Real-time monitoring, fault prediction and health management for offshore wind turbine systems. *Renewable Energy, 218*, 119258. DOI: 10.1016/j.renene.2023.119258

Haghshenas, A., Hasan, A., Osen, O., & Mikalsen, E. T. (2023). Predictive digital twin for offshore wind farms. *Energy Informatics, 6*(1), 1. DOI: 10.1186/s42162-023-00257-4

Han, T., Xie, W., & Pei, Z. (2023). Semi-supervised adversarial discriminative learning approach for intelligent fault diagnosis of wind turbine. *Information Sciences, 648*, 119496. DOI: 10.1016/j.ins.2023.119496

Heydari, A., Garcia, D. A., Fekih, A., Keynia, F., Tjernberg, L. B., & De Santoli, L. (2021). A hybrid intelligent model for the condition monitoring and diagnostics of wind turbines gearbox. *IEEE Access : Practical Innovations, Open Solutions, 9*, 89878–89890. DOI: 10.1109/ACCESS.2021.3090434

Hu, W., Jiao, Q., Liu, H., Wang, K., Jiang, Z., Wu, J., Cong, F., & Hao, G. (2024). A transferable diagnosis method with incipient fault detection for a digital twin of wind turbine. *Digital Engineering, 1*, 100001. DOI: 10.1016/j.dte.2024.100001

Huang, H., Lin, Y., Zhou, Y., Zhao, Y., Zhang, P., & Fan, L. (2023). Data-driven modeling of power system dynamics: Challenges, state of the art, and future work. iEnergy, 2(3), 200–221.

Hussain, M., Mirjat, N. H., Shaikh, F., Dhirani, L. L., Kumar, L., & Sleiti, A. K. (2024). Condition monitoring and fault diagnosis of wind turbine: A systematic literature review. *IEEE Access : Practical Innovations, Open Solutions, 12*, 190220–190239. DOI: 10.1109/ACCESS.2024.3514747

Ibrahim, M., Rassõlkin, A., Vaimann, T., Kallaste, A., Zakis, J., Hyunh, V. K., & Pomarnacki, R. (2023). Digital twin as a virtual sensor for wind turbine applications. *Energies*, *16*(17), 6246. DOI: 10.3390/en16176246

Inturi, V., Shreyas, N., Chetti, K., & Sabareesh, G. R. (2021). Comprehensive fault diagnostics of wind turbine gearbox through adaptive condition monitoring scheme. *Applied Acoustics*, *174*, 107738. DOI: 10.1016/j.apacoust.2020.107738

Kamarzarrin, M., Refan, M. H., Amiri, P., & Dameshghi, A. (2022). A new intelligent fault diagnosis and prognosis method for wind turbine doubly-fed induction generator. *Wind Engineering*, *46*(1), 308–340. DOI: 10.1177/0309524X211027808

Karad, S., & Thakur, R. (2021). Efficient monitoring and control of wind energy conversion systems using Internet of Things (IoT): A comprehensive review. *Environment, Development and Sustainability*, *23*(10), 14197–14214. DOI: 10.1007/s10668-021-01267-6

Karthik Selva Kumar, K., Brintha, R., Aruna Devi, K., Balaji, P. S., & Jaiswal, K. K., Sangmesh, ..., & Krishna, B. V. V. K. (2023, December). AI-enabled condition monitoring for a wind energy conversion system—Vertical Axis Wind Turbine (VAWT). In Conference on Fluid Mechanics and Fluid Power (pp. 705–714). Springer Nature Singapore.

Kathirvel, A., Gobalakrishnan, N., & Naren, A. K. (2025a). Cyber security in industrial automation using AI. In *AI-enhanced cybersecurity for industrial automation*. IGI Global.

Kathirvel, A., Gobinath, V. M., & Naren, A. K. (2025e). AI-based healthcare systems for pets and birds. In *Cases on AI-driven solutions to environmental challenges* (pp. 419–447). IGI Global.

Kathirvel, A., Gobinath, V. M., Naren, A. K., & Maheswaran, C. P. (2025d). AI-based automated disease detection based on symptoms healthcare for pets and birds. In *AI and the revival of big data* (pp. 311–345). IGI Global.

Kathirvel, A., Gobinath, V. M., Sathya, V., Naren, A. K., & Raj, A. A. (2025). The role of robotics in smart manufacturing: Increasing efficiency and reducing costs. In *The quantum AI era of neuromarketing*. IGI Global.

Khan, P. W., & Byun, Y. C. (2024). A review of machine learning techniques for wind turbine's fault detection, diagnosis, and prognosis. *International Journal of Green Energy*, *21*(4), 771–786. DOI: 10.1080/15435075.2023.2217901

Khan, P. W., & Byun, Y. C. (2024). Fault prognosis of wind turbines using multimodal machine learning. *Energy Reports*, *12*, 4938–4952. DOI: 10.1016/j.egyr.2024.10.043

Li, Y., & Shen, X. (2021). A novel wind speed-sensing methodology for wind turbines based on digital twin technology. *IEEE Transactions on Instrumentation and Measurement*, *71*, 1–13.

Liu, S., Ren, S., & Jiang, H. (2023). Predictive maintenance of wind turbines based on digital twin technology. *Energy Reports*, *9*, 1344–1352. DOI: 10.1016/j.egyr.2023.05.052

Madani, S. S., Shabeer, Y., Fowler, M., Panchal, S., Chaoui, H., Mekhilef, S., Dou, S. X., & See, K. (2025). Artificial intelligence and digital twin technologies for intelligent lithium-ion battery management systems: A comprehensive review. *Batteries*, *11*(8), 298. DOI: 10.3390/batteries11080298

Mahmoud, M., Semeraro, C., Abdelkareem, M. A., & Olabi, A. G. (2024). Designing and prototyping the architecture of a digital twin for wind turbine. *International Journal of Thermofluids*, *22*, 100622. DOI: 10.1016/j.ijft.2024.100622

Maron, J., Anagnostos, D., Brodbeck, B., & Meyer, A. (2022). Artificial intelligence-based condition monitoring and predictive maintenance framework for wind turbines. *Journal of Physics: Conference Series*, *2151*(1), 012007. DOI: 10.1088/1742-6596/2151/1/012007

Matsui, T., Yamamoto, K., & Ogata, J. (2022). Anomaly detection using a SCADA feature extractor and machine learning to detect lightning damage on wind turbine blades. *IEEJ Transactions on Electrical and Electronic Engineering*, *17*(6), 945–951. DOI: 10.1002/tee.23599

Mehlan, F. C., Nejad, A. R., & Gao, Z. (2022). Digital twin-based virtual sensor for online fatigue damage monitoring in offshore wind turbine drivetrains. *Journal of Offshore Mechanics and Arctic Engineering*, *144*(6), 060901. DOI: 10.1115/1.4055551

Mikołajewska, E., Mikołajewski, D., Mikołajczyk, T., & Paczkowski, T. (2025). Generative AI in AI-based digital twins for fault diagnosis for predictive maintenance in Industry 4.0/5.0. *Applied Sciences (Basel, Switzerland)*, *15*(6), 3166. DOI: 10.3390/app15063166

Ng, E. Y. K., & Lim, J. T. (2022). Machine learning on fault diagnosis in wind turbines. *Fluids (Basel, Switzerland)*, *7*(12), 371. DOI: 10.3390/fluids7120371

Olatunji, O. O., Adedeji, P. A., Madushele, N., & Jen, T. C. (2021, May). Overview of digital twin technology in wind turbine fault diagnosis and condition monitoring. In 2021 IEEE 12th International Conference on Mechanical and Intelligent Manufacturing Technologies (ICMIMT) (pp. 201–207). IEEE. DOI: 10.1109/ICMIMT52186.2021.9476186

Pandit, R., Astolfi, D., Hong, J., Infield, D., & Santos, M. (2023). SCADA data for wind turbine data-driven condition/performance monitoring: A review on state-of-art, challenges and future trends. *Wind Engineering*, *47*(2), 422–441. DOI: 10.1177/0309524X221124031

Pandit, R., Santos, M., & Sierra-García, J. E. (2024). Comparative analysis of novel data-driven techniques for remaining useful life estimation of wind turbine high-speed shaft bearings. *Energy Science & Engineering*, *12*(10), 4613–4623. DOI: 10.1002/ese3.1911

Pereira, A., Hovstø, A., Rodrigues, E., Silvia, F., & Suciu, G., AL-Naday, M., ... & Atmojo, U. D. (2024). Edge driven digital twins in distributed energy systems: Role and opportunities for hybrid data driven solutions. *Applied Sciences (Basel, Switzerland)*.

Prasath, C. A. (2025). AI-enabled digital twin framework for predictive maintenance in smart urban infrastructure. *Journal of Smart Infrastructure and Environmental Sustainability*, *2*(1), 1–10.

Rithik, G., Kathirvel, A., & Naren, A. K. (2024). The IoT of robotics: The frontier of automation. In *Artificial intelligence-enabled digital twin for smart manufacturing*. Wiley–Scrivener Publishing LLC. DOI: 10.1002/9781394303601.ch13

Rojas, L., Peña, Á., & Garcia, J. (2025). AI-driven predictive maintenance in mining: A systematic literature review on fault detection, digital twins, and intelligent asset management. *Applied Sciences (Basel, Switzerland)*, *15*(6), 3337. DOI: 10.3390/app15063337

Rositani, M. (2023). Physics-informed learning and data-driven control of an Airborne Wind Energy system.

Sarić, A. A., Transtrum, M. K., Sarić, A. T., & Stanković, A. M. (2022). Integration of physics- and data-driven power system models in transient analysis after major disturbances. *IEEE Systems Journal*, *17*(1), 479–490. DOI: 10.1109/JSYST.2022.3150237

Sasinthiran, A., Gnanasekaran, S., & Ragala, R. (2024). A review of artificial intelligence applications in wind turbine health monitoring. *International Journal of Sustainable Energy*, *43*(1), 2326296. DOI: 10.1080/14786451.2024.2326296

Shah, S. S., Daoliang, T., & Kumar, S. C. (2024). RUL forecasting for wind turbine predictive maintenance based on deep learning. *Heliyon*, *10*(20), e39268. DOI: 10.1016/j.heliyon.2024.e39268 PMID: 39678167

Shao, K., He, Y., Xing, Z., & Du, B. (2023). Detecting wind turbine anomalies using nonlinear dynamic parameters-assisted machine learning with normal samples. *Reliability Engineering & System Safety*, *233*, 109092. DOI: 10.1016/j.ress.2023.109092

Sifat, M. M. H., & Das, S. K. (2024). Proactive and reactive maintenance strategies for self-healing digital twin islanded microgrids using fuzzy logic controllers and machine learning techniques. *IEEE Transactions on Power Systems*, *40*(1), 178–187. DOI: 10.1109/TPWRS.2024.3408096

Sun, S., Wang, T., Yang, H., & Chu, F. (2022). Condition monitoring of wind turbine blades based on self-supervised health representation learning. *Applied Energy*, *313*, 118882. DOI: 10.1016/j.apenergy.2022.118882

Sun, T., Yu, G., Gao, M., Zhao, L., Bai, C., & Yang, W. (2021). Fault diagnosis methods based on machine learning and its applications for wind turbines: A review. *IEEE Access : Practical Innovations, Open Solutions*, *9*, 147481–147511. DOI: 10.1109/ACCESS.2021.3124025

Tang, M., Zhao, Q., Wu, H., Wang, Z., Meng, C., & Wang, Y. (2021). Review and perspectives of machine learning methods for wind turbine fault diagnosis. *Frontiers in Energy Research*, *9*, 751066. DOI: 10.3389/fenrg.2021.751066

V. R., & K. A. (2025). Improving EEG signal classification in brain–computer interfaces via Grey Wolf Optimizer-based deep learning. In *Proceedings of the 3rd International Conference on Sustainable Computing and Data Communication Systems (ICSCDS)* (pp. 1312–1318). IEEE.

Wang, N., Chen, Q., Zhu, L., & Sun, H. (2022). Integration of data-driven and physics-based modeling of wind waves in a shallow estuary. *Ocean Modelling (Oxford)*, *172*, 101978. DOI: 10.1016/j.ocemod.2022.101978

Wang, Y., Sun, W., Liu, L., Wang, B., Bao, S., & Jiang, R. (2023). Fault diagnosis of wind turbine planetary gear based on a digital twin. *Applied Sciences (Basel, Switzerland)*, *13*(8), 4776. DOI: 10.3390/app13084776

Zhang, K., Cao, J., & Zhang, Y. (2021). Adaptive digital twin and multiagent deep reinforcement learning for vehicular edge computing and networks. *IEEE Transactions on Industrial Informatics*, *18*(2), 1405–1413. DOI: 10.1109/TII.2021.3088407

Zhang, L., Zhang, H., & Cai, G. (2022). The multiclass fault diagnosis of wind turbine bearing based on multisource signal fusion and deep learning generative model. *IEEE Transactions on Instrumentation and Measurement*, *71*, 1–12. DOI: 10.1109/TIM.2022.3178483

Zhou, J., Yang, M., Zhan, Y., & Xu, L. (2023). Digital twin application for reinforcement learning-based optimal scheduling and reliability management enhancement of systems. *Solar Energy*, *252*, 29–38. DOI: 10.1016/j.solener.2023.01.042

Zhou, Y., Zhou, J., Cui, Q., Wen, J., & Fei, X. (2024). Digital twin-driven online intelligent assessment of wind turbine gearbox. *Wind Energy (Chichester, England)*, *27*(8), 797–815. DOI: 10.1002/we.2912

Zhu, Z., Zhang, J., Zhu, S., & Yang, J. (2023). Digital twin technology for wind turbine towers based on joint load–response estimation: A laboratory experimental study. *Applied Energy*, *352*, 121953. DOI: 10.1016/j.apenergy.2023.121953

Chapter 7
Digital Twin Frameworks for AI-Driven Wind Turbine Monitoring and Predictive Maintenance

Nguyen Duc Thuan
http://orcid.org/0000-0002-5749-5409
Hanoi University of Science and Technology, Vietnam

ABSTRACT

The rapid digitalization of renewable energy systems has made digital twin technology a transformative approach for modeling, monitoring, and optimizing wind turbines. This chapter presents a framework for an AI-powered digital twin that integrates multi-physics modeling, machine learning, and real-time data synchronization. It develops detailed aerodynamic, mechanical, electrical, structural, and thermal models forming a hybrid simulation grounded in physics. Artificial intelligence methods, from machine learning to physics-informed neural networks, enhance fault detection, anomaly diagnosis, and life prediction. A small-scale turbine simulation validates the framework, showing real-time operation, improved torque prediction, and precise anomaly detection. The discussion highlights future directions in multi-agent wind farm twins, uncertainty quantification, explainable AI, and self-evolving models, outlining a roadmap toward autonomous and trustworthy digital energy ecosystems.

1. INTRODUCTION

The global transition toward renewable energy has brought wind power to the forefront as one of the most sustainable and rapidly expanding sources of clean

DOI: 10.4018/979-8-3373-4159-0.ch007

electricity (Gielen et al., 2019). Modern wind turbines have evolved into highly complex electromechanical systems that integrate aerodynamic, mechanical, electrical, and control subsystems (Yousefzadeh & Ahmadi Kamarposhti, 2024). As the scale of deployment grows, from individual turbines to geographically distributed wind farms, maintaining reliability and maximizing energy yield have become crucial (McKenna et al., 2016). The increasing penetration of wind energy in power grids also imposes stringent requirements on predictive maintenance, fault diagnosis, and operational optimization (R. K. Pandit et al., 2023). Achieving these goals demands a level of system observability, modeling, and intelligence far beyond that of traditional supervisory control and data acquisition (SCADA) systems.

As wind farms expand in both scale and geographical dispersion, the operational landscape becomes increasingly complex. Offshore wind installations, now exceeding 60-story structures and located tens of kilometers from shore, operate under harsh conditions involving salt corrosion, humidity, severe turbulence, and difficult accessibility (Hassani & Dackermann, 2023). These constraints significantly increase the cost of corrective maintenance, making proactive health assessment essential. Moreover, the variability of wind resources and the intermittency of renewable generation pose additional challenges for grid stability, further motivating the need for intelligent monitoring systems capable of anticipating failures before they propagate into power-quality disruptions or grid imbalance events (Kull et al., 2025; Saleh et al., 2024). The convergence of these industrial pressures underscores why modern wind energy systems require monitoring paradigms far more advanced than the legacy SCADA-based frameworks.

Conventional monitoring frameworks, typically based on SCADA and condition monitoring systems (CMS), provide valuable but limited insights into turbine health. They operate primarily through threshold-based alarms derived from aggregated signals such as rotor speed, wind velocity, temperature, or vibration RMS (R. Pandit et al., 2023). However, these systems often suffer from insufficient temporal resolution, restricted sensor coverage, and weak adaptability to complex degradation dynamics (Tautz-Weinert & Watson, 2017). As a result, they struggle to detect early-stage faults, handle sensor drift, or adapt to changing environmental conditions. The limitations of such rule-based systems are especially evident when dealing with nonlinear and coupled phenomena, such as gearbox wear under fluctuating wind loads or blade fatigue influenced by turbulence intensity (Da Silva et al., 2025).

Beyond their sensing limitations, SCADA and conventional CMS architectures are fundamentally constrained by their one-directional data pipelines and low-level processing capabilities. They lack the architectural flexibility needed to integrate physics-based models, high-rate vibration streams, or edge-level analytics in real time (Wang et al., 2026). As wind turbines increase in rating, moving toward 10–20 MW offshore units, the resulting data ecosystem becomes too rich and complex for traditional systems, which were never designed for multi-physics simulation coupling or machine-learning–driven inference (Z. Zhang et al., 2023). This architectural gap directly motivates the transition toward digital twin–based solutions that unify data, models, and control in a single cyber–physical ecosystem.

To overcome these challenges, researchers and practitioners have increasingly turned to digital twin (DT) technology, an emerging paradigm that tightly integrates physical assets and their virtual counterparts through continuous data exchange. Originally conceptualized by NASA for spacecraft system simulation and later formalized in industrial contexts by Grieves and Vickers, the digital twin represents a dynamic digital replica that evolves concurrently with the physical system (Grieves & Vickers, 2017). In wind energy applications, the DT enables real-time synchronization between sensor data, simulation models, and analytical algorithms, providing an unprecedented platform for monitoring, diagnosis, and optimization (Ibrahim et al., 2023a; Kandemir et al., 2024).

The rapid adoption of digital twin architectures in wind energy is also driven by broader trends in industrial digitalization, including the proliferation of IoT-enabled sensors, advancements in edge computing, and the increasing affordability of high-fidelity finite-element and computational fluid dynamics (CFD) simulations (Hasan & Crawford, 2025; Mbasso et al., 2025). These technological advancements enable the creation of multi-layer twins that integrate structural models, aerodynamic solvers, control-system emulators, and data-driven performance models into a unified virtual environment. Importantly, digital twins provide a scalable mechanism to incorporate domain knowledge, such as drivetrain stiffness matrices, blade modal shapes, or generator loss maps, into the monitoring pipeline. This allows the virtual model not only to mirror the physical system but also to simulate "what-if" scenarios involving varied wind inflow, component aging, or extreme events such as gusts and grid faults.

The essence of a digital twin lies in its bidirectional information flow between the real and virtual domains (Jiang et al., 2021). Physical sensors mounted on blades, gearboxes, generators, and towers continuously stream data into the virtual model, which in turn processes, analyzes, and feeds back actionable insights to guide operation or maintenance (Maksimović et al., 2025). This closed feedback loop allows predictive maintenance to transition from reactive or scheduled strategies to condition-based and adaptive strategies. For instance, instead of replacing a gearbox after a fixed interval, the digital twin can predict the remaining useful life (RUL) based on vibration or acoustic emission (AE) patterns, enabling maintenance only when necessary.

The recent digitalization of wind energy infrastructure has further accelerated the shift toward intelligent monitoring solutions. High-speed edge computing platforms, standardized communication protocols, and cloud-based supervisory systems now enable continuous synchronization between physical assets and advanced analytical engines (Amin et al., 2025; Ismail et al., n.d.). At the same time, the maturation of AI, particularly in time-series modeling, deep representation learning, and uncertainty quantification, provides the computational foundation for building adaptive and interpretable digital twins (Juarez et al., 2025). These technological advances create the ideal environment for AI-driven DTs to move from conceptual prototypes to large-scale industrial deployment.

AI, particularly machine learning (ML) and deep learning (DL), has emerged as a powerful enabler of data-driven understanding. When integrated into the digital twin framework, AI methods can automatically extract complex patterns from heterogeneous sensor data, identify anomalous behaviors, and predict potential failures (Rojas et al., 2025). The synergy between physics-based modeling (the backbone of digital twins) and data-driven AI analytics (the cognitive layer) defines a new generation of intelligent monitoring systems. Such hybrid AI-DT frameworks not only improve prediction accuracy but also enhance interpretability by embedding physical constraints into learning models.

In the context of wind turbines, this synergy is especially valuable. The aerodynamic behavior of the rotor, the torsional dynamics of the drivetrain, and the electrical performance of the generator are governed by coupled nonlinear equations. Yet, real-world operation introduces stochastic disturbances from turbulence, icing, or component wear. A pure physics model may not fully capture these stochasticities, while a pure data-driven model lacks physical grounding. Hence, a hybrid digital twin, in which the physical model provides a reliable baseline and AI models adaptively correct deviations, offers the most balanced and effective approach. Such an integrated framework forms the foundation of AI-driven predictive maintenance, capable of detecting anomalies early, estimating degradation trends, and optimizing control actions in real time.

Beyond performance prediction, digital twins also support resilience by serving as platforms for quantifying uncertainty and evaluating operational risks (Dui et al., 2025). Wind turbines operate in environments where measurement noise, sensor degradation, and unpredictable wind turbulence are unavoidable. By integrating probabilistic models such as stochastic load simulations, Monte Carlo surrogate models, or Bayesian estimators, the digital twin can provide uncertainty-aware forecasts of loads, stresses, and component deterioration (Jorgensen et al., 2023). Such probabilistic insights are crucial for decision-making in curtailment strategies, extreme-weather response, and life-extension assessments. As the future grid demands increasingly firm and dispatchable renewable power, uncertainty-aware digital twins will become indispensable for ensuring stable and safe turbine operation.

This chapter aims to provide a comprehensive overview and framework for developing AI-enhanced digital twins for wind turbine monitoring and predictive maintenance. It bridges the gap between theoretical modeling and real-time implementation by integrating multi-domain physical models, AI algorithms, and data communication architectures. The remainder of the chapter is organized as follows. Section 2 introduces the fundamentals of digital twin technology, its conceptual hierarchy, core components, and communication standards. Section 3 develops detailed system models for wind turbines, encompassing aerodynamic, mechanical, electrical, structural, and thermal domains. Section 4 presents AI-enhanced analytics, including machine learning, deep learning, and physics-informed methods for fault detection and control optimization. Section 5 describes real-time data integration architectures, emphasizing IoT, edge computing, and cyber-security. Section 6 showcases an implementation case study on a small-scale wind turbine digital twin. Section 7 discusses broader implications, scalability, and future research perspectives. Section 8 concludes the chapter by summarizing the key insights and technological roadmap.

2. FUNDAMENTALS OF DIGITAL TWIN TECHNOLOGY

Digital twin technology represents a fundamental shift in how engineered systems are modeled, monitored, and optimized. A digital twin is not merely a digital replica of a physical object but a living, continuously synchronized virtual entity that mirrors the state, behavior, and performance of its physical counterpart throughout its lifecycle (Rasheed et al., 2020). The twin integrates multi-domain modeling, real-time data acquisition, artificial intelligence, and bidirectional communication, forming the backbone of intelligent and self-adaptive systems in modern industry (Rahman et al., 2025).

The conceptual depth of DT lies in its ability to connect physical reality with cyber intelligence through continuous feedback loops. In the context of wind turbines, the digital twin serves as a high-fidelity simulation and decision-support system that evolves with the physical turbine, constantly updating parameters based on sensor data, predicting future conditions, and generating optimized control commands. Understanding the definition, structure, and operating principles of digital twins is therefore essential before delving into their application in predictive maintenance.

2.1. Definition and Conceptual Hierarchy

The evolution toward the digital twin can be understood as a three-tier hierarchy that gradually increases in data integration and autonomy: digital model, digital shadow, and digital twin (Figure 1) (R. Zhang et al., 2025).

Digital model: A digital model is a static or semi-static virtual representation of a physical asset. It is typically used for design or analysis but has no automatic connection to real-time operational data. Examples include CAD geometry or finite element models of turbine blades. While accurate within their design context, these models remain isolated from real-world dynamics.

Digital shadow: The digital shadow introduces one-way data flow, from the physical asset to its virtual representation. Sensor data updates the model periodically, allowing engineers to observe degradation or performance trends. However, no feedback is returned to the physical system. The digital shadow thus functions as a diagnostic tool rather than an autonomous decision-making system.

Digital twin: The digital twin completes the loop through bidirectional data exchange. Real-time measurements continuously update the virtual model, while the model's predictions, insights, or control decisions are transmitted back to the physical system. This feedback enables self-optimization, predictive maintenance, and adaptive control. The digital twin thus embodies real–virtual co-evolution, where both domains mutually inform and refine each other.

Formally, the twin can be expressed as a coupled system:

$$\mathscr{T} = \{\mathscr{P}, \mathscr{V}, \mathscr{D}, \mathscr{C}\} \tag{1}$$

where $\mathscr{P}$ denotes the physical asset, $\mathscr{V}$ the virtual model, $\mathscr{D}$ the data exchange mechanism, and $\mathscr{C}$ the control or decision feedback. The twin's fidelity evolves over time as $\mathscr{V}$ adapts to changes observed in $\mathscr{P}$, thus maintaining dynamic consistency.

Figure 1. Conceptual hierarchy of digital model, digital shadow, and digital twin, showing increasing levels of connectivity and autonomy.

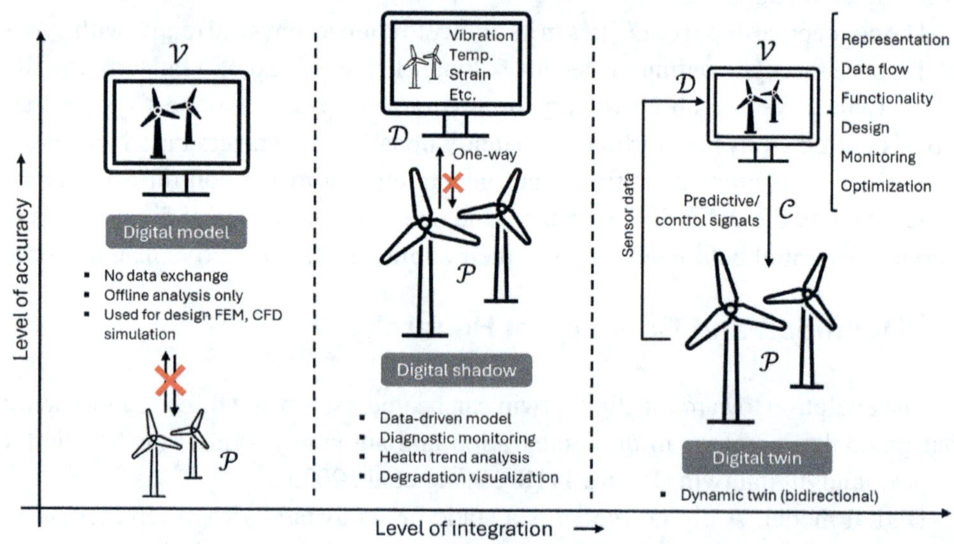

2.2 Key Components of a Digital Twin

A complete digital twin consists of three tightly integrated components: the physical asset, the virtual model, and the data connection (Figure 2) (Wanasinghe et al., 2020). Each plays a distinct yet interdependent role in sustaining the twin's real-time intelligence.

The physical asset refers to the actual wind turbine or subsystem equipped with sensors, actuators, and controllers. It generates multi-source data: rotor speed, pitch angle, torque, vibration, acoustic emission, temperature, and weather conditions, collected via a data acquisition (DAQ) system. These measurements capture both the operational state and environmental context of the turbine. Modern turbines may include hundreds of sensors distributed across blades, nacelle, gearbox, generator, and tower, forming the foundation of the cyber-physical interface.

The virtual model is the computational representation of the turbine's physical and operational behavior. Depending on fidelity requirements, it may comprise: physics-based models, such as aerodynamic or multi-body dynamic simulations; data-driven models, constructed using machine learning algorithms trained on sensor data; or hybrid models, which combine both approaches to balance interpretability and adaptability. In practice, hybrid models often outperform pure physical or data-driven approaches by capturing nonlinearities and uncertainties while respecting physical laws.

The data connection is the communication infrastructure that synchronizes the physical and virtual spaces. It includes data transmission protocols (OPC-UA, MQTT, REST API), cloud/edge gateways, and middleware for message queuing, storage, and time alignment. This layer ensures real-time synchronization and bidirectional information flow, allowing both observation (from physical to virtual) and feedback (from virtual to physical).

Figure 2. Core components of a digital twin: physical asset, virtual model, and data connection.

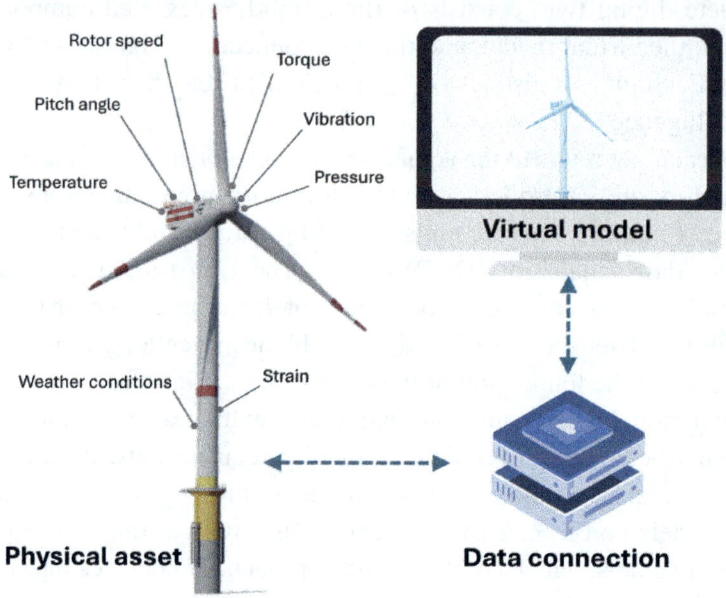

2.3. Functional Loop

The functional mechanism of a digital twin operates as a closed-loop system, as depicted in Figure 3. The loop consists of four interrelated stages: data acquisition; modeling and analysis; feedback and control; and optimization (Fei et al., 2025; Leon-Medina et al., 2025).

Sensor networks and data acquisition systems continuously collect signals from the physical asset. These include SCADA data (low-frequency operational parameters) and high-frequency condition-monitoring data (vibration, AE, strain). Data quality, timestamp accuracy, and synchronization across channels are critical to ensure fidelity.

The acquired data feed into the virtual model, which executes real-time simulations or predictions. Depending on the objective, this stage may involve finite element analysis, dynamic simulations, or AI-based regression/classification models. The model estimates latent states such as stresses, fatigue damage, or efficiency losses.

Insights derived from the virtual model, such as torque deviation, bearing wear rate, or anomaly probability, are transmitted back to the physical system. Controllers or maintenance schedulers act upon these recommendations, enabling adaptive operation and predictive maintenance.

The feedback loop can include optimization routines (e.g., reinforcement learning or multi-objective optimization) that adjust control parameters, pitch angle, yaw alignment, or torque limits, to maximize energy output and minimize stress or vibration.

Mathematically, this closed-loop interaction can be expressed as:

$$\mathcal{V}_{t+1} = f(\mathcal{V}_t, \mathcal{D}_t); \; \mathcal{P}_{t+1} = g(\mathcal{P}_t, \mathcal{C}_t) \quad (2)$$

where $\mathcal{D}_t$ represents incoming data from sensors, and $\mathcal{C}_t$ denotes control feedback from the virtual model to the physical asset.

Figure 3. Functional loop of a digital twin showing continuous data flow and feedback between physical and virtual domains.

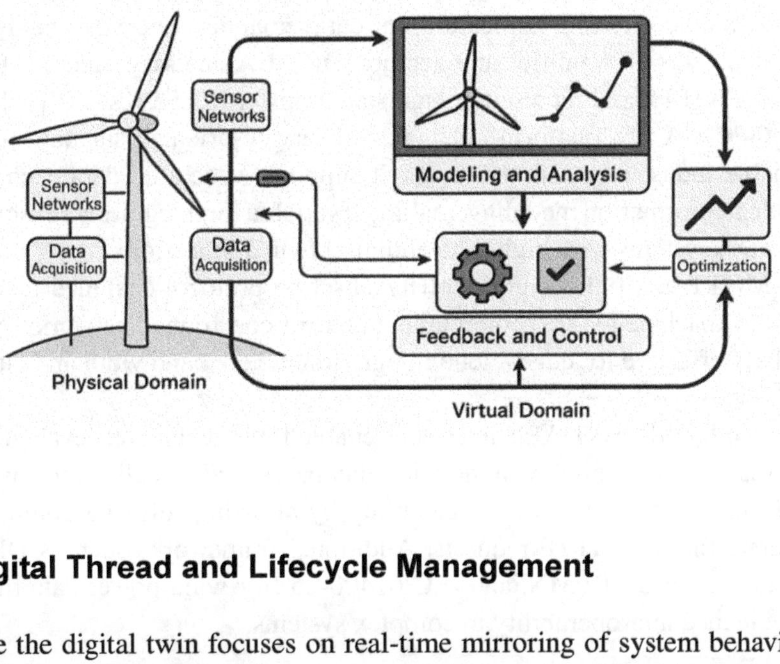

2.4. Digital Thread and Lifecycle Management

While the digital twin focuses on real-time mirroring of system behavior, the digital thread provides continuity across the entire product lifecycle, from concept design and manufacturing to operation, maintenance, and decommissioning. It acts as the chronological data backbone that links all digital artifacts associated with a physical asset.

In wind turbine systems, the digital thread ensures that design parameters, simulation data, field measurements, and maintenance logs are interlinked and traceable. For instance, a blade's digital thread may contain its CAD geometry, material certificates, finite element meshes, strain measurements, and fatigue history. This comprehensive data lineage allows engineers to perform lifecycle analytics, such as correlating manufacturing tolerances with operational failures.

Key functionalities of the digital thread include: Traceability – maintaining version control and provenance of data across design iterations and operational updates; Configurability – adapting the twin to reflect hardware or software modifications; and Lifecycle feedback – using operational data to inform future designs, enhancing reliability and manufacturability. Thus, the digital thread complements the digital twin by ensuring that every piece of information, whether generated in simulation or measured in the field, is persistently linked within a unified data framework.

2.5. Standards and Interoperability

A crucial enabler of digital twin ecosystems is interoperability, the ability of heterogeneous devices, software, and networks to communicate and collaborate seamlessly. For wind turbine digital twins, interoperability covers hardware-level protocols, middleware communication, and data semantics. The most widely adopted standards include: OPC unified architecture (OPC-UA), message queuing telemetry transport (MQTT), and representational state transfer (REST) APIs (Table 1).

The OPC-UA is a platform-independent, service-oriented standard for secure and reliable industrial communication. It supports structured data exchange and hierarchical information modeling, making it suitable for integrating sensors, PLCs, and SCADA systems with higher-level digital twin applications.

The MQTT is a lightweight publish/subscribe protocol designed for efficient machine-to-machine (M2M) communication over constrained networks. MQTT is particularly effective for edge–cloud integration, where bandwidth and latency are critical factors.

The web-based REST APIs interfaces enable flexible data retrieval and control commands between digital twin modules, databases, and visualization dashboards. REST facilitates modularity and scalability by allowing software components to interoperate through HTTP requests. Additional frameworks such as DDS (data distribution service), BPMN, and IEC 61400-25 (for wind power data modeling) further enhance interoperability in complex systems.

Table 1. Comparison of major interoperability protocols used in digital twin architecture.

Protocol	Architecture Type	Typical Use	Advantages	Limitations
OPC-UA	Client–Server	Industrial automation, SCADA	Secure, structured data model	Heavier bandwidth
MQTT	Publish–Subscribe	Edge IoT, sensor networks	Lightweight, low latency	Limited data semantics
REST API	Web services	Application integration	Flexible, widely supported	Less suited for high-frequency streaming
DDS	Peer-to-Peer	Distributed control	Real-time, deterministic	Complex configuration

3. SYSTEM MODELING OF WIND TURBINES

The development of a digital twin for wind turbines begins with constructing an accurate system model that captures the key physical interactions among aerodynamic, mechanical, electrical, structural, and thermal domains. Each subsystem contributes uniquely to the overall dynamics, and together they determine the turbine's power conversion efficiency, stability, and lifetime. In a digital-twin context, these models serve as the physics backbone that guides data assimilation, anomaly detection, and control optimization.

3.1. Aerodynamic Modeling

The aerodynamic subsystem converts the kinetic energy of wind into mechanical torque on the rotor shaft. Its modeling provides the foundation for all subsequent mechanical and electrical analyses. The power extracted from the wind can be expressed as:

$$P = \frac{1}{2}\rho A\, C_p(\lambda,\beta)\, v^3 \tag{2}$$

where ρ is air density (kg/m^3), $A = \pi R^2$ is swept area of the rotor (m^2), v is free-stream wind speed (m/s), $C_p(\lambda,\beta)$ is power coefficient dependent on tip-speed ratio λ and pitch angle β. The aerodynamic torque applied to the low-speed shaft is then:

$$T_a = \frac{P}{\omega_t} = \frac{1}{2}\rho A \frac{C_p(\lambda,\beta)}{\lambda} R\, v^2 \tag{3}$$

where ω_t is the rotor angular speed and $R$the rotor radius. The tip-speed ratio is defined by:

$$\lambda = \frac{\omega_t R}{v} \tag{4}$$

and determines the operating point of the turbine. For each pitch angle β, there exists an optimal λ_{opt} at which C_p reaches its maximum $C_{p,\max}$ (Figure 4).

Empirical formulations are often used to approximate $C_p(\lambda,\beta)$. One popular expression is:

$$C_p(\lambda,\beta) = C_1\left(\frac{C_2}{\lambda_i} - C_3\beta - C_4\right)e^{-C_5/\lambda_i} + C_6\lambda \tag{5}$$

with:

$$\frac{1}{\lambda_i} = \frac{1}{\lambda + 0.08\beta} - \frac{0.035}{\beta^3 + 1} \tag{6}$$

where $C_1, C_2, \ldots, C_6$ are empirically identified constants. This relation enables the twin to simulate aerodynamic torque under variable pitch and wind conditions in real time.

Figure 4. Typical $C_p - \lambda$ curves for various pitch angles, illustrating the maximum aerodynamic efficiency region (Abdul-Ruhman & Hawas, 2020).

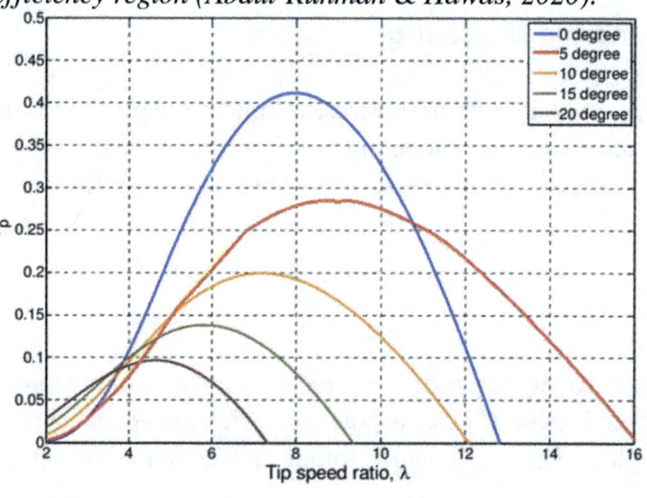

3.2. Mechanical Drive-Train Modeling

The mechanical drive-train transmits aerodynamic torque to the generator and includes the hub, low-speed shaft, gearbox, and high-speed shaft. Dynamic interactions among these components dominate vibration and fatigue behaviors.

A widely accepted representation is the two-mass torsional model (Figure 5), where the turbine side (rotor + low-speed shaft) and generator side are connected by a flexible shaft with stiffness k_s and damping c_s:

$$\begin{aligned} J_t\ddot{\theta}_t + c_s(\dot{\theta}_t - \dot{\theta}_g) + k_s(\theta_t - \theta_g) &= T_a \\ J_g\ddot{\theta}_g - c_s(\dot{\theta}_t - \dot{\theta}_g) - k_s(\theta_t - \theta_g) &= -T_e \end{aligned} \tag{7}$$

where J_t, J_g are moments of inertia of turbine and generator, T_a is aerodynamic torque, T_e is electromagnetic torque. These coupled differential equations capture torsional resonance phenomena and are crucial for condition monitoring of gearboxes and shafts.

The gearbox amplifies torque and speed according to the ratio $n_g = \frac{\omega_g}{\omega_t}$. However, backlash, gear-mesh stiffness variation, and bearing friction introduce nonlinearities. For higher fidelity, multi-stage gear dynamics can be modeled using lumped-parameter methods or finite-element submodels integrated within the twin.

Figure 5. Two-mass torsional model of a wind-turbine drive-train.

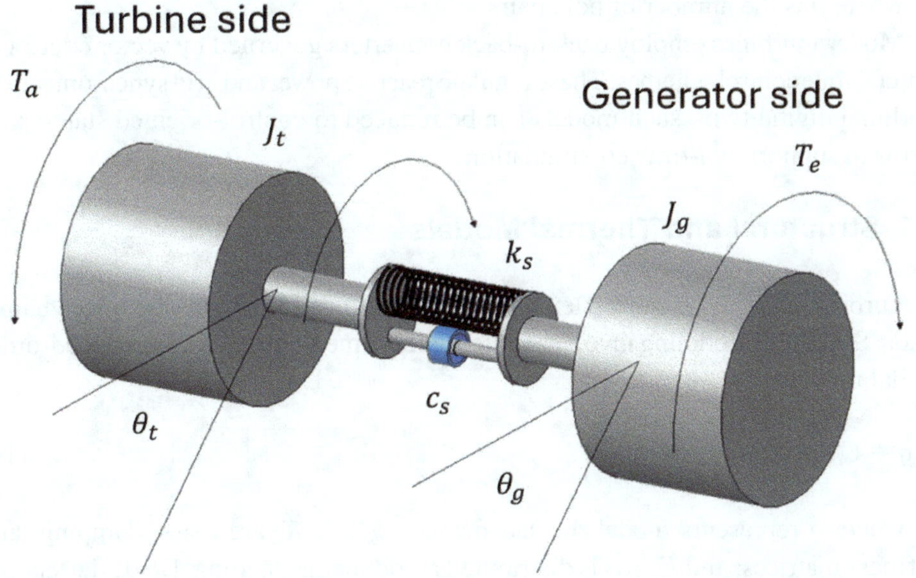

3.3. Electrical and Generator Modeling

The electrical subsystem converts mechanical power to electrical energy and regulates grid interaction. Two common architectures are the doubly-fed induction generator (DFIG) and the permanent-magnet synchronous generator (PMSG).

In the dq reference frame, the stator and rotor voltage equations of a DFIG are:

$$\begin{aligned} v_{sd} &= R_s i_{sd} + \frac{d\psi_{sd}}{dt} - \omega_s \psi_{sq} \\ v_{sq} &= R_s i_{sq} + \frac{d\psi_{sq}}{dt} + \omega_s \psi_{sd} \\ v_{rd} &= R_r i_{rd} + \frac{d\psi_{rd}}{dt} - (\omega_s - \omega_r)\psi_{rq} \\ v_{rq} &= R_r i_{rq} + \frac{d\psi_{rq}}{dt} + (\omega_s - \omega_r)\psi_{rd} \end{aligned} \tag{8}$$

with flux linkages:

$$\psi_{sd} = L_s i_{sd} + L_m i_{rd}, \psi_{rd} = L_m i_{sd} + L_r i_{rd} \tag{9}$$

and similar for q-components. The electromagnetic torque is obtained as:

$$T_e = \frac{3}{2} p L_m \left(i_{sq} i_{rd} - i_{sd} i_{rq} \right) \tag{10}$$

where p is the number of pole pairs.

Modern turbines employ back-to-back converters governed by vector control or direct torque control schemes. These regulate reactive power and grid synchronization. Within a digital twin, such models can be reduced to control-oriented state-space forms to support real-time co-simulation.

3.4. Structural and Thermal Models

Turbine blades are slender, flexible structures exposed to stochastic aerodynamic loads. Structural modeling involves finite-element methods (FEM) or reduced-order modal models:

$$M_b \ddot{q} + C_b \dot{q} + K_b q = F_{aero}(t) \tag{11}$$

where q represents modal displacements; M_b, C_b, K_b are mass, damping, and stiffness matrices; and $F_{aero}(t)$ is distributed aerodynamic loading. Blade deflection, bending moment, and strain distributions derived from these models provide inputs for fatigue and health assessment (Figure 6).

Figure 6. Mode-shape illustration and deflection profile of a wind-turbine blade (Sheibani & Akbari, 2015).

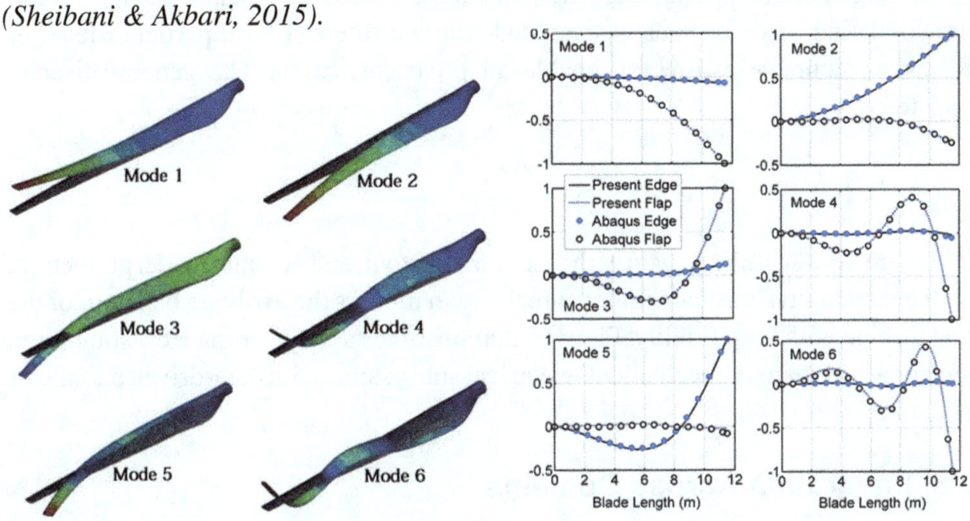

Thermal stresses critically affect bearings, generators, and power electronics. The lumped-parameter heat-balance equation is:

$$C_{th}\frac{dT}{dt} = P_{loss} - hA(T - T_{amb}) \tag{12}$$

where C_{th} is thermal capacitance, P_{loss} is power losses (friction + electrical), $hA(T - T_{amb})$ is heat dissipation to the environment. Coupling thermal models with mechanical and electrical subsystems allows the twin to estimate temperature-dependent efficiency and derating behavior.

3.5. Model Calibration and Parameter Identification

For a digital twin to remain faithful over time, model parameters must be continually aligned with real operational data, a process known as calibration or parameter identification.

Parameters such as stiffness, damping, and aerodynamic coefficients are estimated from experimental data or high-fidelity simulations using techniques like least-squares fitting or genetic optimization. For example, minimizing the squared torque-error between measured T_a^{meas} and modeled T_a^{model}:

$$\min_{\theta} J(\theta) = \sum_{k=1}^{N} \left[T_a^{meas}(k) - T_a^{model}(k,\theta)\right]^2 \quad (13)$$

yields the parameter vector $\theta = [C_1, \ldots, C_6]$ or other coefficients.

During operation, parameters drift due to wear or environmental changes. Real-time updating methods such as extended Kalman filters (EKF), particle filters, or Bayesian recursive estimation enable adaptive correction. The general discrete update is:

$$\theta_{k+1} = \theta_k + K_k\left[y_k - h(x_k, \theta_k)\right] \quad (14)$$

where K_k is the Kalman gain, y_k the measured output, and $h(\cdot)$ the model prediction.

These adaptive mechanisms ensure the twin mirrors the evolving behavior of the real turbine with high fidelity. Figure 7 demonstrates a multi-domain co-simulation loop of aerodynamic–mechanical–electrical subsystems with data-driven parameter updating.

3.6. Integration Across Domains

To realize a full digital twin, the above subsystem models are integrated within a co-simulation environment as shown in Figure 7, enabling energy and signal exchange among domains. Common frameworks include MATLAB/Simulink + Python, FMI/FMU-based coupling, or Modelica environments. Time-synchronized integration (Table 2) ensures consistent causality between fast electrical dynamics (millisecond scale) and slower mechanical or aerodynamic responses (second scale). Through such integration, the digital twin becomes capable of multi-rate simulation, fault propagation analysis, and real-time predictive control.

While multi-domain physical models provide interpretability and structural consistency, they remain limited in representing stochastic and time-varying operational conditions. This motivates the introduction of AI-based analytics, which complement physics models by learning nonlinear patterns directly from data. The next section therefore integrates these two perspectives and presents the AI layer of the digital twin.

Figure 7. Multi-domain co-simulation loop of aerodynamic–mechanical–electrical subsystems with data-driven parameter updating.

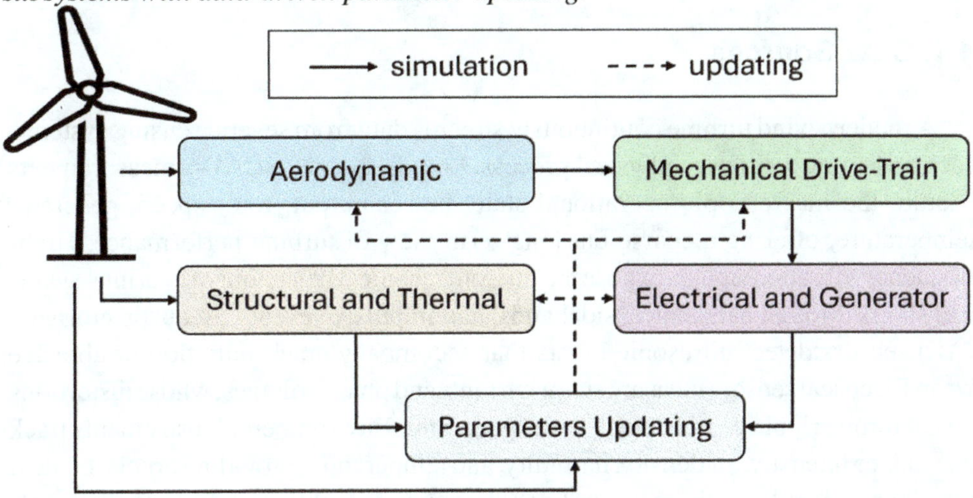

Table 2. Example of model domains and typical time-scales.

Domain	Governing Variables	Typical Time Step	Coupling Direction
Aerodynamic	Wind speed, torque	0.1–1 s	→ Mechanical
Mechanical	Angular speed, stress	1–10 ms	↔ Electrical
Electrical	Currents, voltage	0.1–1 ms	↔ Control
Thermal	Temperature	1–10 s	← All

4. AI-ENHANCED ANALYTICS FOR WIND TURBINE MONITORING

Physical models provide structure and interpretability for digital twins, yet real-world wind turbines operate in environments where aerodynamics, mechanical loads, and electrical dynamics interact in ways that cannot be expressed entirely through first-principles equations. Turbulence, thermal variation, component fatigue, and stochastic grid conditions introduce nonlinear behaviors that evolve over time. Artificial Intelligence (AI) complements traditional modeling by enabling the digital twin to learn from operational data, adapt to changing conditions, and infer patterns that remain hidden in analytical formulations. Within the DT framework, AI acts as a cognitive layer that transforms raw sensor streams into diagnostic knowledge, predictive indicators, and control actions.

4.1. Data Sources

A modern wind turbine continuously streams data from several sensing systems, each reflecting a different physical process. Low-frequency SCADA measurements capture the macroscopic operational state, power output, rotor speed, generator temperature, offering a coarse but long-term view of turbine performance. High-frequency vibration signals expose the micromechanical behaviour of bearings, gears, and shafts through harmonics, sidebands, and impulsive events. Acoustic emission (AE) sensors detect ultrasonic bursts that accompany crack initiation or abrasive wear. Electrical sensors measure stator currents and phase voltages, whose distortions reveal torque ripple or imbalance in the generator. Environmental instruments track turbulence intensity, air density, humidity, and temperature, providing crucial context that shapes aerodynamic loading. Each data type in Table 3 and Figure 8 reveals distinct aspects of system behavior. Fusing these modalities enables comprehensive state estimation and cross-domain correlation analysis.

Table 3. Representative data sources in wind-turbine monitoring.

Sensor Type	Variable	Sampling Rate	Unit
SCADA	Power, speed, temp.	1 Hz	—
Accelerometer	Vibration	1–20 kHz	m/s^2
AE Sensor	RMS, energy, counts	50–500 kHz	—
Electrical	Current, voltage	10–50 kHz	A, V
Environmental	Wind speed, temp.	1 Hz	m/s, °C

Figure 8. Multimodal data streams feeding the AI-enhanced digital twin.

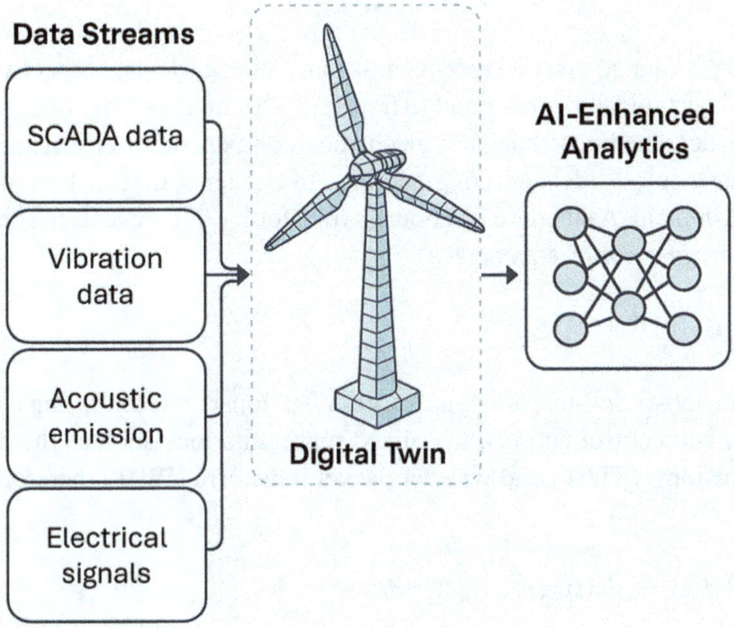

4.2. Data Preprocessing and Feature Extraction

The preprocessing pipeline generally begins with filtering to remove high-frequency noise, electrical interference, or harmonic contamination. Butterworth low-pass filters are widely used for their flat magnitude response, whereas Kalman filters provide adaptive noise suppression for dynamic environments such as wind turbines.

Because sensors often operate at different sampling rates, including accelerometers, strain gauges, AE sensors, and controller logs, resampling and time alignment are required to create synchronized data windows. Finally, normalization to zero mean and unit variance stabilizes machine-learning training and prevents features with large magnitudes from dominating the learning process.

Time-domain features provide compact descriptions of signal amplitude, energy, and impulsiveness. The root-mean-square (RMS), for instance, reflects overall energy and is a key indicator of bearing and gearbox degradation. Similarly, kurtosis highlights impulsive events caused by defects such as cracks or spalls:

$$x_{\text{RMS}} = \sqrt{\frac{1}{N}\sum_{i=1}^{N} x_i^2}, K = \frac{\frac{1}{N}\sum(x_i - \bar{x})^4}{(\frac{1}{N}\sum(x_i - \bar{x})^2)^2} \tag{15}$$

where x_{RMS} and K are the root-mean-square and kurtosis, respectively. High kurtosis values typically correspond to transient AE bursts or early bearing defects.

In frequency and time–frequency, mechanical components often generate periodic imprints, harmonics, sidebands, or blade-pass frequencies, that are best observed in the spectral domain. A standard fast Fourier transform (FFT) reveals the distribution of signal energy across frequencies.:

$$X(f) = \text{FFT}[x(t)] \tag{16}$$

However, most wind-turbine signals are non-stationary due to varying wind speed, turbulence, and control actions. Localized representations such as the short-time Fourier transform (STFT) and wavelet packet transform (WPT) therefore become essential:

$$STFT\{x(t)\}(f,\tau) = \int x(t)w(t-\tau)e^{-j2\pi ft}\,dt \tag{17}$$

Wavelet-based features, such as band-energy or wavelet entropy, capture transient and impulsive signatures typical in AE or early-stage mechanical faults. Combined, these features form high-dimensional vectors that represent the system's operational state and serve as inputs for machine-learning and deep-learning models within the digital twin.

4.3. Machine Learning for Anomaly Detection

Machine learning enables a digital twin to identify deviations from normal operating behavior, even when no explicit fault labels exist. This capability is especially valuable in wind turbines, where failures are rare, but operational variability is high.

When labeled fault data are scarce, unsupervised models are preferred because they learn the statistical characteristics of healthy operation. Principal Component Analysis (PCA) reduces feature dimensionality and projects data onto orthogonal components. Anomalies appear as samples with large reconstruction error or unusual projections:

$$e_i = \| x_i - \hat{x}_i \|_2 \tag{18}$$

Autoencoders (AE) use a neural network to reconstruct inputs. Deviations from normal patterns increase the reconstruction loss:

$$L = \frac{1}{N}\sum_{i=1}^{N} \| x_i - \hat{x}_i \|^2 \tag{19}$$

Clustering algorithms such as K-Means or DBSCAN group similar operating states, where outliers represent abnormal conditions. These approaches detect early degradation trends without requiring annotated fault datasets.

When labeled examples of faults are available, classical supervised algorithms such as support vector machines (SVM) or random forests (RF) learn decision boundaries separating different classes of mechanical health. Their performance is typically quantified by:

$$P = \frac{TP}{TP + FP}, R = \frac{TP}{TP + FN} \tag{20}$$

and the F1-score, which jointly measures classification accuracy and robustness. Such metrics are essential for benchmarking digital-twin diagnostic modules across varying wind conditions and turbine configurations.

4.4. Deep Learning for Fault Classification

Deep learning enhances digital-twin intelligence by learning discriminative representations directly from raw or lightly processed signals, thereby reducing reliance on handcrafted features. Spectrograms, scalograms, or raw waveforms can be fed into convolutional neural networks (CNNs). Convolutional layers automatically learn filters that detect harmonics, sidebands, and fault-related modulations:

$$h_{ij} = \sigma\left(\sum w_{mn} S_{i+m,j+n} + b\right) \tag{21}$$

This parameter sharing allows CNNs to effectively capture localized, repeatable patterns in vibration or AE data, even under noisy field conditions.

Wind-turbine faults often evolve gradually. Long short-term memory (LSTM) networks are designed to model such temporal dependencies:

$$h_t = f\left(W_x x_t + W_h h_{t-1} + b\right) \tag{22}$$

where the cell state tracks long-term degradation trends such as increasing vibration activity or rising blade-pass modulation.

Combining CNNs and LSTMs allows the model to extract both spatial and temporal features. CNN layers act as powerful encoders of spectral or waveform patterns, while LSTM layers capture temporal progression, achieving high robustness against environmental variability, rotor-speed changes, and turbulent inflow conditions. Such deep-learning architectures form the core of advanced digital-twin analytics, enabling continuous monitoring, early fault detection, and long-term degradation modeling.

4.5. Physics-Informed Neural Networks (PINNs) for Hybrid Modeling

Despite their accuracy, purely data-driven models often lack physical interpretability. Physics-Informed Neural Networks (PINNs) bridge this gap by embedding governing equations into the loss function, allowing the model to respect physical constraints. Let $y = f_\theta(x, t)$ be a neural-network approximation of a physical variable (e.g., torque). The total loss function combines data and physics terms:

$$\mathcal{L}_{total} = \mathcal{L}_{data} + \lambda \mathcal{L}_{physics} \tag{23}$$

where:

$$\mathcal{L}_{data} = \frac{1}{N}\sum_i (y_i - y_i^{obs})^2, \mathcal{L}_{physics} = \frac{1}{M}\sum_j \left(\frac{dy_j}{dt} - g(y_j, u_j)\right)^2 \tag{24}$$

Here $g(\cdot)$represents the governing physical law (e.g., aerodynamic torque balance). The hyperparameter λtunes the physics-to-data weighting.

For example, for torque prediction (Figure 9):

$$T_a = \frac{1}{2}\rho A C_p(\lambda, \beta) v^3 / \omega_t \tag{25}$$

Figure 9. Illustration of physics-informed training loop within the digital-twin environment.

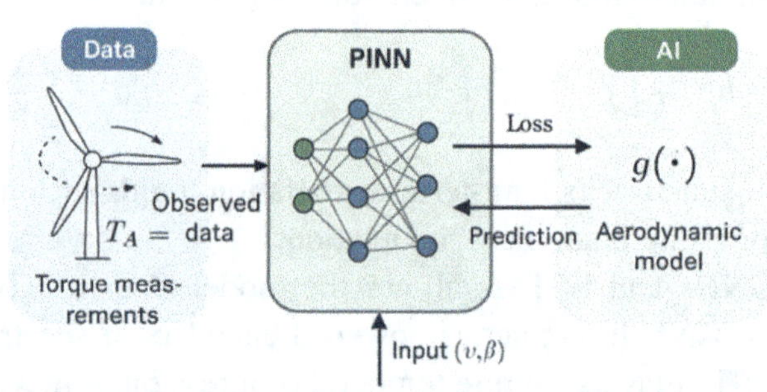

The PINN enforces this aerodynamic relationship during training, improving generalization under unseen wind conditions. Compared to black-box neural nets, PINNs achieve lower extrapolation error and enhanced physical plausibility.

4.6. Reinforcement Learning for Control Optimization

Beyond fault detection, AI can actively optimize turbine control through Reinforcement Learning (RL), where an agent learns control actions that maximize cumulative rewards. The state–action–reward structure is:

$$s_t = [v_t, \omega_t, \beta_t, P_t], a_t = [\Delta\beta_t, \Delta\gamma_t]$$

$$r_t = \alpha P_t - \beta\sigma_{stress} - \gamma\sigma_{vib} \tag{26}$$

Where P_t is instantaneous power, σ_{stress} is structural stress indicator, σ_{vib} is vibration amplitude. The agent updates its policy $\pi(a \mid s)$ to maximize expected cumulative reward $\mathbb{E}\left[\sum_t r_t\right]$. Common algorithms are: Q-Learning / Deep Q-Network (DQN) for discrete actions (e.g., pitch up/down); Policy Gradient / Actor–Critic for continuous control (pitch angle, yaw torque). Training within the digital-twin simulator avoids physical risk; the resulting controller is deployed to the real turbine via the feedback channel. RL thus enables adaptive control, automatically balancing energy capture against fatigue loads and extending component lifetime.

Deploying AI models in practice requires a robust real-time data pipeline that bridges the physical and virtual layers. Section 5 therefore details the communication architecture enabling edge inference, synchronization, and reliable data exchange.

5. REAL-TIME DATA INTEGRATION AND COMMUNICATION ARCHITECTURE

A digital twin's intelligence depends not only on accurate models and AI analytics but also on the real-time exchange of data between the physical turbine and its virtual representation. In a wind-energy environment, this exchange must be reliable under harsh outdoor conditions, bandwidth limitations, and cybersecurity constraints. The communication architecture therefore forms the nervous system of the digital twin, linking sensors, controllers, databases, and analytical engines in a closed, low-latency loop.

5.1. IoT and Edge Computing in Wind-Turbine Monitoring

Modern wind turbines are essentially cyber-physical systems embedded with Internet of Things (IoT) nodes. These nodes comprise intelligent sensors, local processors, and communication modules capable of pre-processing and transmitting data.

However, continuous high-frequency sensing (vibration, AE, current harmonics) can generate gigabytes of data per day, making cloud-only architectures inefficient. Edge computing resolves this challenge by bringing computation closer to the data source. Edge devices located in the turbine nacelle or tower perform local filtering, feature extraction, and event detection before transmitting summarized information to the cloud twin (Figure 10).

Advantages of this distributed approach include:

- Reduced network latency and bandwidth usage,
- Enhanced reliability in remote or offshore installations,
- Increased data privacy through localized storage and encryption,
- Real-time decision support for safety-critical operations.

Figure 10. Hierarchical edge–cloud architecture for wind-turbine digital twin, showing local analytics at the edge and global coordination at the cloud.

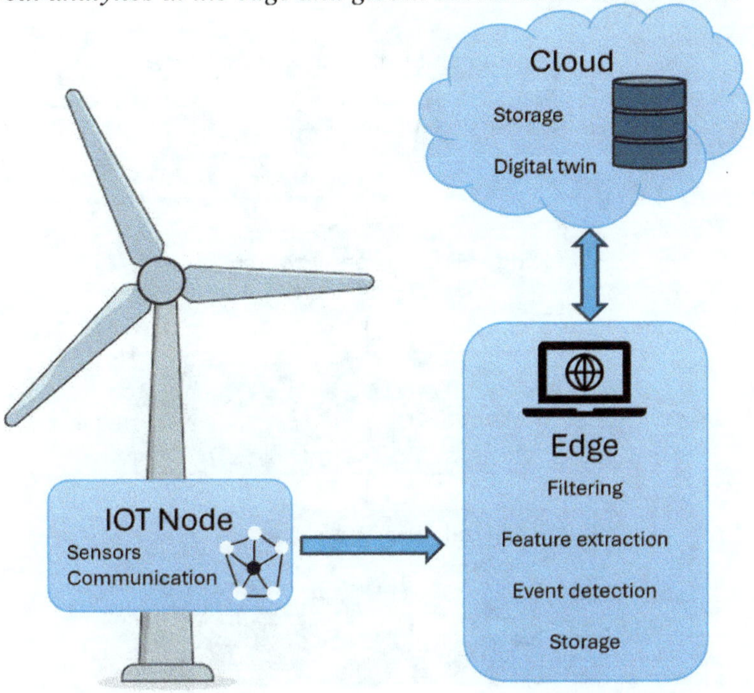

5.2. Communication Stack

A standard communication stack in digital-twin-enabled monitoring systems can be divided into four layers (Figure 11):

1. Sensing and Data Acquisition (DAQ) Layer: Sensors (accelerometers, AE, current probes, temperature) generate raw analog signals that are digitized by DAQ boards. Sampling rates vary from 1 Hz for SCADA to > 100 kHz for AE.
2. Edge Gateway and Protocol Translation Layer: Data packets are encapsulated using lightweight communication protocols such as MQTT (Message Queuing Telemetry Transport) or OPC-UA and forwarded to an edge gateway. This layer handles buffering, compression, and metadata tagging (timestamp, sensor ID).

Figure 11. Simplified communication stack connecting sensors, DAQ, MQTT broker, and digital-twin engine.

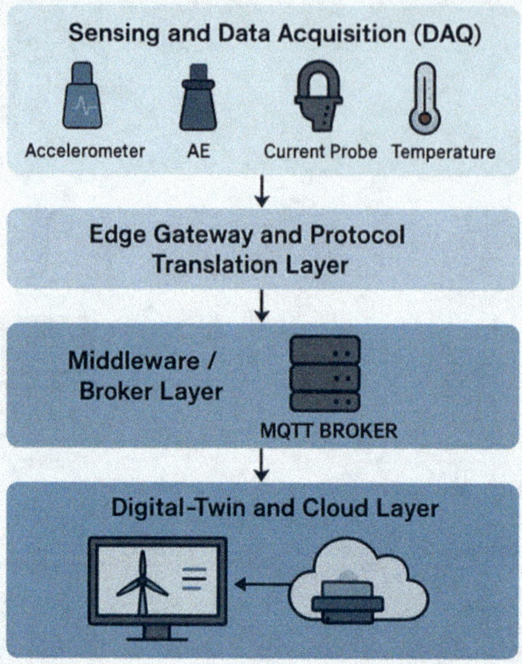

3. Middleware / Broker Layer: The MQTT broker acts as a publish–subscribe hub. Each sensor topic publishes data (e.g., /turbine1/vibration) while subscribers, such as digital-twin analytics or visualization dashboards, receive updates asynchronously. This decouples producers and consumers, ensuring scalability.
4. Digital-Twin and Cloud Layer: Incoming messages are streamed into databases and simulation engines. The twin performs inference or model updates and optionally sends control commands back through the broker.

Reducing each component is crucial to maintaining near real-time synchronization between the physical turbine and its virtual twin.

5.3. Real-Time Synchronization Mechanisms

When data originate from heterogeneous devices, timestamp misalignment can lead to phase errors in vibration or power-signal analysis. Synchronization protocols such as NTP (Network Time Protocol) or IEEE 1588 Precision Time Protocol (PTP) align clocks within microseconds, ensuring coherent correlation across channels.

In real-time analytics, signals are processed in sliding windows of size W and step S. If the sampling rate is f_s, then each window represents a time interval W/f_s.

Streaming frameworks (e.g., Apache Kafka, RabbitMQ) handle buffering and delivery guarantees, allowing the digital twin to operate continuously despite network jitter.

To achieve both responsiveness and computational depth, tasks are divided between edge and cloud:

$$T_{total} = T_{edge}^{proc} + T_{transmit} + T_{cloud}^{proc} \quad (27)$$

Optimization aims to minimize T_{total} subject to resource constraints. Edge processors handle time-critical inference (e.g., anomaly alarms), while the cloud executes heavier simulations and long-term learning.

In real deployments, communication bottlenecks often arise from limited bandwidth in offshore wind farms, packet loss under harsh weather, or inconsistent synchronization between high-frequency vibration data and low-frequency SCADA signals.

5.4. Data Storage and Management

Digital-twin environments require structured data storage systems capable of handling high-volume, high-velocity time-series data. Specialized time-series databases (TSDBs) such as InfluxDB, TimescaleDB, or OpenTSDB are optimized for this purpose.

In Time-Series Database (TSDB) architecture, data are stored as key–value pairs indexed by timestamp and tag: "Record: "{"time","sensor_id","value","tags"}. Compression algorithms (e.g., Gorilla encoding) reduce storage size by exploiting temporal correlation. Retention policies automatically down-sample historical data, preserving trends while saving space.

All sensor readings, model states, and control commands are linked to the turbine's digital thread. Each event carries a globally unique identifier (GUID), ensuring traceability across lifecycle phases. Versioning tools like Git-LFS or DVC (Data Version Control) allow roll-back and model provenance.

Table 4. Comparison of time-series databases used in digital-twin applications.

Database	Storage Engine	Query Language	Typical Use	Strengths
InfluxDB	TSM	InfluxQL	Edge gateway logging	Lightweight, open source
TimescaleDB	PostgreSQL-based	SQL	Cloud analytics	Relational + time-series flexibility
OpenTSDB	HBase	REST	Large-scale monitoring	Scalability to billions of points

5.5. Cyber-Security and Reliability Considerations

With the growing digitalization of wind farms, cyber-security has become a prerequisite for trustworthy digital-twin operation. Key aspects include data confidentiality and encryption, integrity and fault tolerance, access control and audit logging, and resilience and health monitoring.

First, all MQTT or REST communications should employ TLS/SSL encryption. Authentication mechanisms (username/password or X.509 certificates) ensure only authorized devices connect to the broker. For additional protection, data payloads can be encrypted at the sensor node using symmetric keys K_s, where:

$$C = E_{K_s}(M), M = \text{sensor message}, C = \text{ciphertext}. \tag{28}$$

Second, checksums (CRC) and message sequencing prevent packet corruption or loss. Redundant brokers or mirrored databases guarantee continuous operation even if individual nodes fail. The twin should implement automatic fail-over to backup servers or local caches.

Third, role-based access control (RBAC) distinguishes between operators, data scientists, and maintenance engineers. Every transaction, data upload, model update, or control command, is recorded in an audit log for traceability and compliance with standards such as ISO 27001.

Finally, network and server health metrics (CPU, memory, packet loss) are continuously monitored. Predictive analytics on the infrastructure itself can forecast failures and trigger self-healing protocols, completing the meta-digital-twin loop—where the twin supervises its own computational environment.

To demonstrate how these components work together, the following section presents a complete implementation of a small-scale wind turbine digital twin.

6. IMPLEMENTATION CASE STUDY: SMALL-SCALE WIND TURBINE DIGITAL TWIN

This section presents a complete case study illustrating how a DT can be implemented, validated, and used as a virtual sensing and monitoring module for a small-scale wind turbine (Ibrahim et al., 2023b). The case study demonstrates the entire development pipeline, from mathematical modeling and test-bench construction to real-time virtual sensing and DT–physical synchronization. Although the example uses a laboratory-scale direct-drive wind turbine generator (WTG), the methodology generalizes directly to full-scale industrial turbines equipped with advanced AI-driven monitoring systems.

6.1. Overview of the Case Study

Wind turbines rely heavily on accurate sensing systems, especially wind speed, rotor speed, electrical outputs, and structural loads. However, physical sensors often degrade due to weathering, icing, electromagnetic interference, or calibration drift. A DT-based virtual sensor provides a complementary (and sometimes alternative) source of measurements by inferring unmeasured quantities from validated physical models and available sensor data.

The implementation showcased here constructs a DT that estimates wind speed from generator electrical measurements and rotor speed, thereby overcoming missing or faulty anemometer readings. This example highlights the capability of a DT to operate as an intelligent virtual instrument tightly integrated with the physical system. Figure 12 shows the overview of the DT structure.

Figure 12. Overview of the DT structure.

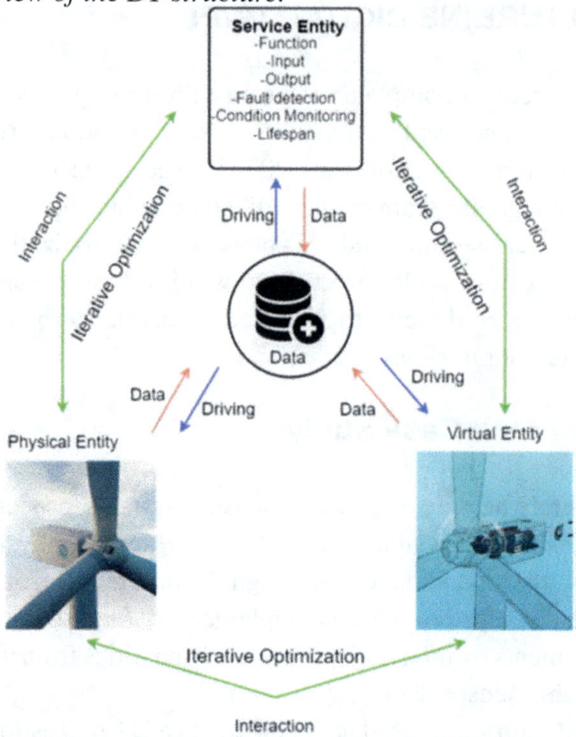

6.2. Physical System

The experimental platform consists of a 600 W direct-drive PMSG-based wind turbine mounted on a controlled test bench (Figure 13). The physical system includes:

- Surface-mounted 8-pole PMSG
- Servo motor as wind emulator, capable of reproducing arbitrary wind profiles
- Power analyzer and multi-channel data acquisition system (DAS)
- Anemometer for ground-truth wind speed

The test bench enables controlled emulation of steady and dynamic wind conditions while directly measuring three-phase voltage, current, electrical power, and rotor speed. These form the input data used to drive and validate the DT. A summary of key mechanical and electrical parameters (e.g., rated speed, rated wind speed, blade length, torque constant) is given in the orignial paper.

Figure 13. Small-Scale WTG Test Bench.

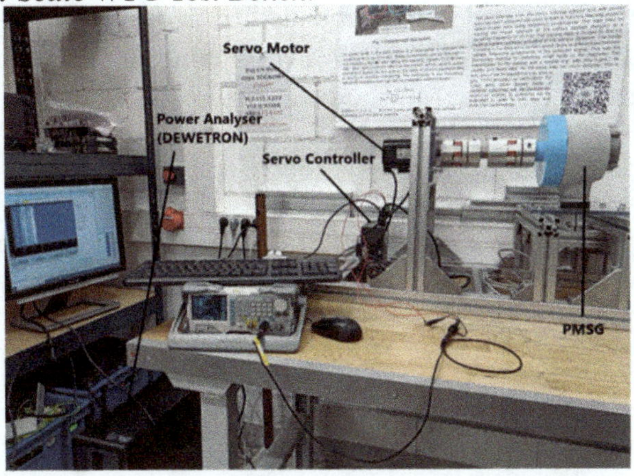

6.3. Virtual System

The virtual turbine is developed in MATLAB/Simulink, combining:

- an aerodynamic wind power model
- a detailed PMSG d–q axis electrical model
- the mechanical rotor dynamics
- a wind-speed estimation block built upon the combined power–torque–TSR relations

Wind power and extracted mechanical power are modeled as:

$$P_{\text{wind}} = \tfrac{1}{2}\rho A V^3 C_p, P_{\text{mech}} = \tfrac{1}{2}\rho A V^3 C_p \eta, T_m = \frac{P_{\text{mech}}}{\omega_m} \tag{29}$$

A standard d–q axis model is implemented:

$$v_d = R i_d + \omega_e \psi_q, v_q = R i_q - \omega_e \psi_d + \omega_m L_d i_d, \; T_e = \tfrac{3}{2} P\left(\psi_d i_q - \psi_q i_d\right) \tag{30}$$

This provides the instantaneous electromagnetic torque, stator flux linkage, and electrical power.

The DT estimates wind speed by solving the inverse mapping:

$$Pe \rightarrow Te \rightarrow \omega m \rightarrow Vest \tag{31}$$

using the generator dynamics and tip-speed ratio relationship: $\omega m = \lambda V/R$.

The estimated wind speed is thus obtained from electrical power, rotor speed, and turbine characteristics. This block (WTG combined model) constitutes the core of the virtual wind-speed sensor as shown in Figure 14.

Figure 14. Wind turbine virtual model in MATLAB/Simulink.

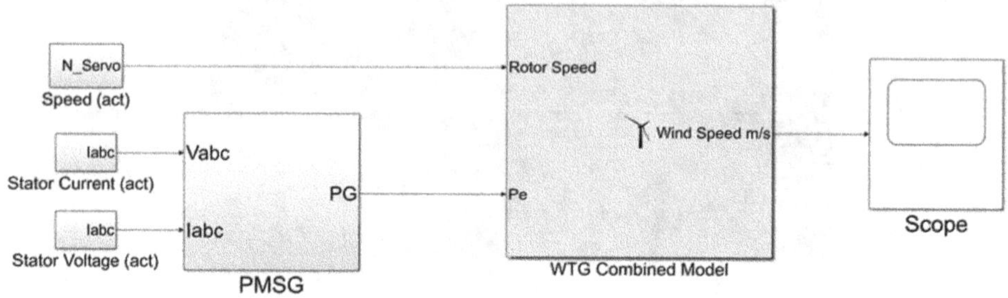

6.4. Real-Time Data Exchange and Synchronization

To achieve continuous synchronization between the physical and virtual turbine, a service unit is developed using an ActiveX server interfaced with the DAS. The communication workflow is:

1. DAS acquires real-time measurements (V, I, rotor speed).
2. ActiveX service retrieves the measurements from hardware buffers.
3. Data is streamed directly into MATLAB/Simulink.
4. The DT executes real-time inference and outputs the estimated quantities.

This architecture ensures that the DT stays synchronized with the physical turbine at all times, enabling both monitoring and future predictive modeling.

6.5. Validation Methodology

Validation ensures that the virtual turbine accurately represents physical behavior. The turbine was first operated under no-load conditions, using the servo motor to produce various rotor speeds corresponding to different wind speeds. The DT processed the measured generator electrical quantities and returned estimated wind speeds, which were compared against anemometer readings.

Next, the turbine was exposed to real variable wind conditions. Wind speed from the anemometer was compared with the DT-estimated wind speed, revealing:

Mean Absolute Error (MAE): 0.26–0.57 m/s; Average estimation accuracy ≈ 83%. The comparison table provided in the source study demonstrates strong agreement between actual and estimated wind speed for various power levels.

6.6. Results and Interpretation

6.6.1. Estimation Accuracy

The DT tracks actual wind speed with good fidelity over a wide operating range (Figure 15). Errors arise mainly due to:

- unmodeled aerodynamic losses
- turbulence and tower–shadow effects
- mechanical friction not perfectly captured in the model
- generator efficiency variations under transient loading

Despite these limitations, the DT consistently exhibits low estimation error and stable behavior, validating its usefulness as an online virtual sensor.

Figure 15. Wind speed estimation results from the proposed DT.

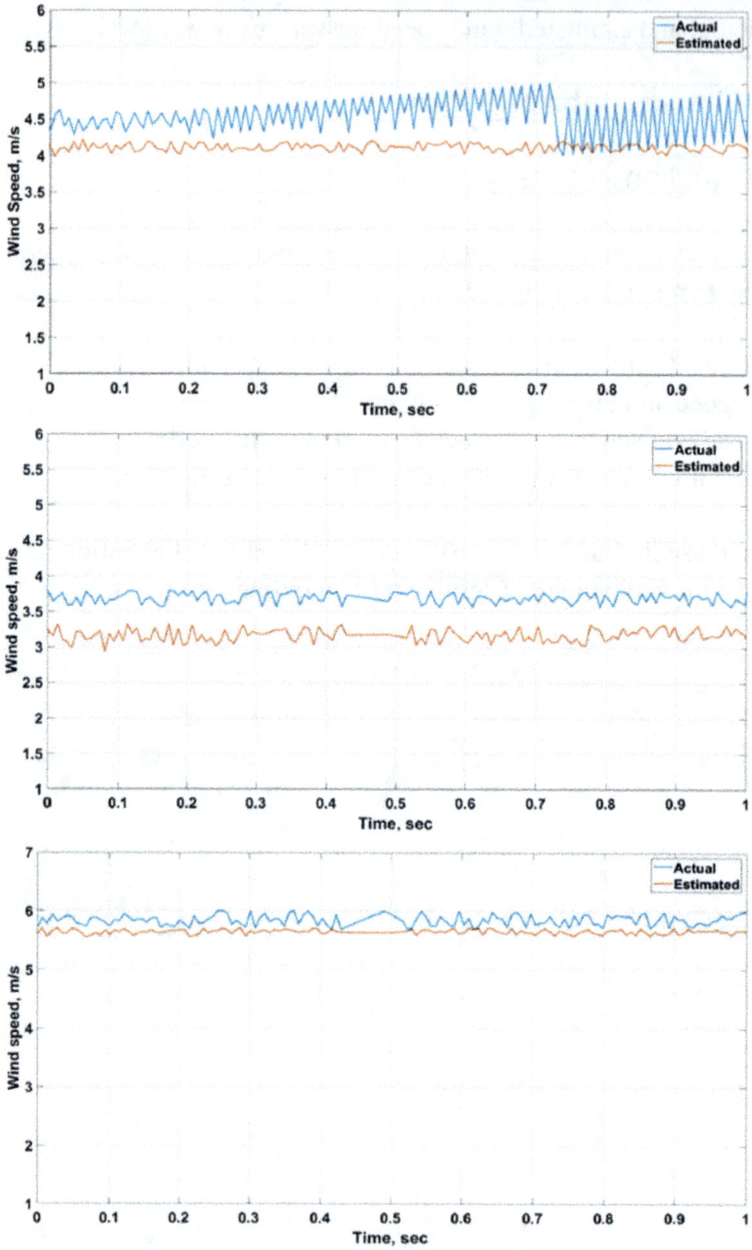

6.6.2. Benefits for AI-Driven Monitoring

The combination of a virtual sensor and a physical turbine enables advanced capabilities:

- Fault tolerance: DT can compensate for missing or faulty anemometer data.
- Enhanced observability: Internal, unmeasured variables (e.g., torque, wind speed, aerodynamic power) become available for AI models.
- Training data augmentation: Synthetic but physically consistent data help train AI models when real data is limited.
- Online anomaly detection: DT residuals (actual vs. estimated) form an anomaly indicator for sensor faults or mechanical degradation.

Although validated under laboratory conditions, deploying the DT in real turbines would require addressing environmental disturbances, electromagnetic noise, and calibration drift.

7. DISCUSSION AND FUTURE PERSPECTIVES

The implementation case study demonstrates that digital twin (DT) technology, when combined with AI-based analytics and real-time connectivity, offers a powerful tool for predictive maintenance and operational optimization of wind turbines. However, the broader implications extend beyond a single prototype. This section discusses how such systems can scale toward wind farms, interact with other energy assets, and evolve toward self-learning and self-verifying ecosystems. The discussion also highlights emerging challenges, data governance, uncertainty quantification, and trustworthiness, that must be addressed before full industrial adoption.

7.1. Scalability from Single-Turbine to Wind-Farm Digital Twins

While a single-turbine digital twin focuses on localized performance and fault diagnosis, wind farms consist of tens to hundreds of interconnected turbines, often distributed across large areas. Scaling the twin architecture from component level to system-of-systems introduces new complexity in communication, synchronization, and collective optimization.

A scalable solution follows a hierarchical architecture:

- Level 1 – Component twins: blades, gearbox, generator, and tower each maintain local models.

- Level 2 – Turbine twin: aggregates component states for global health estimation.
- Level 3 – Wind-farm twin: coordinates turbines through shared weather data, power-flow optimization, and wake modeling.

This hierarchy supports federated learning, where each turbine trains local AI models using its own data but contributes to a shared global model through aggregated gradients. Such federated digital twins preserve data privacy while enabling collective intelligence across geographically distributed assets.

At the wind-farm level, the DT can incorporate wake interactions between turbines using simplified flow-field models such as Jensen or dynamic wake meandering. Optimization objectives extend from maximizing total power output to minimizing fatigue loads and balancing maintenance schedules.

Mathematically, this multi-objective problem can be expressed as:

$$\max_{\mathbf{u}} \left[P_{farm}(\mathbf{u}) - \alpha L_{fatigue}(\mathbf{u})\right] \tag{32}$$

where control vector $\mathbf{u} = [\beta_i, \gamma_i, \omega_i]$ (pitch, yaw, and rotational speed of each turbine) and αweighs power versus fatigue. Reinforcement-learning-based agents deployed across turbines can learn cooperative control policies, giving rise to self-optimizing wind farms.

7.2. Integration with Microgrid and Energy Storage Twins

The next frontier in renewable energy management involves coupling wind-turbine twins with microgrid and battery energy storage twins to form multi-energy digital ecosystems. In a cyber-physical energy network, the network interconnects the following digital twins:

- Wind farm twin: predicts short-term power generation.
- Battery twin: models charge/discharge dynamics, degradation, and state-of-health (SoH).
- Grid twin: simulates load demand and network stability.

Information exchange between these entities enables coordinated scheduling. For example, when the wind twin forecasts a power drop, the storage twin pre-emptively discharges to stabilize supply.

In a coupled modeling framework, energy balance in such integrated systems can be represented as:

$$P_{grid}(t) = P_{wind}(t) + P_{battery}(t) - P_{load}(t) \tag{33}$$

subject to battery constraints:

$$SOC_{min} \leq SOC(t) \leq SOC_{max}, SOC(t+1) = SOC(t) + \eta_{ch} P_{ch} \Delta t - \frac{P_{dis} \Delta t}{\eta_{dis}} \tag{34}$$

Within the DT, AI algorithms perform predictive dispatch, while physical models ensure compliance with electrical and thermal limits. This synergy extends predictive maintenance from individual assets to entire renewable clusters.

7.3. Communication Bottlenecks and Integration

A core challenge in deploying AI-enabled DT systems for wind turbine monitoring lies in ensuring reliable, low-latency communication across heterogeneous components, turbine-level sensors, edge devices, supervisory controllers, and cloud platforms. Modern wind farms generate high-volume, high-frequency data streams, including SCADA measurements (1–10 s), vibration and acoustics (kHz), and environmental data from distributed meteorological sensors. When these multimodal streams are transmitted through shared wireless links or constrained backhaul networks, bandwidth saturation can occur, resulting in packet loss, increased latency, and degraded synchronization between the physical turbine and its virtual counterpart. Such desynchronization directly reduces DT fidelity and impairs AI model performance, especially for time-sensitive tasks such as anomaly detection, early fault localization, and remaining useful life (RUL) prediction.

Communication constraints also emerge from heterogeneous protocol stacks and vendor-specific interfaces that make data integration non-trivial. Turbines may use mixed communication standards, Modbus TCP, OPC-UA, CAN bus, proprietary vibration acquisition interfaces, creating interoperability gaps when constructing a unified DT data pipeline. The lack of standardized semantic models for turbine components further complicates the ingestion of condition-monitoring data into cloud- or edge-hosted analytics engines. As a result, additional middleware layers or protocol translators are often required, adding latency and introducing potential points of failure.

To mitigate these bottlenecks, emerging architectures emphasize edge-centric integration, where initial signal processing, event filtering, and AI inference are executed on intelligent edge nodes co-located with the turbine. This reduces the volume of raw high-frequency data transmitted upstream and enables near-real-time decisions, even under intermittent connectivity. Techniques such as adaptive sampling, compressive sensing, and federated learning frameworks can further alleviate bandwidth demands by minimizing data transmissions without sacrificing model fidelity. At the integration layer, the adoption of interoperable standards, OPC-UA with information modeling, MQTT for lightweight messaging, and digital twin ontology frameworks, allows more seamless communication between turbines, farm-level controllers, and cloud-based DT platforms. Ultimately, addressing communication bottlenecks requires co-designing the DT architecture with network constraints in mind, ensuring that the physical assets, communication infrastructure, and AI/DT modules form a coherent, resilient, and scalable ecosystem for predictive maintenance.

7.4. Uncertainty Quantification and Explainable AI

As digital twins grow in autonomy, ensuring trust and transparency becomes critical. Predictions from AI models must be accompanied by uncertainty estimates and interpretable reasoning. Digital twins can incorporate probabilistic frameworks such as Bayesian Neural Networks (BNNs) or Gaussian Process Regression (GPR) to quantify uncertainty in predictions. For a model output $y = f(x) + \epsilon$, with $\epsilon \sim \mathcal{N}(0, \sigma^2)$, the predictive distribution becomes:

$$p(y_* \mid x_*, \mathcal{D}) = \int p(y_* \mid x_*, \theta) p(\theta \mid \mathcal{D}) \, d\theta \tag{35}$$

where $p(\theta \mid \mathcal{D})$ captures parameter uncertainty learned from data. Such probabilistic modeling enables maintenance decisions to consider confidence intervals rather than single deterministic outputs.

Interpretability methods like SHAP (Shapley Additive Explanations) and LIME (Local Interpretable Model-agnostic Explanations) allow engineers to understand which features (e.g., vibration RMS, AE energy, temperature) most influence fault predictions. Visualizing these feature attributions within the DT dashboard builds operator trust and facilitates root-cause diagnosis.

The following example is about uncertainty-aware gearbox bearing fault prediction. Consider a wind turbine DT equipped with a hybrid AI model that combines physics-informed GPR with SCADA and high-frequency vibration data to monitor the gearbox high-speed shaft (HSS) bearing. The DT continuously processes features such as:

- Vibration spectral kurtosis around 1–5 kHz
- AE burst rate during transient torque events
- Oil temperature and debris sensor output
- Generator torque ripple as an indirect load indicator

Suppose the model outputs a predicted bearing health index h_t, along with a 95% predictive confidence interval derived from the GPR posterior. During nominal operation, the health index remains within the safe band with narrow uncertainty, reflecting consistent model confidence.

However, after a period of operation under turbulent wind conditions, the DT detects a rising trend in spectral kurtosis and intermittent AE bursts. The GPR model predicts a degradation signal: h_{t+1}=0.42, 95% CI=[0.29,0.58].

Even though the point estimate (0.42) does not yet cross the fault threshold (e.g., 0.50), the upper bound of the confidence interval (0.58) exceeds it. As a result, the DT classifies the turbine as entering an early-warning regime. This allows the operations and maintenance (O&M) team to pre-schedule inspection for the next low-wind window, reducing the risk of costly unplanned downtime.

To further diagnose the cause, SHAP analysis is applied to the same prediction. The top contributing features reveal that:

- AE energy and spectral kurtosis contribute +0.11 and +0.07 toward the anomaly score,
- Torque ripple contributes +0.05,
- Oil temperature provides a small stabilizing (negative) contribution.

These attributions suggest that the anomaly arises from impulsive mechanical events—consistent with early-stage bearing pitting rather than lubrication issues or load imbalance. Such physics-aligned explanations strengthen confidence in the DT's recommendation and help O&M teams deploy the correct maintenance action.

7.5. Cyber-Physical Trust and Model Verification

A critical challenge for industrial adoption is establishing trustworthiness across the cyber-physical boundary. Verification ensures the digital twin's implementation correctly represents its design, while validation confirms its predictions match reality. Standard frameworks such as ASME V&V 40 and ISO 23247 define criteria for fidelity and credibility in digital twins.

The twin's performance can be validated through ground-truth comparison:

$$E_{val} = \frac{1}{N}\sum_{i=1}^{N} | y_i^{meas} - y_i^{pred} | \qquad (36)$$

with thresholds defining acceptable deviations. Automated continuous validation, where the twin self-assesses its accuracy and triggers recalibration when E_val>E_thr, represents an emerging trend known as self-verifying digital twins.

As discussed in Section 5, secure communication and authentication are essential. Future digital twins will embed blockchain-based ledgers to ensure tamper-proof data provenance and trustable maintenance records. Each update, sensor reading, model revision, or AI retraining, can be hashed and stored, creating a traceable digital audit trail.

7.6. Future Research

The ultimate vision for digital twin evolution involves autonomy, adaptability, and creativity, enabling systems that not only mirror but also anticipate and redesign themselves.

By integrating Generative AI models (e.g., Variational Autoencoders or Generative Adversarial Networks), digital twins could simulate multiple virtual scenarios or synthetic fault cases to improve model robustness. Generative twins can automatically produce diverse operating datasets, accelerating training of predictive models without physical experimentation.

Emerging research explores self-evolving twins, where meta-learning algorithms adjust both model structures and hyperparameters over time. Such meta-optimization allows the digital twin to autonomously adapt to aging effects or environmental drift, gradually improving its own fidelity and intelligence.

In the long term, wind-turbine twins will operate as cognitive agents, communicating, negotiating, and learning collectively across fleets. Federated frameworks will allow twins to share insights without sharing raw data, enabling global learning while preserving privacy. This evolution parallels biological ecosystems, giving rise to self-organizing energy networks.

8. CONCLUSION

The convergence of Digital Twin (DT) technology and Artificial Intelligence (AI) represents a transformative paradigm for the wind energy sector, redefining how turbines are designed, monitored, and maintained. Through this chapter, we have traversed the conceptual, theoretical, and practical dimensions of building AI-driven digital twins for wind turbine monitoring and predictive maintenance, from their foundational principles to real-world implementation.

At its core, the digital twin serves as a dynamic mirror of reality, continuously reflecting the physical turbine's operational state in cyberspace. By integrating multi-physics models, encompassing aerodynamic, mechanical, electrical, structural, and thermal domains, the DT provides an interpretable, physics-grounded foundation for understanding system behavior. On top of this, AI algorithms introduce adaptability and intelligence, enabling the twin to recognize patterns, detect anomalies, and forecast degradation well before failures occur.

In summary, the digital twin redefines predictive maintenance from a reactive, schedule-based process into a proactive, intelligent lifecycle management strategy. It bridges the gap between physics-based understanding and data-driven adaptability, offering a unified platform for simulation, analysis, and control. When integrated across fleets and coupled with other energy assets, digital twins will enable holistic, resilient, and sustainable operations for the next generation of smart wind-energy systems.

Looking ahead, research should focus on three converging directions: scalable architecture; trustworthy AI; self-evolving intelligence. The synergy between AI and digital twin technology thus heralds a new era of cognitive wind energy systems, where data, physics, and intelligence coalesce to realize the long-sought goals of reliability, sustainability, and autonomy in renewable energy.

REFERENCES

Abdul-Ruhman, M. S., & Hawas, M. N. (2020). Optimal Pitch Angle Control for Wind Turbine Using Intelligent Controller. *IOP Conference Series. Materials Science and Engineering*, *745*(1), 012017. DOI: 10.1088/1757-899X/745/1/012017

Amin, U., Kim, D., Ahmed, F. N., Ahmad, G., & Hossain, M. J. (2025). Digital twins for smart asset management in the energy industry: State-of-the-art. *Expert Systems with Applications*, *289*, 128358. DOI: 10.1016/j.eswa.2025.128358

Da Silva, T. C., Da Silva Antunes, F. A., Teixeira, J. C., & Contreras, R. C. (2025). Correlation Between Wind Turbine Failures and Environmental Conditions: A Machine Learning Approach. *IEEE Access : Practical Innovations, Open Solutions*, *13*, 50043–50058. DOI: 10.1109/ACCESS.2025.3551241

Dui, H., Cao, T., & Wang, F. (2025). Digital twin-based resilience evaluation and intelligent strategies of smart urban water distribution networks for emergency management. *Resilient Cities and Structures*, *4*(1), 41–52. DOI: 10.1016/j.rcns.2025.02.001

Fei, X., Jiang, H., Zhou, J., & Zhou, Y. (2025). Digital twin modeling method for wind turbine generator system based on operational mechanism and scenario fusion. The International Journal of Advanced Manufacturing Technology 2025 139:3, 139(3), 1979–2001. DOI: 10.1007/s00170-025-15892-2

Gielen, D., Boshell, F., Saygin, D., Bazilian, M. D., Wagner, N., & Gorini, R. (2019). The role of renewable energy in the global energy transformation. *Energy Strategy Reviews*, *24*, 38–50. DOI: 10.1016/j.esr.2019.01.006

Grieves, M., & Vickers, J. (2017). Digital Twin: Mitigating Unpredictable, Undesirable Emergent Behavior in Complex Systems. Transdisciplinary Perspectives on Complex Systems: New Findings and Approaches, 85–113. DOI: 10.1007/978-3-319-38756-7_4

Hasan, S. K. K., & Crawford, C. W. (2025). *A new era for digital twins: progress and industry adoption*. Digital Twin., DOI: 10.1080/27525783.2025.2555877

Hassani, S., & Dackermann, U. (2023). A Systematic Review of Advanced Sensor Technologies for Non-Destructive Testing and Structural Health Monitoring. Sensors 2023, Vol. 23, Page 2204, 23(4), 2204. DOI: 10.3390/s23042204

Ibrahim, M., Rassõlkin, A., Vaimann, T., Kallaste, A., Zakis, J., Hyunh, V. K., & Pomarnacki, R. (2023a). Digital Twin as a Virtual Sensor for Wind Turbine Applications. Energies 2023, Vol. 16, Page 6246, 16(17), 6246. DOI: 10.3390/en16176246

Ibrahim, M., Rassõlkin, A., Vaimann, T., Kallaste, A., Zakis, J., Hyunh, V. K., & Pomarnacki, R. (2023b). Digital Twin as a Virtual Sensor for Wind Turbine Applications. Energies 2023, Vol. 16, Page 6246, 16(17), 6246. DOI: 10.3390/en16176246

Ismail, L., Abdelmoti, A., Basu, A., Berini, D. E., & Naouss, M. (n.d.). A Systematic Review of Digital Twin-Driven Predictive Maintenance in Industrial Engineering: Taxonomy, Architectural Elements, and Future Research Directions. Retrieved December 2, 2025, from https://arxiv.org/pdf/2509.24443v1

Jiang, H., Qin, S., Fu, J., Zhang, J., & Ding, G. (2021). How to model and implement connections between physical and virtual models for digital twin application. *Journal of Manufacturing Systems*, *58*, 36–51. DOI: 10.1016/j.jmsy.2020.05.012

Jorgensen, J., Hodkiewicz, M., Cripps, E., & Hassan, G. M. (2023). Requirements for the application of the Digital Twin Paradigm to offshore wind turbine structures for uncertain fatigue analysis. *Computers in Industry*, *145*, 103806. DOI: 10.1016/j.compind.2022.103806

Juarez, M. G. J., Giret, A., & Botti, V. (2025). Semantic and modular orchestration of AI-driven digital twins for industrial interoperability and optimization. *Journal of Industrial Information Integration*, *48*, 100959. DOI: 10.1016/j.jii.2025.100959

Kandemir, E., Hasan, A., Kvamsdal, T., & Abdel-Afou Alaliyat, S. (2024). Predictive digital twin for wind energy systems: a literature review. Energy Informatics 2024 7:1, 7(1), 1–36. DOI: 10.1186/s42162-024-00373-9

Kull, K., Asad, B., Khan, M. A., Naseer, M. U., Kallaste, A., & Vaimann, T. (2025). Faults, Failures, Reliability, and Predictive Maintenance of Grid-Connected Solar Systems: A Comprehensive Review. Applied Sciences 2025, Vol. 15, Page 11461, 15(21), 11461. DOI: 10.3390/app152111461

Leon-Medina, J. X., Tibaduiza, D. A., Parés, N., & Pozo, F. (2025). Digital twin technology in wind turbine components: A review. *Intelligent Systems with Applications*, *26*, 200535. DOI: 10.1016/j.iswa.2025.200535

Maksimović, M., Jokić, S., & Bošković, M. (2025). Innovative Horizons for Sustainable Smart Energy: Exploring the Synergy of 5G and Digital Twin Technologies. Process Integration and Optimization for Sustainability 2025 9:2, 9(2), 431–470. DOI: 10.1007/s41660-024-00478-4

Mbasso, W. F., Harrison, A., Dagal, I., Jangir, P., Khishe, M., Kotb, H., Shaikh, M. S., Smerat, A., Fendzi Donfack, E., & Kumar, R. (2025). Digital twins in renewable energy systems: A comprehensive review of concepts, applications, and future directions. *Energy Strategy Reviews*, *61*, 101814. DOI: 10.1016/j.esr.2025.101814

McKenna, R., Ostman, P., & Fichtner, W. (2016). Key challenges and prospects for large wind turbines. *Renewable & Sustainable Energy Reviews*, *53*, 1212–1221. DOI: 10.1016/j.rser.2015.09.080

Pandit, R., Astolfi, D., Hong, J., Infield, D., & Santos, M. (2023). SCADA data for wind turbine data-driven condition/performance monitoring: A review on state-of-art, challenges and future trends. *Wind Engineering*, *47*(2), 422–441. DOI: 10.1177/0309524X221124031

Pandit, R. K., Astolfi, D., & Durazo Cardenas, I. (2023). A Review of Predictive Techniques Used to Support Decision Making for Maintenance Operations of Wind Turbines. Energies 2023, Vol. 16, Page 1654, 16(4), 1654. DOI: 10.3390/en16041654

Rahman, M. A., Shahrior, M. F., Iqbal, K., & Abushaiba, A. A. (2025). Enabling Intelligent Industrial Automation: A Review of Machine Learning Applications with Digital Twin and Edge AI Integration. Automation 2025, Vol. 6, Page 37, 6(3), 37. DOI: 10.3390/automation6030037

Rasheed, A., San, O., & Kvamsdal, T. (2020). Digital twin: Values, challenges and enablers from a modeling perspective. *IEEE Access : Practical Innovations, Open Solutions*, *8*, 21980–22012. DOI: 10.1109/ACCESS.2020.2970143

Rojas, L., Peña, Á., & Garcia, J. (2025). AI-Driven Predictive Maintenance in Mining: A Systematic Literature Review on Fault Detection, Digital Twins, and Intelligent Asset Management. Applied Sciences 2025, Vol. 15, Page 3337, 15(6), 3337. DOI: 10.3390/app15063337

Saleh, A. M., István, V., Khan, M. A., Waseem, M., & Ali Ahmed, A. N. (2024). Power system stability in the Era of energy Transition: Importance, Opportunities, Challenges, and future directions. Energy Conversion and Management: X, 24, 100820. DOI: 10.1016/j.ecmx.2024.100820

Sheibani, M., & Akbari, A. A. (2015). Finite element modeling of a wind turbine blade. *Journal of Vibroengineering*, *17*(7), 3774–3791. https://www.extrica.com/article/16033

Tautz-Weinert, J., & Watson, S. J. (2017). Using SCADA data for wind turbine condition monitoring - A review. *IET Renewable Power Generation*, *11*(4), 382–394. DOI: 10.1049/iet-rpg.2016.0248

Wanasinghe, T. R., Wroblewski, L., Petersen, B. K., Gosine, R. G., James, L. A., De Silva, O., Mann, G. K. I., & Warrian, P. J. (2020). Digital Twin for the Oil and Gas Industry: Overview, Research Trends, Opportunities, and Challenges. *IEEE Access : Practical Innovations, Open Solutions*, *8*, 104175–104197. DOI: 10.1109/ACCESS.2020.2998723

Wang, S., Vidal, Y., & Pozo, F. (2026). Recent advances in wind turbine condition monitoring using SCADA data: A state-of-the-art review. *Reliability Engineering & System Safety*, *267*, 111838. DOI: 10.1016/j.ress.2025.111838

Yousefzadeh, M., & Ahmadi Kamarposhti, M. (2024). Wind Energy Conversion Systems: A Review on Aerodynamic, Electrical and Control Aspects, Recent Trends, Comparisons and Insights. *Green Energy and Technology*, (Part F2899), 25–60. DOI: 10.1007/978-3-031-18389-8_2

Zhang, R., Zhu, H., Chang, Q., & Mao, Q. (2025). A Comprehensive Review of Digital Twins Technology in Agriculture. Agriculture 2025, Vol. 15, Page 903, 15(9), 903. DOI: 10.3390/agriculture15090903

Zhang, Z., Liu, X., Zhao, D., Post, S., & Chen, J. (2023). Overview of the development and application of wind energy in New Zealand. *Energy and Built Environment*, *4*(6), 725–742. DOI: 10.1016/j.enbenv.2022.06.009

Chapter 8
Condition Monitoring and Fault Detection With AI and Digital Twin Technologies

R. N. Ravikumar
http://orcid.org/0009-0009-3705-1681
Marwadi University, Rajkot, India

S. Aarthi
http://orcid.org/0009-0006-9064-2091
Marwadi University, Rajkot, India

Shakhboz Meylikulov
Termez University of Economics and Service, Termez, Uzbekistan

C. Navamani
Nandha Engineering College, Erode, India

Bekzod Madaminov
Mamun University, Khiva, Uzbekistan

T. M. Saravanan
http://orcid.org/0000-0001-8746-952X
Kongu Engineering College, Erode, India

ABSTRACT

As the world moves toward renewable energy, wind power stands out as a key sustainable source. However, wind turbines face challenges in reliability due to harsh conditions, mechanical wear, and electrical stress. Traditional maintenance falls short for large wind farms. This chapter explores how Artificial Intelligence (AI) and Digital Twin technologies enable real-time monitoring of blades, gearboxes, bearings, generators, and converters. AI models like Neural Networks, Support Vector Machines, Convolutional Neural Network (CNN), and Long Short-Term Memory (LSTM) help detect faults and predict failures early. Digital twins enhance diagnostics by simulating turbine behavior using real-time data. The chapter also covers hybrid modeling, anomaly detection, and federated learning. Case studies include offshore

DOI: 10.4018/979-8-3373-4159-0.ch008

wind AI, blade crack detection, and converter diagnostics. Challenges include data quality, cybersecurity, and scalability. Future trends point to edge AI, cloud-based diagnostics, and blockchain for secure monitoring and predictive maintenance.

INTRODUCTION

The importance of wind energy has shifted in importance in response to the adoption of clean energy generation by different countries across the globe. This has resulted in the construction of wind turbines that are slap on growing in size globally. These wind turbines are also getting complex and are becoming harder to service because they are increasingly being built offshore. These matters are primarily operational in nature and are proving to be the primary challenge. The problems are also emerging in the context of serviceable dependency and effectiveness. The combination of scheduled fault prevention approach and the reactive maintenance strategy, leads to unplanned downtime and even safety risks. More so, there are unplanned and unaccounted safety risks. The operational use of AI and the rest of the advanced Fourth Industrial revolution technologies such as Digital twins have proven useful in wind energy apparatus optimization. Condition monitoring that drives AI and Digital Twins based strategies plays a critical role by assessing the condition of blades, generators and gearboxes. Blades, gearboxes and generators are known to be some of the most problematic reactive. The turbines are also reactive, meaning the condition of the turbine must also be evaluated. Otherwise, there are significant passive monitoring adversarial repercussions. Anomalies as well as the downward trends of the various units are corrected through the application of IoT and advanced AI models before they occur. The digital twin also augments the process by developing dynamically changing virtual models of turbines under predictive simulations and performance improvement optimization. The lifetime operational risk and cost of a machine are decreased considerably when there is a shift from reactive to predictive maintenance, which is made possible through the application of AI and digital twins technology. This chapter lays the groundwork to grasp the focus of these new technologies innovations towards integrating the monitoring and diagnostics of wind turbine and energy system sustainability (Bassey, 2022b).

Importance of Condition Monitoring in Wind Energy

Keeping track of all these conditions is integrated in new wind technology through monitoring, which helps in these new systems functioning safely, reliably and in a cost-effective manner. The blades, gearboxes, bearings, and generators of a wind turbine are put through extreme fluctuations and rough environmental conditions, which results in extreme, fatigue, wear and breakdowns. Sensors that monitor vibration, temperature, pressure, and noise will identify soft problems that proactive maintenance can deal with. Repairs and identification of minor problems will minimize turbine downtime and maximize turbine life. In wind farms, monitoring has replaced expensive manual inspections which helps in efficient resource planning and reducing wasted energy. In conclusion, condition monitoring provides asset managers with data-based information and helps relay long term goals, reducing operating expenses, and maximizing energy generation (Benbouzid et al., 2021).

Limitations of Conventional Fault Detection Methods

Routine inspections and maintenance checks, fixed limit maintenance, and reactionary maintenance are all forms of technological 'fault' detection of wind turbines. While these 'techniques' are simple and easy to understand, they do not align with the complexity of modern turbines and their wider use and energy demands. 'These techniques' also tend to neglect covert failures that develop over time, which are extremely damaging. There is a lack of machine learning and flexibility within the system to understand and 'recognize' changes in decline that are 'formulated' from a set of historical data. There is also a lack of understanding of the massive volume of data that comes from large wind farms. There is risk of false alarms or faults being missed in the rule-based methods under the influence of noise and interactions within the complex systems, and unnecessary maintenance is performed or maintenance is delayed (Bindingsbø et al., 2023). That is why data-driven, intelligent, adaptive diagnostic models are needed.

Role of AI and Digital Twins in Next-Gen Wind System Reliability

The combination of Artificial Intelligence (AI) and Digital Twin technologies is transforming the future of wind energy systems and changing the strategies towards reliability and maintenance. AI boosts state detection via sophisticated data sorting out, sequence finding, and forecasting. With AI, historical and real-time sensor data can be trained to machine learning and deep learning models to detect fault before it occurs and classify failures, and predict performance, in real-time with no static thresholds or manual fault diagnosis. Digital twins, computerized representations of actual turbines, replicate actual-life performance and conduct behaviour in various circumstances based on live data of Supervisory Control and Data Acquisition (SCADA) schemes and sensor systems. These dynamic models support in real-time performance-based comparisons, fault-based simulations and optimisation of maintenance with minimum downtime interruption. AI and Digital Twins can be used to implement predictive maintenance, adaptive control, and continuous learning mechanisms to turbine supervision, and leave turbine operation paradigm as being reactive to being proactive (Bisu et al., 2024). Such synergy yields better cost-efficiency, energy production and system transparency, and leads to long-term dependability, safety, scalability and sustainability of the global wind energy infrastructure.

FUNDAMENTALS OF CONDITION MONITORING IN WIND TURBINES

Wind turbine condition monitoring is the process of monitoring of critical parts of the turbine- gearboxes, generators, blades, bearings, and converters to identify them at an early stage of failure and facilitate the adoption of predictive practices. Over time mechanical stress, fatigue, corrosion, and misalignment are some of the issues that these components encounter. Sensors sense vibrations, temperature, pressures, rotation frequency, and acoustics, to check the health of a turbine (Bosch, 2022). As an example, imbalance is detected in gearboxes with the presence of vibration sensors and generator heat is monitored with the presence of thermocouples. Early cracks in blade can be detected using acoustic sensor. Raw sensor data, which is frequently incomplete or noisy, should be processed, e.g. by filtering, normalization and interpolation. These actions make data reliable and enable AI-based fault detection in order to reduce downtime and improve the efficiency of wind turbines.

Key Components and Failure Modes (Gearbox, Generator, Blades, etc.)

The wind turbines have major parts that can fail because of constant environmental and mechanical strains. Loss of insulation is a very real consequence of overheating a generator, whilst the gearbox is prone to excessive wear, lubricant depletion and misalignment. Cracks may form on the edges of blades due to fatigue, lightning, or the accumulation of ice. Rotational forces can spin prop shafts and bearing blocks, while the power converter is prone to overheating, nacelle, and yaw fault. The yaw fault is utmost rotatable down tower nacelle part, core and nacelle. Their distinct behavior and possible failure modes serve to tailor target advanced predictive maintenance aids aimed at boosting turbine uptime. In contrast, Table 1 depicts constituent elements, such as the gearbox and generator, in correlation to their failure modes, such as wear and overheating. With a focus on the edges, the table assists readers in making the connection between the most vulnerable components, explaining the need for condition monitoring.

Table 1. Common Wind Turbine Components and Failure Modes

Component	Typical Failure Mode	Impact on System
Gearbox	Gear wear, lubrication failure	Reduced torque transmission, overheating
Generator	Insulation degradation, thermal stress	Reduced power output, electrical failure
Blades	Cracks, delamination	Loss of aerodynamic efficiency, safety hazard
Bearings	Fatigue, corrosion	Increased vibration, system misalignment
Power Converter	Switching faults, overheating	Grid incompatibility, shutdown

Sensor Types and Data Acquisition (Vibration, Temperature, Acoustic)

The monitoring of wind turbine condition is based on strategically positioned sensor network that provides real time data of the operational process. The most common vibration sensors (also known as accelerometers) predict faults such as imbalance, misalignment, or bearing wear, and are usually conveyed on gearboxes, generators, and shafts. Heat, humidity sensors are used to monitor heat accumulation, friction or cooling problems in generators or electronic equipment. Acoustic emission sensors detect cracking of the blade by detecting sound of high frequency due to micro-fractures (Casallas et al., 2023). Torque sensors and strain gauges are used to detect the variations in stress and load of rotating components. Data is sent over SCADA or edge systems, and the acquisition must be synch-ed with high resolutions. Precise choice of sensors, location and calibration is key to proper diagnostics and machinery condition monitoring. The Table 2 Summarizes sensor types (vibration, temperature, acoustic, strain) with their placement, the components they monitor, signal characteristics, and typical fault types detected.

Table 2. Sensor Types and Their Monitoring Applications

Sensor Type	Monitored Component	Measurement Parameter	Detected Faults
Vibration Sensor	Gearbox, Bearings	Acceleration, Frequency	Imbalance, Misalignment, Bearing Wear
Temperature Sensor	Generator, Converter	Heat	Overheating, Friction, Insulation Breakdown
Acoustic Sensor	Blades	Sound Waves	Crack Propagation, Delamination
Strain Gauge	Blades, Shaft	Deformation	Structural Stress, Fatigue
Current Sensor	Generator	Current Flow	Harmonic Distortion, Load Imbalance

Data Challenges: Noise, Imbalance, and Missing Values

Data-driven condition monitoring in wind turbines is encountered to have key challenges notwithstanding the improvement of sensor technology. Noise in an environment, electrical and mechanical vibrations may cause the signal to be distorted resulting in false positive or negative results. Supervised learning is skewed towards an unbalanced data, where only healthy states are present, and early defects are omitted (Dai et al., 2018). The absence of values caused by the lack of communication or sensor failure, and other issues ruins continuity and makes diagnostics hard. The inaccuracy of modelling is influenced by discrepancies in scale, units, and frequency in between turbines. The only solution will be robust preprocessing filtering, normalization, smoothing, and augmentation. It is of prime importance that good data handling is necessary to build quality yet scalable AI models to monitor the condition of wind turbines and predict maintenance.

AI TECHNIQUES FOR FAULT DETECTION

Artificial Intelligence (AI) introduced adaptive and data-driven diagnostics, and it has changed wind turbine fault detection. In contrast to the rule-based approaches, AI models process historical and real-time measurements as well as identify sophisticated patterns of degradation and failures to predict their occurrence with high precision. Other methods can be based on machine learning, such as Support Vector Machine (SVM), Decision Trees, Random Forests, and Gradient Boosting of normal and faulty states based on vibration, temperature, and acoustic characteristics. The deep learning technologies, including Convolutional Neural Network (CNNs) and Long Short-Term Memory (LSTMs), demonstrate high performance when dealing with time-series data to capture spatial and temporal features. In unlabelled identification of faults, autoencoders and clustering are useful. Such advanced methods as transfer and online training help extend the model ability, increasing reliability and minimizing off-line times of the turbines. The Table 3 Compares AI methods (SVM, Decision Trees, CNN, LSTM, Autoencoders) in terms of input type, learning method, use case (classification/anomaly detection), advantages, limitations, and suitability for wind turbine diagnostics.

Table 3. Comparison of AI Techniques for Fault Detection

AI Method	Input Type	Use Case	Strengths	Limitations
SVM	Feature Vectors	Binary/Multi-class Classification	High accuracy for small datasets, good generalization	Sensitive to feature scaling, less effective with noise
Decision Trees	Feature Vectors	Classification/Regression	Easy to interpret, low computational cost	Prone to overfitting with complex data
Random Forest	Feature Vectors	Fault Classification	Robust to noise, handles feature interactions well	Less interpretable, may require more computation
CNN	Images/Spectrograms	Pattern Recognition in Signals	Automatically learns features, high accuracy	Requires large datasets, high computational cost
LSTM	Time-Series Data	Sequential Pattern Analysis	Captures temporal dependencies, suitable for trends	Complex architecture, slow training
Autoencoder	Time-Series/High-Dim Data	Anomaly Detection	Learns compressed representations, unsupervised	Hard to tune thresholds, interpretability is limited

Figure 1. Performance Comparison of AI Models for Fault Detection

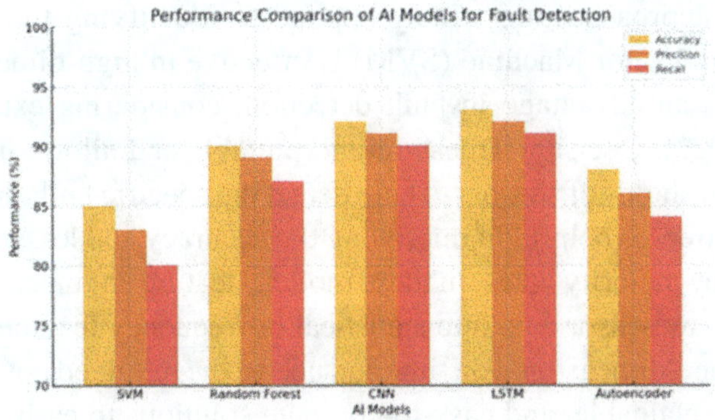

Figure 2. Fault Prediction Accuracy Over Time (Model Comparison)

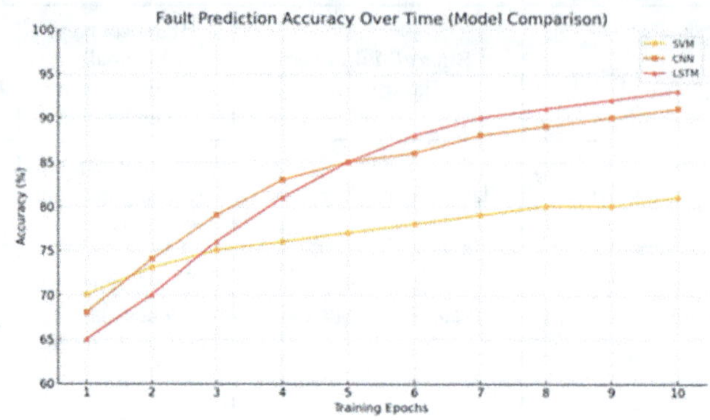

The above Figure 1 Compares the accuracy, precision, and recall of different AI models (e.g., SVM, Random Forest, CNN, LSTM, Autoencoder) in wind turbine fault classification tasks. The above Figure 2 Tracks how different AI models (SVM, CNN, LSTM) improve in fault prediction accuracy over multiple training iterations or as more sensor data is collected.

Machine Learning Methods (SVM, Decision Trees, Random Forests)

Due to the possibility to identify patterns by using historical data, the Machine Learning (ML) approaches are widely applied to identifying the faults in wind turbines. Support Vector Machine (SVM) is effective in high-dimensional binary learning and has an advantage in fault detection, considering extracted features (Garan et al., 2022). Decision Trees are interpretable and allows input data to be divided using thresholds of sensors to help detect the specific fault. Random Forest, ensemble of the trees is helpful in improving the accuracy and lessening overfitting that is achieved by majority votes. In these models, feature engineering is necessary in transforming raw sensor data into statistical or frequency features. Even though ML models are one-dimensional on time and need to be retrained now and then, they offer scalable, explainable, and easy-to-compute solutions to early fault diagnosis in SCADA or edge systems.

Deep Learning Applications (CNN, LSTM for Time-Series Data)

Deep learning models have already had a huge impact on wind turbine fault diagnosis, as the type of feature learning required to diagnose the fault can be done automatically using raw sensor data. CNNs are also successful in analyzing vibration spectrograms, thermal images, and time-frequency representations to extract spatial patterns and local anomalies, e.g. blade or gear faults, even in noisy environments (González et al., 2023). Recurrent Neural Network (RNN) Long Short-Term Memory (LSTM) networks are appropriate in modelling rotating components time-series data due to its capability to identify slow-developing faults and predict anomalies. Compared to classic ML, deep learning does not involve any manual feature engineering, but the inputs must be high-dimensional and the computation intensive while hyperparameters have to be optimized with care. Nonetheless, with multi-sensor, high-resolution, and multimodal data, CNNs and LSTMs can be counted among the methods of present-day real-time monitoring and predictive maintenance within wind energy systems.

Anomaly Detection and Classification

The ability to detect anomalies in real-time data plays an important role in the fault detection of wind turbine, particularly in cases where labelled failure data is scarce or inaccessible. Such methods detect abnormal trends in sensor data which is outside the scope of normal operations. Unsupervised-based approaches such as K-means clustering, Isolation Forests, and One-Class SVM will model the normal data and alert them of any outliers as possible anomalies. Deep learning Autoencoders rebuild input data and mark instances with large reconstruction error as anomaly (Hwang & Yoo, 2021). They perform prominently in high-dimensional and time-series data and are able to pick on minor deviations which indicate early errors. When anomalies have been detected it can use supervised classifiers in case labelled data has been provided. The hybrid method that uses both anomaly and classification have the advantages of high sensitivity and specificity, thus suitable in real-life wind turbine applications where the timely detection of faults becomes a priority but the false alarms and the needless maintenance should be reduced to a minimum.

Transfer Learning and Online Learning Models

Transfer learning and online learning could provide great improvements in fault detection in wind energy systems, where the data on labelled faults is scarce and where turbine operating regimes may differ. The transfer learning enables models trained on one on a particular turbine or dataset to be applied to another using previously gained knowledge to enhance its performance with only little additional fresh data. As an example, a deep learning model developed on a gearbox vibration of one wind park can be tuned to another turbine working in other conditions (Mahaveera & Pasuk, 2021). Online Learning means that constant model updates can be made upon the new appearance of data, which is ideal to work in the changing environment. Real-time changing algorithms such as Online Gradient Descent and Adaptive Boosting are algorithms that do not change accuracy of detection. These learning paradigms, which are coupled with edge computing and SCADA integration, enables scalable, responsive and low latency to monitor the dynamic nature of distributed wind farms.

DIGITAL TWIN TECHNOLOGY

Digital Twin technology allows making a virtual copy of wind turbines functioning in real time which is used to simulate and control performance and optimize it. It combines real-time data in the sensors with computational models that represent the current state of the turbine as well as its future behavior and possible failures (Masoumi, 2023). It simulates the stress, dynamic vibration and component degradation through the coupling of the physical properties, environmental conditions, and operational data. Real and simulated sensor values can be used to identify faults early on which has the effect of maintaining predictive maintenance and minimizing denouncement. Superior digital twins can involve multi-physics modelling, SCADA integration, and AI-powered activities such as virtual fault injection, root cause analysis, and performance optimization of onshore and offshore wind farms. The Figure 3 showing the structure of a digital twin, including the physical system, data layer (sensors/SCADA), AI models, simulation layer, and feedback loop.

Figure 3. Digital Twin Architecture for Wind Turbines

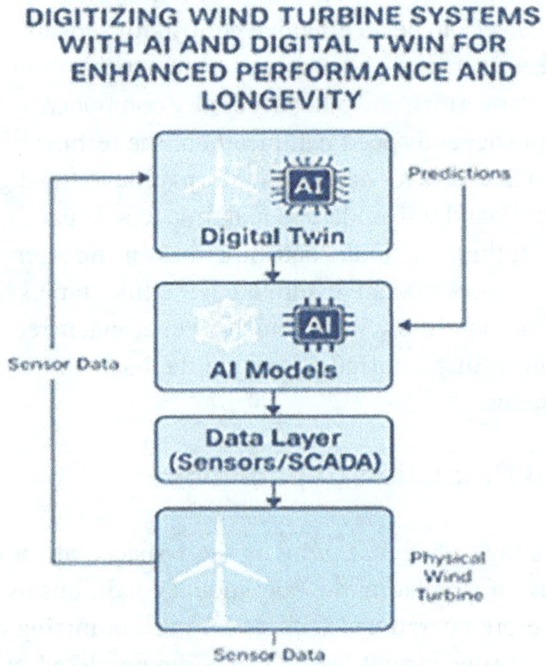

Concept and Architecture of Digital Twins

A digital twin is an actual representation, in a virtual environment in real-time, of how a physical wind turbine looks, acts and performs. The components of its architecture are generally taken to be three-layered, comprising a physical layer (the turbine components and the sensors), a virtual model layer (physics-based and data-driven, simulations), and a data exchange layer (real-time data connectivity) (Prabhavathi et al., 2023). This hierarchical arrangement would allow a smooth synchronization of the actual turbine and the computer representation. The twin itself is constantly being developed as it consumes operating data like that of vibration, temperature, torque and wind speed as it becomes optimized to the current conditions of the turbine (Sethi et al., 2023). Its support to lifecycle tracking, risk analysis and modular diagnostics is where its core lies, as the basis of predictive maintenance and smart wind farm management.

Integration With Physical Systems and Real-Time Data

In order to have an effective combination of digital twin and physical wind turbine systems, there needs to be an effective network of sensors and communication infrastructure. The sensors must be mounted on the primary components of turbine gearboxes, blades, converters, and other accessory components to monitor temperature, vibration, acoustic, and speed parameters of the turbine blades. The SCADA systems or the IoT Gateways forward such information to the digital twin which, in turn, is able to modify and self-update in real time to account for the present health and behavior of the turbine. APIs and data pipelines ensure interoperability between edge or cloud platforms. Associative blinking with time stamps helps in monitoring the history and operation of the system, which is beneficial in recognizing anomalies or pre alert warnings. Such configurations are the basis for anticipatory, flexible, and intelligent systems.

Simulation and Prediction Capabilities

The main advantage of digital twins is the f spacecraft, a revolutionary tech feature, the promise of simulating the operational conditions using for eliminating system response before operational failures for pre-optimizing conditions. Digital twins simulating real-time virtual reality in a computer, like Load increase, system Load increase, Wat, system burn out, and Breather Gush, speed set the gauges to cool the controls. These kinds of, Finite Element Dynamic Analysis of components to keto components (Sikdar et al., 2024). The AI and machine learning algorithms increase predictive capabilities by studying the performance trends and predicting the Remaining Useful Life (RUL) of the critical components. The virtual environment enables engineers to introduce faults, test design interventions or optimize control strategies in a way that it leaves physical assets intact. Digital twins enable proactive preventative schedules, energy production optimisation, and cost-effective asset management by predicting faults and performance losses prior to their occurrence. Digital twins can be a potent tool in lowering downtime and increasing system reliability due to the possibility to merge live data and physics-based modelling.

Multi-Physics and Hybrid Modelling Approaches

High-fidelity digital twins are realistic since it involves multi-physics models to simulate mechanical, thermal, electrical and aerodynamic interactions within the components of the wind turbines. As an example, a gearbox twin can be at the same time simulation of gear friction (mechanical), lubricant temperature (thermal) and vibration dynamics. Such simulations are realistic with respect to loads and behavior (da Silva et al., 2025). To be more accurate, hybrid modelling also improves accuracy by integrating physics-based models with data-driven models such as machine learning. An example is that a physics-based torque model can be dynamically tuned with the use of AI to suit actual performance. Through this integration, the twin can learn, adapt and improve over a long interval, and can capture nonlinear behaviors and inter-domain effects as well as rare faults that tend to be overlooked by a standalone model. The multi-physics and hybrid digital twin are therefore the means to perform more realistic simulations, smart diagnostics, and prognosis of the wind turbine system.

AI + Digital Twin Integration for Predictive Maintenance

Powerful Wind power predictive maintenance can be achieved using AI and Digital Twin integration. Digital twins are real-time models of turbines, and AI learns models over historical and streaming sensor data to diagnose anomalies, predict which fuel cell components will fail, and optimize consumables. LSTM, autoencoder, and ensemble classifier are some of the techniques used to predict a failure as well as calculate the Remaining Useful Life (RUL) of critical parts of the vehicle like gearboxes and bearings (Udo et al., 2023). This synergy facilitates injection of virtual fault and modelling of behavior of the machine without operation disturbances. The constant feedback loop between simulations, AI insights, and physical data allows the maintenance needs to be met in time and provides cost-effective maintenance service and scalable, cloud-based monitoring of wind farms. The figure 4 that visualizes how AI models interact with digital twins for tasks like fault prediction, virtual simulations, and triggering predictive maintenance. The Figure 5 Compares maintenance-related downtime in turbines using traditional methods vs. digital twin-enabled predictive maintenance over months or years.

Figure 4. AI and Digital Twin Integration Workflow

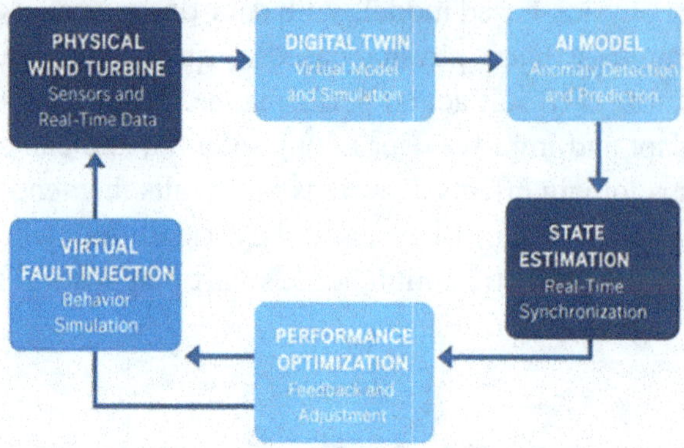

Figure 5. Digital Twin vs. Traditional Maintenance – Downtime Trend

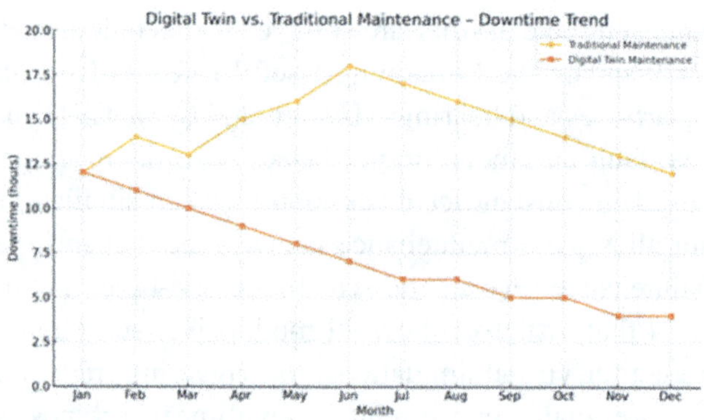

Real-Time Synchronization and State Estimation

To be able to maintain an effective predictive maintenance, real-time synchronization between the digital twin and the wind turbine is essential. It is accomplished by non-stop covering data of in-built sensors that allow the digital twin to recreate the work reality of the turn-out in near-real-time. With the application of AI-based state estimation approaches, the system may draw conclusions about the hidden or partially observable states that cannot be directly measured with its sensors, including internal gear wear or thermal stress. In this estimation, the Kalman filters, neural networks, or Bayesian models are used, which can absorb the data to anticipate the present and upcoming conditions of the turbine. Real-time synchronization allows stretching the digital twin dynamically, and it can evolve to degradation trends and anomalies in operations in real-time (Uhanto et al., 2024). Reasonably accurate state estimation also enables the prediction of faults early, and it facilitates maintenance decisions using risk principles. This integration is therefore a main driving force towards developing proactive and affordable maintenance strategies of wind farms systems because it makes their system models more reliable due to their boosted predictions.

Virtual Fault Injection and Behavior Simulation

One of the most impressive uses of digital twin augmented by AI, virtual fault injection enables engineers to model the behavior of a wind turbine across a range of conditions under which the turbine may fail without bringing the real system to jeopardy. Artificially injecting modelled faults (e.g., bearing misalignment, gearbox tooth cracks) into the digital twin, one can see how it responds to stress, shift vibrations and thermal change and react to control manipulation in a virtual risk-free environment (Kong et al., 2023). AI models can assist in using data on the past failures to build up the fault realistic signatures, which are in turn conducted to assess the robustness of operational strategies and break point of the sensors. Prediction of this simulation contributes in detecting the initial symptoms of certain failures and optimized detection thresholds. Also, it is possible to prove new maintenance procedures in virtual realities, check changes in designs, or even teach fault recognition programs in very many cases. In the end, such virtual fault injection has increased fault readiness, decreased the risk in operations, and allowed smart model optimization in the wind turbine maintenance processes.

Feedback Loops for Performance Optimization

AI-based digital twins will create the loops of feedback that will perpetually improve both the performance and decisions made regarding the maintenance of turbines. As soon as the real-time sensor data goes through AI models and analysis, its results, i.e., health scores, the estimates of faults, or best choices of control effort, are then fed back into the digital twin and the physical control frameworks. This feedback loop can be used to tenably and dynamically tune the performance (by changing the operational parameters such as yaw angle, blade pitch) or perform maintenance alerts on the basis of any degradation trends that are arising. The AI can enhance its algorithms and gradually learn through any discrepancies between what was predicted and what it actually happened by comparing predicted behavior with the actual behavior of the turbine (Lebranchu et al., 2019). Such self-learning feedback systems will be necessary in autonomous or semi-autonomous maintenance systems. Furthermore, applied to a wind park with many turbines, combined knowledge can be utilized to balance loads, prioritise maintenance and optimise energy production at a large scale. This leads to an iterative improvement system that minimizes service interruptions, maximizes master use of resources, and allows long term sustainability objectives in wind energy infrastructure.

IMPLEMENTATION FRAMEWORK

The usage of digital twins' system on wind turbine maintenance using AI needs orderly hierarchical design of combining sensing, communication, computation, and learning sequent. At the hardware level, sensors over the critical turbine parts with high frequencies (vibration, temperature, acoustic, strain) are installed and provide information in SCADA in real-time. This information is communicated on standardised, low-latency protocols (e.g. MQTT, OPC-UA) to edge or cloud computing units. The distributed cloud will be able to analyze models and updates, the long-term storage process, and the edge AI will be able to process the local rapid process (Odgaard & Stoustrup, 2013). The AI engine reads information, defines anomalies, forecasts fault and facilitates digital twin simulations. Intelligent scheduling of maintenance can be sent by integrated feedback control. The architecture provides scalability, cybersecurity, interoperability and good lifecycle management in wind farms. The Figure 6 showing data flow from sensors → SCADA → edge/cloud → AI modules → decision-making → turbine feedback. Emphasizes real-time data processing and learning.

Figure 6. Implementation Framework for AI-Powered Condition Monitoring

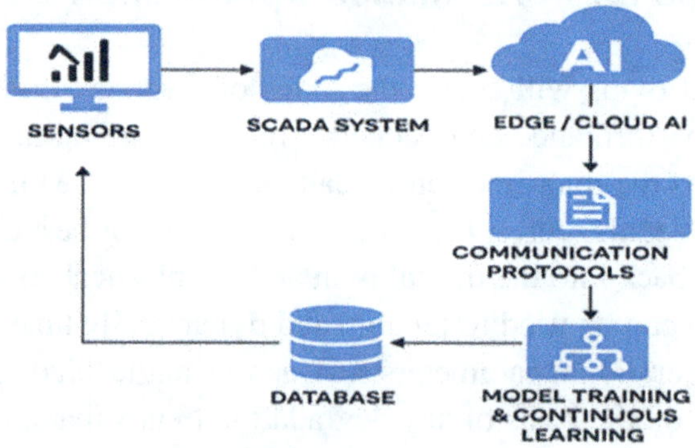

System Architecture: Sensors, SCADA, Edge/Cloud AI Modules

The AI-accommodated digital twin frameworks to maintain wind turbines rely on a solid system structure. The architecture starts with the sensor layer and the architectural blocks: the vibration, temperature, strain, and acoustic sensors and are installed on a critical part such as blades, gear boxes, and generators. The data collected on these sensors is in real-time and can be integrated by the SCADA (Supervisory Control and Data Acquisition) system which can be considered the central point of collecting the operational data and monitoring remotely. The Figure 7 illustrates the data flow from turbine-mounted sensors to a SCADA system, which interfaces with edge and cloud AI modules for real-time monitoring and decision-making. It highlights the layered structure essential for implementing AI-driven condition monitoring in wind energy systems.

Figure 7. System Architecture: Sensors, SCADA, Edge/Cloud AI Modules

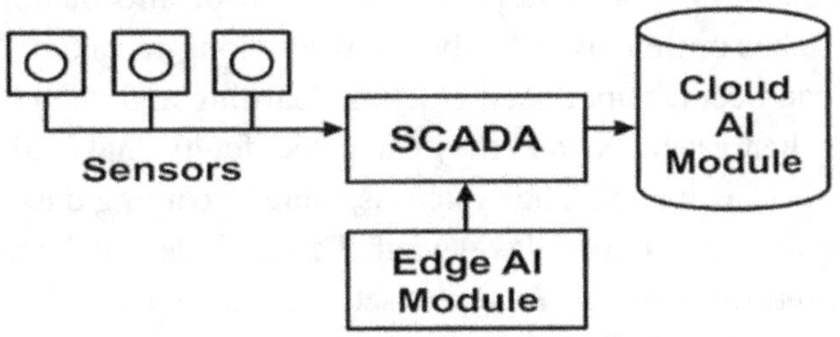

Anomaly alerts and fault notifications with minimal delays is facilitated because SCADA data is brought close to a turbine and relayed to Edge AI modules. If a turbine is configured to serve a demanding load and train deep learning modules or simulate digital twins, serverless Edge AI capabilities in the field can reach EU-based cloud AI modules seamlessly. Thanks to this edge-cloud heterogeneous design, instantaneous response to on-site diagnosis is possible, while extensive cloud-based diagnostics on the entire fleet can still be performed (Sanchez et al., 2015). The distributed arrangement enhances resilience of the system, minimizes bandwidth expenses and facilitates scale predictive maintenance.

Communication Protocols and Interoperability

Data communication is needed for the syncing of the turbine twins, physical and digital. System uses general communication protocols. like MQTT, OPC-UA, Modbus and SCADA, to send sensor, SCADA and edge or cloud data to the analytics modules. Such systems can deliver data for real-time analysis with very low latency, and during data spikes which are normal for multi sensor wind turbines. Interoperability is a core design principle aimed to facilitate seamless functionality of the diverse hardware and software systems. IoT gateways or middleware platforms enable the translation across different systems and devices. This will ensure the turbine supplier, sensor, and cloud provider will work together (Shi et al., 2024). The presence of networks of interoperable APIs and modular microservices-based structures facilitates the application of AI models on diverse wind plants with minimal reconfiguration. They also possess robust cybersecurity controls like access control and end-to-end slip fire cryptography to defend valuable operation data. Standardized and interoperable set of communication provides systems with scalability, flexibility, and secure system-wide cooperation.

Data Storage, Model Training, and Continuous Learning

Data storage and processing pipelines have an essential role in training accurate AI models and operations of digital twins. Raw sensor information is archived to large scale data lakes or time over databases, where long-term trending and system benchmarking can occur. Supervised machine learning and deep learning models are trained using historical data and they can detect faults, make failure predictions and simulate turbine behavior. The system is using streaming data to execute continuous learning to keep the models relevant. This includes the learning algorithms in the online or retraining of the model at set intervals, so that the AI will keep to the changes in the turbine behavior or the changes in the environment. Lightweight real-time inference happens at the edge, and high throughput data analytics and retraining flows and model versioning occur in the cloud (Takanashi et al., 2021). Feeds back in the field operations as e.g. logs on maintenance or a confirmation of a failure is fed back to the learning cycle to improve the accuracy of the learning. Such a feedback loop learning mechanism does not only increase the diagnostic accuracy but also develops the digital twin over time enhancing its intelligence and contextuality.

PRACTICAL IMPLEMENTATION CONSIDERATIONS

Effective implementation of AI and Digital Twin systems to monitor wind turbines is a crucial issue that cannot be overlooked on technical and operational levels. On the hardware side, there is a choice of sensors (vibration, temperature, acoustic) to be installed and positioned in an optimal manner to capture the right data. Software integration should be compatible with the current SCADA systems, should be able to integrate their APIs strongly, and should be able to synchronize edge and cloud platforms to keep in real-time process. Secure authentication and encryption capability should be used to secure operational data. On the financial side, initial investment, and the estimated savings in terms of downtime, maintenance expenditures, and energy losses should be evaluated by means of a cost benefit analysis. The workers must also carefully be prepared they must learn how to respond to AI driven recommendations and alerts promptly. Finally, the development should be able to extend to multiple turbines and sites without overly excessive Medi-engineering, leading to adjustable, durable, and elongate operations in the future.

CASE STUDIES AND APPLICATIONS

Wind turbine maintenance that integrates AI and digital twins is efficient, per case studies performed in the real world. With AI, predictive maintenance has been accomplished for offshore turbines gearboxes using vibration and torque data, thus minimizing unplanned downtimes. In addition, blades with cracks are detected using deep learning classifiers alongside acoustic emission sensors. Generators and converters digital twins have made it possible to detect thermal peculiarities and electrical maladies at an early stage (Ulmer et al., 2020). Also, twins integrated with cloud have been applied in optimization of energy yield, providing real-time performance details as well as control variations. Such applications illustrate the advantage of AI-powered digital twins in enhancing reliability, cost reduction and operational efficiency in the wind farms. The Table 4 Provides an overview of real-world case studies, including application area (e.g., gearbox, blades), techniques used (AI + Digital Twin), turbine type (onshore/offshore), outcome achieved (e.g., RUL prediction, failure detection).

Table 4. Case Studies of AI and Digital Twin Applications in Wind Energy

Application Area	Technique Used	Turbine Type	Outcome Achieved
Gearbox Failure Prediction	Vibration Sensors + SVM	Offshore	Early detection of gear wear; reduced unscheduled downtime
Blade Crack Detection	Acoustic Sensors + CNN	Onshore/ Offshore	Accurate identification of micro-cracks; improved safety
Generator Fault Diagnostics	SCADA + Autoencoder	Onshore	Detection of overheating and harmonic distortion
Converter Monitoring	LSTM + Digital Twin Simulation	Offshore	Forecast of converter failures; thermal anomaly detection
Energy Yield Optimization	Digital Twin + Reinforcement Learning	Onshore/ Offshore	Real-time performance tuning; increased energy output

Figure 8. Impact of AI Applications Across Wind Components

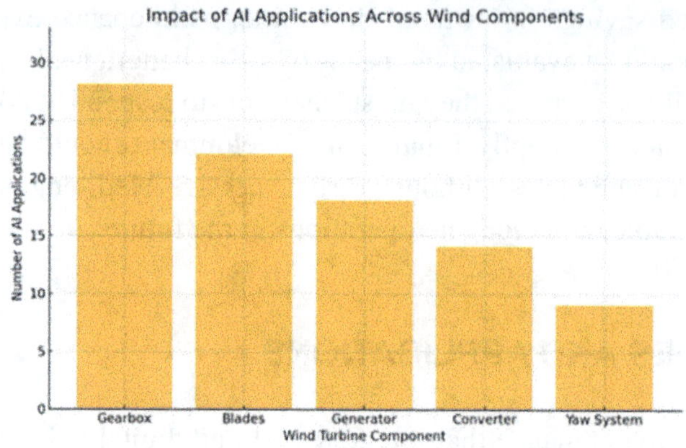

The Figure 8 Visualizes the number of successful real-world applications per component (e.g., gearbox, blades, generator, converter) using AI and digital twin technologies.

Figure 9. Energy Output Before and After AI Optimization Deployment

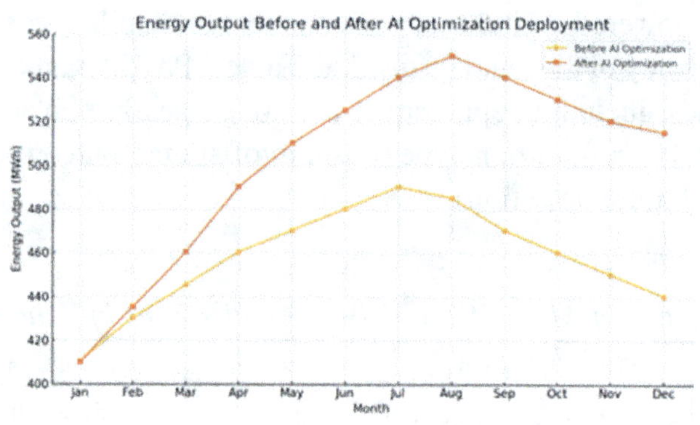

The Figure 9 Plots the monthly or yearly energy output of a turbine/farm before and after implementing AI-based optimization strategies (e.g., pitch/yaw control tuning).

Gearbox Failure Prediction in Offshore Turbines

High humidity, exposure to salts, and varying load on the wind turbines put offshore wind turbines under extremely adverse environmental conditions, and gearbox failure prediction is among the highest priorities. One of the most failure prone components is the gear box, because of its complexity and mechanical nature and constant loading fluctuations. In order to counteract this, AI algorithms have been used on the high-frequency vibrations data and SCADA parameters of offshore turbine gear boxes (Vives, 2022). Random Forests and SVM are trained machine learning models that are used to discover the early evidence of wear on gears, breakage of lubricants, and rotting of bearings.

Combined with digital twins, these models allow making predictions about the behavior of the gearbox during its operation with different loads. Instant synchronization between the physical turbine and the twin allows estimation of stress, torque fluctuation and remaining useful life (RUL) to be accurate. With such configuration, it becomes possible to do predictive maintenance which can reduce unplanned downtimes and huge costs of replacements. This was seen in practice where operators claimed up to 30% in increased ability for fault detection accuracy and far less in emergency maintenance actions. Such proactive ability augments not just reliability but also directly impacts the availability of the offshore turbines which is crucial in situations where physical access to the component is access limited and expensive (Wang et al., 2024). In the context of growing offshore wind farm scales, these increasingly predictive systems are indispensable in environmentally sustainable and cost-efficient operations. Vibration data from an offshore wind turbine gearbox is analyzed in the SVM (Support Vector Machine) model featured in Figure 10 to determine likelihood of mechanical failure. The SVM model is a subcomponent of the framework that facilitates advanced predictive maintenance and early fault detection.

Figure 10. Gearbox Failure Prediction in Offshore Turbines

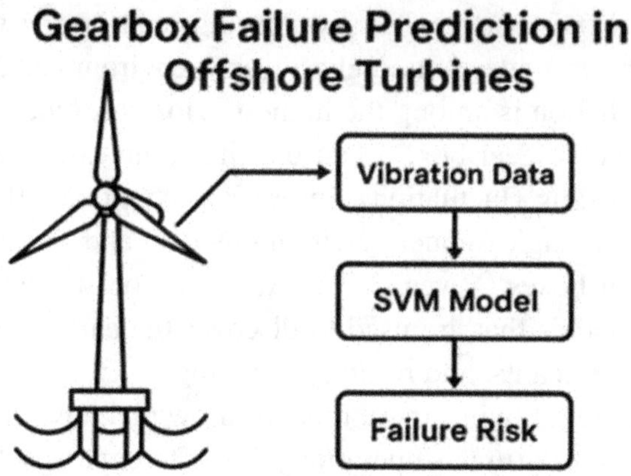

Blade Crack Detection Using Acoustic Analysis

The permanent aerodynamic loading on the blades of wind turbines causes progressive wear out and material fatigue as well as structural cracks. The necessity to detect blade cracks early is essential since they may affect an alarming degree on the incident of aerodynamic efficiency and it may cause tragic failures in case it is not addressed shortly. Acoustic emission (AE) analysis with AI has proven an effective method to monitor health of blades in real-time and without the use of invasive measures. High-frequency acoustics that are produced by micro-cracks or delamination events are recorded by measuring the sensors attached to the blade surface. These cues are usually weak and get lost in the ambient clatter and thus are hard to detect manually (Xiao et al., 2021).

Convolutional Neural Networks (CNNs), or any deep learning model, is trained using labelled AE dataset in order to discriminate between normal structural noise and cracks-induced abnormalities. Spatial and frequency patterns can be learned by these models through raw waveforms, and this is how an accurate prediction on the degree of severity and location of the crack can be achieved. Combined with a digital twin of the blade, the system would allow simulating structural stress distribution and modelling how detected faults would develop under different wind circumstances. The maintenance teams can then be alerted in advance and it can carry out scheduled interventions instead of emergency shutdown. The technique will greatly cut down on the costs of inspection, enhance turbine availability, and safety of operations mostly of offshore and onshore turbines in remote and hard to reach locations. The figure 11 illustrates the process of detecting cracks in wind turbine blades using acoustic sensors. Sound signals captured from the blade surface are processed through a Convolutional Neural Network (CNN), which classifies the signal to identify structural cracks, enabling early fault detection and preventive maintenance.

Figure 11. Blade Crack Detection Using Acoustic Analysis

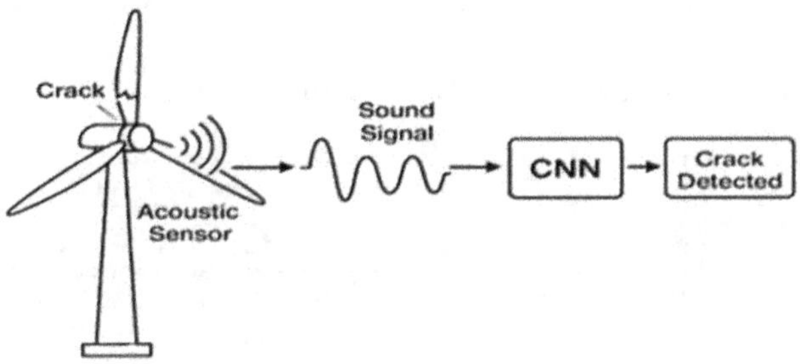

Generator and Converter Fault Diagnostics

A generator and a power converter wind up the major components of a wind turbine responsible of converting mechanical energy to electric and conditioning it to the appropriate grid format. These components are particularly vulnerable to thermal overstress, insulation breakdown, voltage rises, and the switching puzzler, significantly affecting efficiencies and culminating to a total breakdown. Such occurrences must be diagnosed early to guarantee uninterrupted energy production warranty (Ahmed & Asamoah, 2024). Based on SCADA data and embedded thermal sensors, AI models, including decision trees, LSTM networks, autoencoders are being used to develop and train to monitor changes in generator temperatures, current harmonics, and converter switching patterns that represent operational abnormalities. Such models are capable of recognizing such abnormal behaviors as the overheating, voltage imbalance, or harmonic distortion before the component fails.

There is also the identification of soft changes in the operation baselines, made possible with unsupervised learning methods, which aid in the indication of undetected or infrequent fault patterns. The generator-converter system has a digital twin that allows the virtual testing of load situations and trans. Simulations can be used to forecast the reaction of a system to external stresses, including gusty winds or grid-impairments, to increase fault tolerance within a system. Through AI-assisted diagnostics, virtual model-based proactive maintenance enables losing or fine-tuning faulty subsystems before the malfunction destroys the function. Such combined method guarantees an increase in the quality of the power, decreases the chance of having grid disconnections and the extension of the lifetime of some important electrical components. The figure 12 illustrates the diagnostic workflow for detecting faults in wind turbine generators and converters. SCADA data is analyzed using an autoencoder model to learn normal operating patterns. Deviations from this learned behavior are flagged as anomalies, enabling early detection of electrical or thermal faults in the generator-converter system.

Figure 12. Generator and Converter Fault Diagnostics

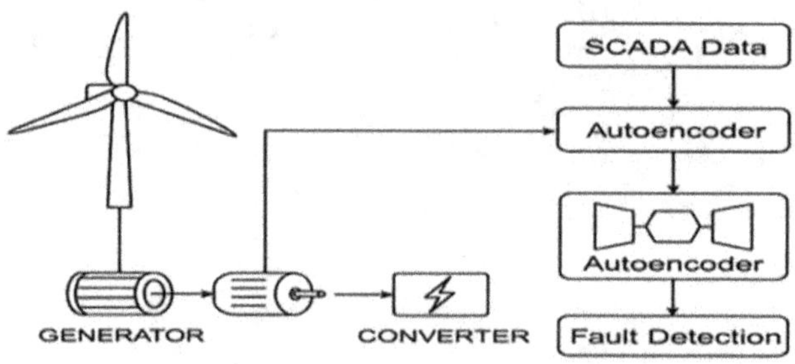

Energy Yield Optimization Through Real-Time Insights

Optimal energy production is an important goal of the operators of wind farms, and digital twins integrated with AI provide an effective technique of continuously scouting wind farms hour by hour to maximise energy output (Avci, 2024). The AI algorithms will detect wind turbine efficiency shortcomings by continuously tracking the wind speed, blade pitch angle, yaw alignment and turbine load. Such inefficiencies tend to emerge because of inappropriate settings of control, incomplete faults or dynamic environmental fluctuations. Digital twins can model the turbine functioning under both theoretical and operational conditions and make a comparison between the two to be able to make changes to it in real-time. The blade pitching and blade yawing mechanisms can be dynamically and fine-tuned with reinforcement learning algorithms or adaptive optimization models to ensure that the turbine is always running at its power coefficient, given the changing wind profile.

In addition, information received by many turbines can be centralized into a central platform to coordinate the whole farm load potential to limit wake losses and thus increasing overall capacity factor. Past records of performance are also used to predict the possibly losses that might be experienced because of seasonal changes or wear and tear. Using performance improvement and optimizing maintenance in tandem would help operators have less downtime while maximizing energy production and making more profit (ROI). This framework follows new goals of sustainable performance where real-time AI and digital twins have operational feedback loops on effectiveness and energy long-term production planning (Bassey, 2022a). Figure 13, for instance, shows how a cloud system analyzes real-time information from a wind turbine for optimization opportunities. These learned optimization strategies are then integrated into the control system which causes the turbine's energy and overall performance to keep getting improved over time.

Figure 13. Energy Yield Optimization Through Real-Time Insights

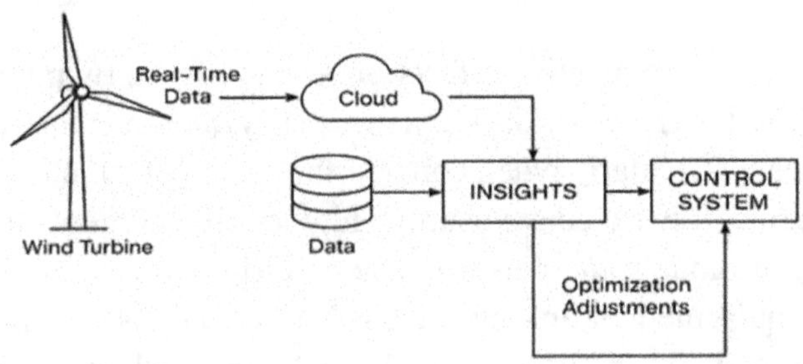

INTEGRATION CHALLENGES AND RISK MITIGATION STRATEGIES

The application of AI and Digital Twin technology in the wind power system poses some integration issues, which would require solutions to ascertain stability and performance. The main concern is associated with the system interoperability since wind farms are likely to use heterogeneous sensors, SCADA platforms, and AI modules of various vendors, and they might not talk to each other efficiently. Gap between interfaces, and between middleware solutions and standardized protocols can be closed by adopting standardized protocols and middleware solutions. The issues of inaccuracy related to noise, missing, or unbalanced data sets pose the problem of data reliability resulting in inaccurate diagnostics. Such problems can be alleviated through robust preprocessing, data augmentation and redundant sensing. Cybersecurity risks such as malicious activity, data corruption, and disruption of their operations must be met with encryption, secure authentication, intrusion detection tools, and logging which is tamper-proof using blockchain. Operational risks include the over-dependence on totally automated systems deprived of the human element; hybrid AI-human decision-making systems offer a counterbalance of preventing false alarms or failing to identify faults. Also, over time, model drift may erode the performance of AI systems that may have changed in the environment or the turbine is operating differently. Being aware of this risk, they can be minimized by constant monitoring, on line training and retraining of models on a schedule. Proactive risk mitigation strategies to solve the mentioned challenges will make AI-Digital Twin integration not only sound in its technical aspect but secure, adaptive, and sustainable to long-term operations of wind energy.

CHALLENGES AND FUTURE TRENDS

Even though there are positive effects of AI and digital twin integration, some challenges still exist. Privacy of data, cybersecurity, and explanation of the model are serious issues, particularly when considering diagnostics in the cloud, and distributed wind farms. Restricted exposure to high quality content labelled fault data hinders training of models and performance on turbine types (Besigomwe, 2025). Moreover, the requirement of linking various heterogeneous systems and maintaining the option of real-time response is still a technical challenge. The trends of the future are edge AI implementation in low-latency processing, blockchain in secure fault logging, and self-evolving models, adaptive learning. New Explainable AI (XAI) and scalable hybrid modelling will improve trust, transparency, and reliability of smart wind energy systems even more. The Table 5 Lists major technical and operational challenges (e.g., data privacy, model explainability) and maps them to proposed solutions and future trends such as blockchain, XAI, edge AI, and adaptive learning models.

Table 5. Key Challenges and Emerging Trends in AI-Powered Wind Monitoring

Challenge Area	Description	Emerging Solution/Trend
Data Privacy & Cybersecurity	Risk of data breaches, unauthorized access, and system manipulation	Blockchain for secure data logging; encrypted SCADA
Model Generalizability	AI models struggle to perform consistently across different turbines or sites	Transfer learning, domain adaptation techniques
Model Explainability	Black-box nature of deep learning limits trust and acceptance	Explainable AI (XAI) to enhance interpretability
Real-Time Processing	Delays due to cloud dependency for analytics and fault detection	Edge AI for low-latency diagnostics at turbine level
Data Quality & Availability	Incomplete, noisy, or imbalanced datasets hinder model training	Data augmentation, synthetic data generation

Figure 14. Top Challenges in AI-Based Wind Monitoring (Industry Survey)

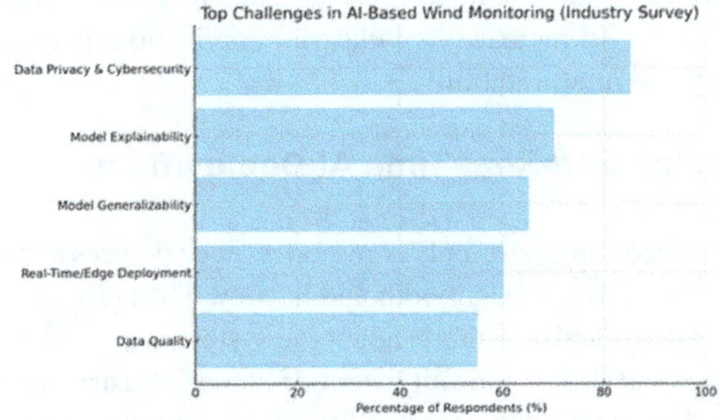

The Figure 14 Shows ranked challenges (e.g., data privacy, explainability, generalizability, real-time deployment) based on frequency or severity reported in recent studies.

Data Privacy and Cybersecurity Concerns

Wind turbines will rely more frequently on sensor networks, edge computing, and cloud computing, so it is reasonable to ensure the safety of operational data. Monitoring and control processes can be disrupted because of cybersecurity threats of unauthorized access, data breaches, or manipulation of SCADA systems. Also, policies regarding data privacy demand that the data about turbines and about operators be treated safely, particularly in farms around the world. It is necessary to implement end to end encryption schemes, secure authentication schemes, and intrusion detection systems (Chiang-Guizar et al., 2025). Adversarial attacks should also not be able to affect AI models. To foster and guarantee the continuity of operations, adequate cyber-physical security architectures are required in facilitating the scalable deployment of AI-based condition monitoring systems.

Model Generalizability and Explainability

AI models specializing in learning a given type of turbine or data set may not perform generally in other kinds of turbines or geographical spreads or operating conditions. This low scalability is caused by the low generalizability, which can lead to wrongful diagnostics. Moreover, most advanced models, especially deep learning networks, are referred to as black boxes, and they are not transparent. This un-explainability is a barrier to trust and regulatory acceptance in safety-critical application (Elgebaly et al., 2025). Explainable AI (XAI) research is starting to fill this gap by creating explainable models and visualizations to explain the way fault predictions are made. To increase its industrial application, it is vital to increase model portability and interpretability.

Edge Computing and Real-Time AI Deployment

AI models are becoming extremely common on edge devices at the point of wind turbines in order to minimize bandwidth and latency. The ability to detect faults in real-time, react rapidly, and constantly adapt the system is possible with edge computing, and it does not utilize the full extent of cloud infrastructure. Nonetheless, there are constraints such as the availability of computational resources, energy efficient functioning, and model updates in distributed edge nodes (Khan, 2024). Model compression, quantization, and federated learning techniques are being studied in order to bring edge AI to reality. With the increase in the size and sophistication of wind Farms, edge computing will play a major role in providing the high-speed, locally based diagnostics and decentralized, scalable maintenance plans.

Blockchain for Secure Fault Logging and Diagnostics

Blockchain technology has the capabilities to provide decentralized, tamper-free data of recording and auditing of fault data in wind energy. It enables transparency, traceability and accountability among the stakeholders by recording diagnostic outputs, maintenance events, and performance measurements on a distributed ledger (Pliuhin et al., 2024). Such is especially useful in the cross-border wind power installations of multi-vendors where reliability and conformity are paramount. Automated facility contract responses to faults detected and intelligent files to trigger maintenance / warranty claims may be automatable as smart contracts. Although integration with real-time AI systems is in its early development stages, blockchain-enabled digital twins show promise of improving data integrity, audibility, and trust in other fields where automation of decision-making is in use.

CONCLUSION

The increased demand around the globe on renewable energy has predisposed wind power as one of the most important pillars in renewable energy structure. Nevertheless, with in-advanced and wide usage of wind turbines, exposed to the dynamic conditions of environment increasing complexity of maintenance, effectiveness and lifetime of operation is achieved. The nature of current wind energy systems and their typical scale, variability and data intensity is exceeding the capabilities of traditional strategies of fault detection and maintenance through a scheduled inspection and repair or reactive repair, or a combination of the two approaches. In such circumstances, using AI integration with Digital Twin technologies is 'magic wand' in addressing such issues with high precision and adaptability, ease of use and ease in managing volume of data. This chapter highlighted how models of artificial intelligence, ranging from basic and fundamental applications of machine learning such as SVM, random forest, to CNN, LSTM and other forms of advanced deep learning, can help in fault detection, failure prediction and learning in operational datasets of a high order of complexity. These approaches can provide low level situational awareness in degrading trends and component level phenomena. Moreover, resilient and adaptive diagnostics can also be engineered with the use of anomaly detection, transfer learning, and online learning, even in the data-barren condition or the high.

An AI-enabled digital replica of a physical wind turbine is referred to as a digital twin. By simulating turbine behavior under various forms of stress and mimicking real-time operational behavior, digital models can offer a complete view of system health and performance. They are capable of applying principles of multi-physics and multi-modal modelling and will achieve exact simulation within the mechanical, electrical, thermal and aero-structural fields. Making use of digital twins enables the positioning of the system as an early warning system, as well as improving efficiency of the maintenance timetable via virtual fault injection and state estimation in real-time, which leads to significant reduction in operational costs due to unplanned shutdowns. The support of artificial intelligence and digital twins streamlines proactive decision-making on Predictive Maintenance the turbine fleet level. The Feedback loop through the physical systems, virtual models, and AI on-the-fly wind farm operations boost energy yield, improve safety, and extend component life. Despite the size of the opportunity, considerable fully integrated into ever-improving systems that will help. Data privacy and cybersecurity, the need for better AI generalization, interpreting 'black box' algorithms, and real-time edge processing form the most important barriers to tackle.

The development and transformation of novel technologies, including edge AI and blockchain, will promote trust and guarantee secure system operation. These novel technologies also focus on Edge AI, Explainable Artificial Intelligence (XAI), and blockchain which promote trust, scalability, and secure system operation. The expansion of our wind turbine operators and, harnessing what the future holds, will result in turbine condition monitoring having advanced autonomy, adaptability, and intelligence. The achievement of secure diagnostics with the use of blockchain technology along with cloud-edge hybrid infrastructures will turn vast wind farms into super-efficient, smart, and robust operations. Both the operational performance as well as the global emissions performance will benefit thanks to the reduced waste, improved reliability, and stability in the energy generation of renewable sources. AI and digital twins give way to technology improvements and form the pillars of the future of wind energy systems. Thanks to this synergy, the research, engineering, and operations stakeholders will be able to sustain wind energy and lower the lifecycle costs for the planet's future.

LIMITATIONS OF THIS STUDY

- **Data Privacy and Cybersecurity Risks –** AI and digital twin systems for wind turbines rely heavily on SCADA, IoT, and cloud/edge communication. This raises concerns about unauthorized access, data breaches, and manipulation of operational data, which could disrupt monitoring and control functions.
- **Limited Model Generalizability –** AI models trained on specific turbine types, locations, or datasets may not perform well when applied to other turbines or different environmental and operational conditions, reducing scalability across diverse wind farms.
- **Lack of Model Explainability –** Many advanced deep learning methods function as "black boxes," making it difficult to interpret fault predictions, which limits trust, regulatory acceptance, and adoption in safety-critical systems.
- **Real-Time Processing Constraints –** Achieving low-latency fault detection at the turbine level is challenging due to the heavy reliance on cloud computing. Edge AI deployment faces limitations in computational resources, power efficiency, and model updating in distributed environments.
- **Data Quality and Availability Issues –** Sensor data can be noisy, incomplete, or imbalanced, especially for fault scenarios, which hinders effective AI model training and reduces prediction reliability.

REFERENCES

Ahmed, I. A., & Asamoah, P. B. (2024). AI-driven predictive maintenance for energy infrastructure. *International Journal of Innovation and Scientific Research, XI*(IX), 507–528. DOI: 10.51244/IJRSI.2024.1109048

Avci, A. B. (2024). AI-driven approaches to enhance energy efficiency in heritage architecture: A review. *6th International Symposium on Innovation in Architecture, Planning and Design Proceedings*, 37–41. DOI: 10.36287/setsci.20.5.037

Bassey, K. E. (2022a). Enhanced design and development simulation and testing. Eng. Sci. Technol. j., 3(2), 18–31. DOI: 10.51594/estj.v3i2.1283

Bassey, K. E. (2022b). Optimizing wind farm performance using machine learning. Eng. Sci. Technol. j., 3(2), 32–44. DOI: 10.51594/estj.v3i2.1284

Benbouzid, M., Berghout, T., Sarma, N., Djurović, S., Wu, Y., & Ma, X. (2021). Intelligent Condition Monitoring of Wind Power Systems: State of the Art Review. *Energies, 14*(18), 5967. DOI: 10.3390/en14185967

Besigomwe, K. (2025). Closed-Loop Manufacturing with AI-Enabled Digital Twin Systems. *Cognizance Journal of Multidisciplinary Studies, 5*(1), 18–38. DOI: 10.47760/cognizance.2025.v05i01.002

Bindingsbø, O. T., Singh, M., Øvsthus, K., & Keprate, A. (2023). Fault detection of a wind turbine generator bearing using interpretable machine learning. *Frontiers in Energy Research, 11*, 1284676. Advance online publication. DOI: 10.3389/fenrg.2023.1284676

Bisu, C., Olaru, A., Olaru, S., Alexei, A., Mihai, N., & Ushaq, H. (2024). Monitoring the Wear Trends in Wind Turbines by Tracking Fourier Vibration Spectra and Density Based Support Vector Machines. *Mathematics, 12*(9), 1307. DOI: 10.3390/math12091307

Bosch, C. (2022, February). Machine Learning for wind turbine fault prediction through the combination of datasets from same type turbines. Floating Offshore Energy Devices. DOI: 10.21741/9781644901731-7

Casallas, J. S., Parra-Peñuela, H. G., & Gaona, E. E. (2023). Wireless real-time data acquisition system for aerodynamic characterization of a wind turbine. Ingenier\'\ia, 28(Suppl), e18860. DOI: 10.14483/23448393.18860

Chiang-Guizar, C. D., Hernandez-Martinez, J. I., Sevilla-Camacho, P. Y., & Solis-Cisneros, H. I. (2025). Artificial intelligence and integrated optimization in the energy sector: Advances in photovoltaic system. Energ\'\ias Renovables, 12(55). DOI: 10.59730/rer.v12n55a5

Dai, J., Yang, W., Cao, J., Liu, D., & Long, X. (2018). Ageing assessment of a wind turbine over time by interpreting wind farm SCADA data. *Renewable Energy*, *116*, 199–208. DOI: 10.1016/j.renene.2017.03.097

Elgebaly, H., Elhariry, B., Noureldin, A., & Stohy, D. (2025). Digital twin for maintenance and smart manufacturing: The mediating role of replacement maintenance in the Saudi industrial sector. *Jlsdgr*, *5*(4), 6107. DOI: 10.47172/2965-730X.SDGsReview.v5.n04.pe06107

Ferreira da Silva, M., Nunes Masson, J. E., Santos, M. F., Rodrigues Silva, W., Wladimir Molina, I., & Martins, G. M. C. (2025). Audible Noise Evaluation in Wind Turbines Through Artificial Intelligence Techniques. *Sensors (Basel)*, *25*(5), 1492. DOI: 10.3390/s25051492 PMID: 40096305

Garan, M., Tidriri, K., & Kovalenko, I. (2022). A Data-Centric Machine Learning Methodology: Application on Predictive Maintenance of Wind Turbines. *Energies*, *15*(3), 826. DOI: 10.3390/en15030826

González, I., Tutivén, C., & Vidal, Y. (2023). Advancing wind turbine reliability: Machine learning-based early detection of rotor imbalance for proactive maintenance strategies. Latin American Workshop on Structural Health Monitoring. December, 5-7 2023. https://www.ndt.net/?id=31544

Hwang, S., & Yoo, C. (2021). Health Monitoring and Diagnosis System for a Small H-Type Darrieus Vertical-Axis Wind Turbine. *Energies*, *14*(21), 7246. DOI: 10.3390/en14217246

Khan, M. R. (2024). Digital twin-driven optimization of bioenergy production from waste materials. *Innovatech Engineering Journal*, *1*(01), 187–204. DOI: 10.70937/itej.v1i01.19

Kong, L., Liang, H., Liu, G., & Liu, S. (2023). Research on Wind Turbine Fault Detection Based on the Fusion of ASL-CatBoost and TtRSA. *Sensors (Basel)*, *23*(15), 6741. DOI: 10.3390/s23156741 PMID: 37571525

Lebranchu, A., Charbonnier, S., Bérenguer, C., & Prévost, F. (2019). A combined mono- and multi-turbine approach for fault indicator synthesis and wind turbine monitoring using SCADA data. *ISA Transactions*, *87*, 272–281. DOI: 10.1016/j.isatra.2018.11.041 PMID: 30545768

Mahaveera, A., & Pasuk, S. (2021). Development of smart monitoring system for wind energy system. International Journal of Systems Applications. *Engineering*, *15*, 132–134. DOI: 10.46300/91015.2021.15.18

Masoumi, M. (2023). Machine learning solutions for offshore wind farms: A review of applications and impacts. *Journal of Marine Science and Engineering*, *11*(10), 1855. DOI: 10.3390/jmse11101855

Odgaard, P. F., & Stoustrup, J. (2013, August). Fault tolerant wind farm control — A benchmark model. *2013 IEEE International Conference on Control (CCA)*. DOI: 10.1109/CCA.2013.6662784

Pliuhin, V., Tsegelnyk, Y., & Slovikovskyi, O. (2024). Digital twins of different types electrical machines. *Lighting Engineering & Power Engineering*, *63*(2), 35–45. DOI: 10.33042/2079-424X.2024.63.2.01

Prabhavathi, K., Prabhakar, P. E., Arunadevi, M., & Shanthi, D. (2023). Design and Implementation of a Machine Learning-Based Wind Turbine Control System. In *E3S Web of Conferences* (Vol. 387, p. 01008). EDP Sciences.

Sanchez, H. E., Escobet, T., Puig, V., & Odgaard, P. F. (2015). Fault diagnosis of advanced wind turbine benchmark using interval-based ARRs and observers. *IEEE Transactions on Industrial Electronics*, *1*, 1. Advance online publication. DOI: 10.1109/TIE.2015.2399401

Sethi, M. R., Sahoo, S., Dhanraj, J. A., & Sugumaran, V. (2023). Vibration signal-based diagnosis of wind turbine blade conditions for improving energy extraction using machine learning approach. *Smart and Sustainable Manufacturing Systems*, *7*(1), 14–40. DOI: 10.1520/SSMS20220023

Shi, P., Jia, L., Yi, S., & Han, D. (2024). Wind turbines fault diagnosis method under variable working conditions based on AMVMD and deep discrimination transfer learning network. *Measurement Science & Technology*, *35*(4), 46120. DOI: 10.1088/1361-6501/ad1fcd

Sikdar, S., Kalgutkar, A. P., & Banerjee, S. (2024). Guided wave based structural health monitoring of composite wind turbine blades. *Proceedings of the Annual British Conference on Non-Destructive*, 2024(1), 1–11. DOI: 10.1784/ndt2024.4b4

Takanashi, M., Sato, S.-I., Indo, K., Nishihara, N., Ichikawa, H., & Watanabe, H. (2021). Anomaly prediction for wind turbines using an autoencoder based on power-curve filtering. IEICE Trans. Inf. Syst., E104.D(9), 1506–1509. DOI: 10.1587/transinf.2020EDL8127

Udo, W. S., Kwakye, J. M., Ekechukwu, D. E., & Ogundipe, O. B. (2023). Optimizing wind Energy Systems using Machine Learning for Predictive Maintenance and Efficiency Enhancement. Comput. Sci. IT Res. j., 4(3), 386–397. DOI: 10.51594/csitrj.v4i3.1398

Uhanto, U., Yandri, E., Hilmi, E., Saiful, R., & Hamja, N. (2024). Predictive maintenance with machine learning: A comparative analysis of wind turbines and PV power plants. *Heca J. Appl. Sci.*, *2*(2), 87–98. DOI: 10.60084/hjas.v2i2.219

Ulmer, M., Jarlskog, E., Pizza, G., & Goren Huber, L. (2020). Cross-turbine training of convolutional neural networks for SCADA-based fault detection in wind turbines. *Proceedings of the Annual Conference of the Prognostics and Health Management Society. Prognostics and Health Management Society. Conference*, *12*(1), 10. DOI: 10.36001/phmconf.2020.v12i1.1205

Vives, J. (2022). Incorporating machine learning into vibration detection for wind turbines. *Modelling and Simulation in Engineering*, *2022*, 1–8. DOI: 10.1155/2022/6572298

Wang, X., Zhou, D., Zhuang, X., Ge, J., & Xiang, J. (2024). *SGG-DGCN: Wind turbine anomaly identification by using deep graph convolutional networks with similarity graph generation strategy. Journal of Dynamics*. Monitoring and Diagnostics., DOI: 10.37965/jdmd.2024.558

Xiao, C., Liu, Z., Zhang, T., & Zhang, X. (2021). Deep learning method for fault detection of wind turbine converter. *Applied Sciences (Basel, Switzerland)*, *11*(3), 1280. DOI: 10.3390/app11031280

KEY TERMS AND DEFINITIONS

Acoustic Emission Sensor: A sensor that captures high-frequency sound waves generated by cracks or delamination in turbine blades.

Artificial Intelligence (AI): Computational techniques enabling machines to learn from data, detect patterns, and make decisions, used here for fault detection and predictive maintenance in wind turbines.

Blockchain: A secure, decentralized ledger technology used for tamper-proof logging of fault data, diagnostics, and maintenance records.

Condition Monitoring: Continuous tracking of turbine component health (e.g., blades, gearbox, generator) through sensors to detect early signs of wear, faults, or inefficiencies.

Digital Twin: A virtual replica of a physical wind turbine that integrates real-time sensor data and simulations to predict performance, detect faults, and optimize maintenance.

Edge AI: Deployment of AI models on local computing devices at or near the turbine to enable low-latency fault detection without relying solely on cloud processing.

Explainable AI (XAI): AI methods designed to provide interpretable and transparent decision-making, increasing trust in automated diagnostics.

Predictive Maintenance: A maintenance strategy that uses AI and real-time monitoring to predict when a component will fail, allowing timely interventions to prevent downtime.

Remaining Useful Life (RUL): The estimated operational time left before a component reaches the end of its functional life.

SCADA (Supervisory Control and Data Acquisition): An industrial control system that collects real-time data from sensors and devices to monitor and control wind turbine operations remotely.

Transfer Learning: A machine learning approach where knowledge gained from one turbine or dataset is adapted to another, improving performance with minimal new data.

Vibration Sensor: A device that measures oscillations in turbine components to detect imbalances, misalignments, or bearing wear.

Chapter 9
Data Privacy and Cybersecurity in Wind Power Systems

Binastya Anggara Sekti
http://orcid.org/0000-0001-5489-4888
Universitas Esa Unggul, Indonesia

ABSTRACT

The growing digitalization of wind power systems has intensified cybersecurity and data privacy challenges across modern renewable energy infrastructures. As AI, digital twins, and cloud-based platforms become integral to turbine monitoring and control, wind farms face expanding attack surfaces and increased vulnerability to malware, sensor spoofing, DDoS attacks, and supply chain breaches. These risks threaten operational continuity, grid stability, and the confidentiality of sensitive operational and geospatial data. Ensuring secure and privacy-compliant wind operations requires multi-layered protection that includes Zero Trust architectures, secure industrial protocols, robust identity and access management, encrypted communication channels, and strong AI governance. A comprehensive understanding of emerging threat vectors and resilient defense mechanisms is essential for safeguarding AI-powered wind energy ecosystems against evolving cyber and privacy risks.

1. INTRODUCTION

Modern wind power systems operate as large-scale cyber-physical infrastructures in which physical assets, operational technology (OT), information technology (IT), cloud platforms, and artificial intelligence models are tightly interconnected. Ensuring cybersecurity and data privacy in such environments requires a holistic architectural

DOI: 10.4018/979-8-3373-4159-0.ch009

approach rather than isolated technical controls. This section presents a layered cybersecurity and data privacy architecture designed specifically for AI-enabled wind farms, integrating protection mechanisms across physical assets, networks, data pipelines, AI systems, and governance layers. Accurate wind power forecasting is one of the most important operations within the economic dispatch problem to increase the performance of power and energy systems (Moayyed, Moradzadeh, Mohammadi-Ivatloo, Aguiar, & Ghorbani, 2022). Table 1 summarizes the layered cybersecurity and data privacy architecture adopted for AI-enabled wind power systems, highlighting how security controls are distributed across physical, digital, and governance domains.

Table 1. Cybersecurity and Data Privacy Architecture for AI-Enabled Wind Power Systems

Architecture Layer	Scope and Security Functions
Physical and Edge Security Layer	Includes turbines, sensors, PLCs, and RTUs. Implements secure boot, cryptographically signed firmware, tamper resistance, device authentication, and edge-level encryption with data minimization to protect data and control signals at the source.
Operational Technology (OT) Network Protection Layer	Secures SCADA and control networks through IEC 62443 zoning, strict network segmentation, industrial firewalls, protocol-aware intrusion detection systems, and encrypted authenticated communications to prevent unauthorized control and lateral movement.
IT, Cloud, and Data Management Layer	Protects cloud platforms, analytics services, and data pipelines using encryption at rest and in transit, identity and access management, secure API gateways, data classification policies, and jurisdiction-aware storage mechanisms.
AI and Digital Twin Security Layer	Governs AI models and digital twins through secure training pipelines, model versioning, integrity verification, explainable AI mechanisms, and safeguards against data poisoning, model drift, and unauthorized model modification.
Governance, Compliance, and Privacy Oversight Layer	Enforces organizational cybersecurity policies, regulatory compliance, data privacy governance, auditing, and incident response across the entire wind farm lifecycle to ensure accountability, transparency, and long-term resilience.

Wind power systems have evolved from standalone mechanical turbines into highly interconnected cyber-physical infrastructures that rely on digital automation, advanced sensing, cloud computing, and AI-driven analytics. Modern wind farms integrate Supervisory Control and Data Acquisition (SCADA) platforms, IoT sensor arrays, wireless telemetry, remote access tools, and predictive maintenance algorithms to achieve unprecedented levels of efficiency and responsiveness. This digital transformation enables operators to optimize turbine performance, reduce downtime, and streamline asset management across geographically distributed wind sites. However, this increased connectivity significantly expands the attack surface for malicious actors and introduces new cybersecurity and data privacy risks that did not exist in earlier generations of wind technology. Cybersecurity for wind energy conversion systems, such as Stand-Alone Wind Energy Conversion Systems (SAWECSs) that use Self-Excited Induction Generators (SEIGs), has been a growing area of focus recently (Demirtas, Sharkh, Gokyildirim, & Calgan, 2025). Figure 1 illustrates a layered cybersecurity and data privacy architecture integrating physical protection, OT security, cloud governance, AI assurance, and regulatory oversight. The architecture demonstrates how protections at each layer reinforce confidentiality, integrity, availability, and trustworthiness across the wind farm lifecycle.

Figure 1. Synergy for Trusted AI-Enabled Wind Power

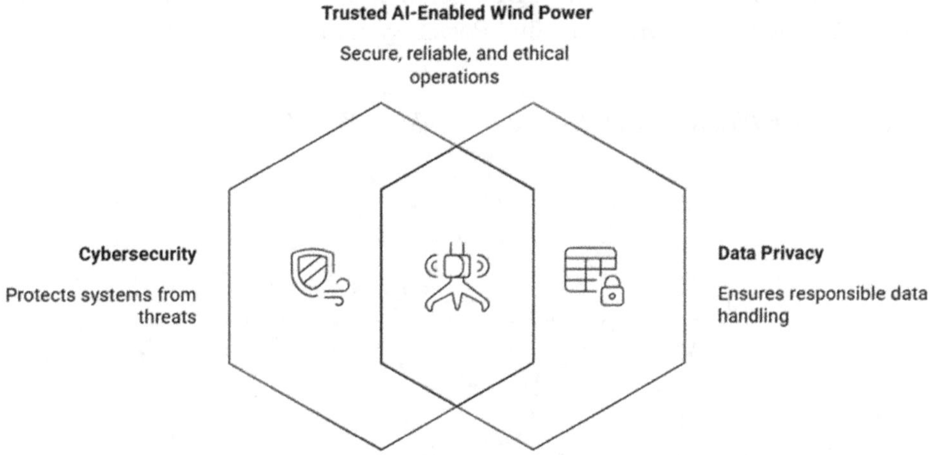

(Biswas et al., 2025)

As wind power becomes a cornerstone of national energy strategies, the consequences of cyberattacks on wind infrastructures have grown more severe. Compromising a single SCADA controller, sensor gateway, or VPN link can disrupt turbine operations, destabilize grid frequency, or expose sensitive operational data. Several publicly reported incidents, including ransomware attacks on wind farm operators in Europe and telecommunication failures affecting remote turbine monitoring, demonstrate how vulnerable these systems can be. Moreover, the reliance on remote diagnostics and third-party maintenance contractors introduces additional governance challenges, as unauthorized access or misconfigured credentials can grant adversaries entry into critical operational networks. The main cybersecurity risks facing the wind energy sector are unauthorized access, data breaches, malware and ransomware, and physical security threats (Shafiee, 2025).

AI-driven automation further intensifies the cybersecurity landscape. While machine learning models enhance predictive maintenance, fault classification, and anomaly detection, they also require large volumes of operational and environmental data. If this data is stolen or manipulated, attackers can degrade AI model accuracy or disrupt decision-making pipelines. Digital twins, which function as virtual replicas of turbines and wind farms, introduce opportunities for simulation but also create new risks of model poisoning if their input data is compromised. As the complexity of wind farm digital ecosystems increases, maintaining confidentiality, integrity, and availability of both operational and analytical data becomes essential. A microgrid is an autonomous local energy system that can work separately from or in conjunction with the main grid, which is necessary for modern energy systems due to the integration of renewable energy sources (RES) such as wind (Hachimi, El Gaabouri, El Bhiri, Motahhir, & Boussetta, 2025). Figure 2 shows the Cyber-Physical Architecture of Modern Wind Power Systems.

Figure 2. Cyber-Physical Architecture of Modern Wind Power Systems

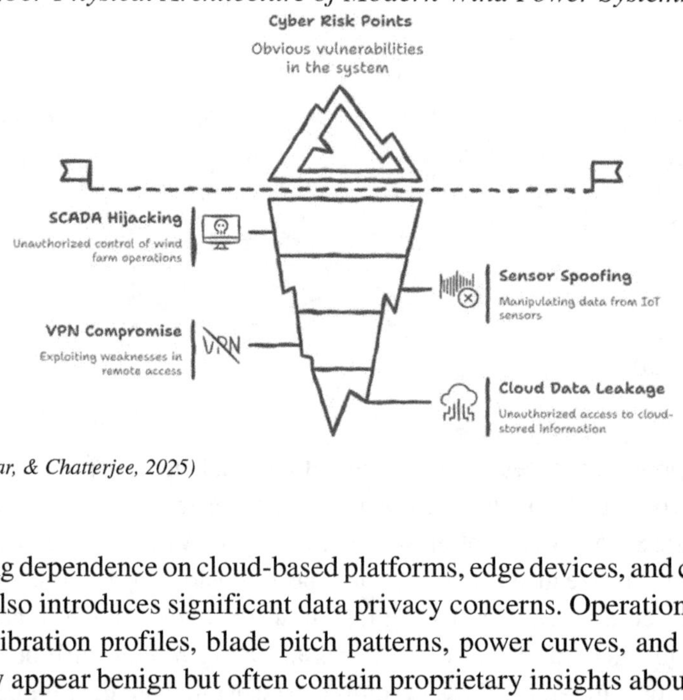

(Paul, Kumar, & Chatterjee, 2025)

The rising dependence on cloud-based platforms, edge devices, and cross-border data flows also introduces significant data privacy concerns. Operational data such as turbine vibration profiles, blade pitch patterns, power curves, and geolocation streams may appear benign but often contain proprietary insights about turbine efficiency, maintenance scheduling, and grid interaction strategies. When transmitted through global cloud providers, this data may fall under regulatory frameworks such as GDPR, requiring strict controls over data storage, retention, and access permissions. The lack of standardized data governance practices across wind operators further amplifies privacy risks and complicates compliance. Renewable energy sector resilience ensures sustainability and reliability together with cybersecurity features for present-day power networks (Ahatsi, Winkler, & Olanrewaju, 2025).

Given these challenges, ensuring cybersecurity and privacy across modern wind power systems requires a holistic and multi-layered approach. Operators must integrate cyber risk assessments, intrusion detection systems, Zero Trust architectures, secure industrial protocols, and strong AI governance practices into every stage of the turbine lifecycle, from design and commissioning to operation and eventual decommissioning. Table 2 summarizes the major forces driving cybersecurity complexity in contemporary wind power infrastructures. The introduction of these digital elements marks a pivotal shift toward viewing wind farms not only as energy-generating assets but as data-intensive critical infrastructure requiring robust protection and governance.

Table 2. Cybersecurity Complexity in Modern Wind Power Systems

Driver	Description	Security Implications
Digitalization of wind farm assets	Integration of SCADA, IoT sensors, and cloud monitoring	Larger attack surface; multi-vector cyber risk
Remote access & third-party maintenance	OEMs and contractors require intermittent access	Credential compromise, VPN abuse
AI-driven operations	Reliance on ML and digital twins for decision-making	Model poisoning, data privacy concerns
Geographic dispersion	Wind farms span remote, hard-to-monitor areas	Difficult physical security and segmentation
Regulatory evolution	GDPR, NERC CIP, IEC 62443 requirements	Compliance pressure and governance complexity

2. CYBER THREAT LANDSCAPE IN WIND POWER SYSTEMS

Wind power systems are increasingly targeted by sophisticated cyber adversaries due to their growing importance in national energy strategies and their reliance on digital automation. As modern wind farms transition into AI-assisted and cloud-connected infrastructures, the number of potential entry points for attackers expands dramatically. Threat actors ranging from cybercriminal groups to state-sponsored units recognize the strategic value of disrupting renewable energy operations, especially as wind power contributes significantly to grid stability in many countries. Cyber intrusions can compromise turbine control behavior, disrupt power delivery, manipulate operational data, or expose sensitive geospatial and performance information. The distributed and remote nature of wind farms, often located in offshore or isolated regions, further exacerbates these vulnerabilities by limiting physical oversight and making continuous monitoring difficult. One of the challenges of modern power systems is the vulnerability of such systems to malicious cyberattacks, among which it dominates false data injection (FDIA) attacks, as it is easier to perform, due to the attack surface introduced by the integration of smart meters and large area monitoring/control assets (Naderi & Asrari, 2025).

Major threat vectors often focus on SCADA systems, IoT sensor networks, edge gateways, programmable logic controllers (PLCs), and cloud interfaces used by OEMs and operators. SCADA systems remain high-value targets because they directly influence turbine pitch, yaw, braking, and power output. Many wind farm incidents in Europe and Asia involved unauthorized SCADA access via outdated VPN credentials or misconfigured remote access portals. Attackers may alter SCADA setpoints, disable safety mechanisms, or induce overspeed events that risk catastrophic mechanical failure. Similarly, PLCs in nacelles or substations are vulnerable due to legacy industrial protocols such as Modbus or DNP3, which often lack encryption and authentication, allowing attackers to modify turbine logic or inject malicious commands over unsecured channels. Supervisory Control and Data Acquisition (SCADA) systems play a crucial role in overseeing and controlling renewable energy sources like solar, wind, hydro, and geothermal resources (Oyucu et al., 2024).

IoT and edge sensors form another critical threat surface in modern wind farms. These sensors provide wind speed, direction, vibration, temperature, and grid frequency measurements necessary for both operational control and AI-driven analytics. Attackers who corrupt sensor values through spoofing, interference, or man-in-the-middle manipulation can trigger incorrect turbine responses, such as unnecessary shutdowns or misaligned yaw positions. In AI-enhanced wind systems, manipulated sensor data can compromise digital twin accuracy, causing predictive maintenance models to misidentify faults or overlook early warning signals. Moreover, as cloud-based dashboards and digital twin environments become standard for remote monitoring, adversaries increasingly target cloud interfaces, API endpoints, and data storage systems. Compromised cloud accounts can lead to total loss of visibility across an entire fleet, enabling attackers to disable turbines, tamper with performance data, or block access through ransomware. Digital Twin (DT) technologies are rapidly transforming the design, operation, and lifecycle management of renewable energy systems (Mbasso et al., 2025).

Supply chain attacks represent one of the most dangerous and rapidly growing threats. Wind operators rely heavily on OEMs, third-party maintenance providers, software vendors, and fleet analytics companies, all of which require periodic access to operational systems. Compromised vendor laptops, malicious firmware updates, insecure VPN tunnels, and infected diagnostic tools can serve as entry points into core turbine controls. Such attacks have been linked to incidents where thousands of turbines lost remote monitoring capabilities after third-party communication modems were compromised. As AI models and digital twins become part of the supply chain, trained, updated, and deployed by external providers, the risks expand to include model poisoning, Trojanized AI algorithms, and compromised training datasets that alter turbine behavior indirectly rather than through direct command injection. Smart inverters are pivotal in modern renewable energy systems, enabling efficient grid integration, stability, and advanced control of distributed energy resources (Awad & Bayoumi, 2025).

Finally, wind farms face large-scale distributed denial-of-service (DDoS) attacks aimed at overwhelming turbine communication networks, remote monitoring portals, or control centers. Loss of communication may force turbines into safe mode or disconnect them from the balancing authority, impacting power availability across large regions. Offshore wind farms are particularly at risk due to reliance on satellite links that are easier to flood or jam compared to terrestrial networks. In addition to DDoS threats, GPS spoofing has been documented as a method of interfering with turbine alignment systems and automated maintenance drones used for blade inspection. Adversaries can distort geolocation signals, causing erratic drone flight paths or inaccurate turbine yaw alignment. Combined with AI-driven operations, such manipulation can propagate errors across energy forecasting models and digital twin simulations, undermining both safety and efficiency. A hybrid machine learning (HML) model that combines the characteristics of Deep Learning (DL) and Long-Short-Term-Memory (LSTM) models can be applied to identify unique patterns of Denial of Service (DoS) and Distributed Denial of Service (DDoS) cyberattacks in the process of power generation, transmission, and distribution (Faheem & Al-Khasawneh, 2024).

2.1 Malware and Ransomware

Malware and ransomware represent some of the most disruptive threats to wind power systems, particularly as wind farms increasingly depend on interconnected SCADA servers, remote access tools, and cloud-based monitoring platforms. Attackers often deploy ransomware to encrypt control room workstations, SCADA databases, or turbine control interfaces, forcing operators to disconnect affected systems from the grid to prevent the spread of infection. In several documented cases across Europe and North America, ransomware groups exploited weak VPN configurations or outdated OS patches to gain a foothold in wind farm networks, ultimately crippling remote monitoring and forcing operators to revert to manual control modes. Such disruptions are especially dangerous in large offshore wind sites where local access is challenging, resulting in prolonged downtime. In AI-enhanced wind operations, ransomware can also target digital twin environments, corrupting model data and halting predictive analytics used for maintenance and grid balancing. The convergence of IT and OT environments increases the risk that malware entering the corporate office network may propagate into turbine control networks, making ransomware mitigation an essential element of wind energy cybersecurity. As the rapid growth of wind energy has positioned wind farm infrastructure as a critical pillar in the global transition to sustainable energy, it is critical to identify the key cybersecurity risks facing the wind energy sector, including unauthorized access, data breaches, malware and ransomware, as well as physical security threats (Shafiee, 2025).

2.2 Denial-of-Service (DoS and DDoS)

Denial-of-Service attacks, including large-scale Distributed Denial-of-Service (DDoS) attacks, present a significant operational risk for wind farms that rely on stable communication links for telemetry, monitoring, and control. Because wind farms often operate in remote or offshore locations, communication infrastructure such as satellite links, wireless relays, or fiber-optic gateways can be relatively bandwidth-constrained. Adversaries exploit this weakness by overwhelming these channels with massive traffic volumes, causing SCADA operators to lose visibility into turbine status and forcing turbines into safe mode due to loss of supervisory signals. DDoS attacks targeting remote OEM portals or cloud dashboards can also block access to diagnostic tools, delaying maintenance and increasing the likelihood of cascading failures. In modern AI-driven wind management systems, these attacks can disrupt the flow of sensor data needed for predictive maintenance algorithms or grid frequency control models, producing incomplete or misleading inputs and degrading model accuracy. These attacks do not always require sophisticated techniques, because even modest DDoS campaigns can significantly impair wind farm operations due to their reliance on uninterrupted telemetry. The emergency response system, water and wastewater treatment facilities, oil and gas pipelines, electrical power transmission and distribution, wind farms, defense systems, and large communication networks are all expanding and becoming more vulnerable (Srivastava, Rawat, & Maurya, 2023).

2.3 SCADA Hijacking

SCADA hijacking is one of the most dangerous forms of cyberattack targeting wind farms, given the direct influence of SCADA systems over turbine pitch, yaw, rotor speed, braking mechanisms, and grid synchronization. Unauthorized modifications to SCADA logic can result in catastrophic outcomes, including turbine overspeed events, yaw misalignment that stresses the nacelle and blades, or shutdown sequences that unintentionally destabilize frequency control in regional grids. The risk is heightened by the fact that many SCADA deployments in wind systems still use legacy industrial protocols such as Modbus or DNP3, which lack strong authentication and encryption. Attackers who gain access through compromised VPN credentials, misconfigured firewalls, or third-party maintenance accounts may manipulate control signals or upload malicious firmware to PLCs. Once SCADA is hijacked, attackers can execute subtle and difficult-to-detect manipulations, such as slowly altering pitch angles to degrade efficiency or accelerating wear on turbine components. In AI-enhanced systems, tampered SCADA inputs can corrupt digital twins, leading to incorrect simulations and misleading operational forecasts. False Data Injection Attacks (FDIAs) pose significant threats to the reliability and safety of wind turbine operations, particularly targeting the Supervisory Control and Data Acquisition (SCADA) systems that monitor critical gearbox parameters (Ramchandani, Vahora, Wagle, & Dhiman, 2025).

2.4 GPS Spoofing and Sensor Manipulation

Wind turbines depend heavily on accurate measurements of wind speed, wind direction, grid frequency, rotor vibration, and environmental conditions to operate safely and efficiently. GPS data plays a critical role in turbine alignment, particularly in large wind farms where yaw systems rely on precise geospatial inputs. GPS spoofing, which involves broadcasting false location or timing signals, can cause turbines to misalign with the wind, leading to unnecessary shutdowns or reduced energy capture. Attackers may also manipulate sensor readings by interfering with anemometers, LIDAR-based wind measurement systems, or grid frequency sensors. In highly digitalized wind environments, corrupted sensor data affects not only real-time control but also produces downstream consequences in AI-driven analytics pipelines. Predictive maintenance models may fail to detect early indications of bearing fatigue, while power forecasting models may generate inaccurate short-term predictions. Sensor manipulation also introduces severe safety risks, as incorrect readings can trigger automatic protective responses that shut down multiple turbines simultaneously and disrupt grid stability. The six dimensions of security in the Identification and analysis of cybersecurity threats of the wind farm industrial control system are Spoofing, Tampering, Repudiation, Information Disclosure, Denial of Service, and Elevation of Privilege (Yang & Zhang, 2022).

2.5 Supply Chain Attacks

Supply chain attacks are among the most challenging threats in wind energy cybersecurity because they exploit trusted relationships between turbine operators, OEMs, subcontractors, IT support vendors, and firmware suppliers. Wind farms depend heavily on external vendors for remote diagnostics, firmware updates, and periodic maintenance, creating numerous trust boundaries that adversaries can infiltrate. Compromised firmware updates, manipulated maintenance laptops, infected USB tools, or insecure vendor VPN access can enable attackers to enter turbine control networks undetected. A widely reported incident in 2022 involved thousands of European wind turbines losing remote communication capability after a third-party satellite modem supplier was compromised — demonstrating how a single vendor can become a systemic point of failure. As AI models and digital twins increasingly rely on third-party cloud services and external datasets, the supply chain threat extends further into the realm of AI poisoning, where attackers compromise training data or insert backdoors into AI models deployed across turbine fleets. Because wind power infrastructures are distributed and operationally interdependent, supply chain attacks can propagate across entire regions and compromise grid reliability.

These diverse and increasingly sophisticated attack vectors demonstrate that modern wind farms operate in a high-risk cyber environment. To understand how these threats materialize in real systems, the next section analyzes the structural vulnerabilities within wind power infrastructures that make such attacks possible.

3. INFRASTRUCTURE VULNERABILITIES IN WIND POWER SYSTEMS

Wind power infrastructures comprise a multilayered ecosystem of turbines, SCADA networks, substations, communication gateways, and cloud-based supervisory platforms. As these components become increasingly digitalized and interconnected, their vulnerability to cyber threats grows substantially. Unlike legacy wind farms that operated primarily as isolated mechanical assets, modern wind turbines depend on networked control systems, IoT sensors, wireless telemetry, and remote maintenance portals. This digital transformation introduces systemic exposure points where faults, misconfigurations, or cyberattacks can propagate throughout the entire wind farm or even across regional power grids. Because wind energy is often integrated into national critical infrastructure, any compromise in availability or integrity can have far-reaching consequences, including instability in grid frequency, loss of power generation, and violation of regulatory compliance. Table 3 shows the Summary of Threats, Vulnerabilities, and Defenses in Wind Power Systems.

Table 3. Summary of Threats, Vulnerabilities, and Defenses in Wind Power Systems

Threat Type	Key Vulnerability	Recommended Defense
Ransomware	Weak VPN, outdated SCADA OS	MFA, network segmentation, backup/DRP
DDoS	Bandwidth-limited offshore networks	Traffic filtering, redundant comms
SCADA Hijacking	Legacy protocols (Modbus/DNP3)	IEC 62351 encryption, Zero Trust
GPS Spoofing	Overreliance on single geolocation source	Multi-sensor fusion, anomaly detection
Supply Chain Attacks	Vendor VPN access, firmware updates	Vendor vetting, signed firmware

One of the most persistent challenges stems from legacy SCADA systems and industrial control protocols that were not originally designed with cybersecurity in mind. Many wind farms still rely on older SCADA implementations using Modbus, IEC 60870-5-104, or DNP3, protocols that lack authentication and encryption by default. Attackers can intercept or manipulate these communications, injecting malicious commands or altering turbine settings. Furthermore, firmware in PLCs and remote terminal units (RTUs) is often outdated and difficult to update, leaving long-standing vulnerabilities unresolved. Once attackers gain access, they can alter yaw or pitch mechanisms, disrupt braking systems, or modify synchronization parameters in ways that strain turbine components and destabilize power output. Figure 3 shows the Attack Surface Diagram for Wind Power Infrastructure.

Figure 3. Attack Surface Diagram for Wind Power Infrastructure

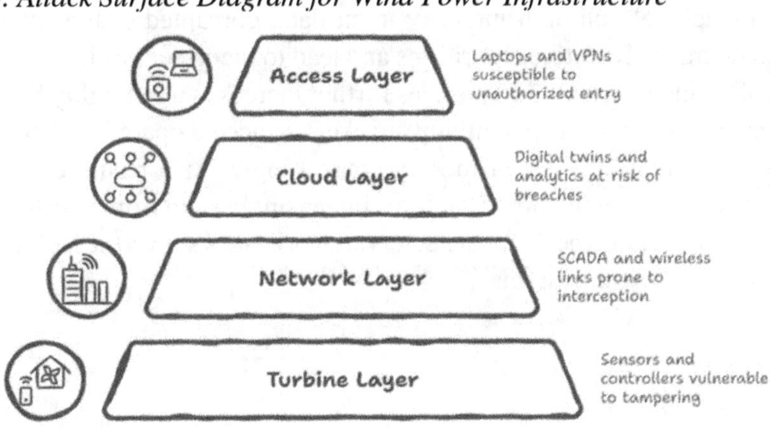

Geographic dispersion also creates unique security challenges in wind power systems. Wind farms often span remote, rural, or offshore locations where physical security controls, such as locked cabinets, surveillance cameras, tamper sensors, or secured substations, may be limited. These locations expose field devices such as edge sensors, power inverters, meteorological masts, and communication relays to physical tampering. In offshore wind facilities, maintenance crews may access equipment through contractor vessels or drones, adding additional uncertainty to access control. When combined with limited on-site personnel, these physical vulnerabilities can enable attackers to connect rogue devices, intercept communication lines, or plant malicious hardware implants that feed false data into SCADA or AI analytics system.

Another major vulnerability arises from the increasing reliance on third-party maintenance contractors, grid operators, OEM service teams, and external analytics providers. Because wind turbines require periodic servicing and diagnostics, maintenance personnel often access operational networks through VPN tunnels or remote access portals. Weak identity and access management (IAM), including shared passwords, outdated credentials, and insufficient audit logging, greatly increases the risk. Compromised contractor accounts have been linked to several attempted intrusions into wind farm SCADA networks in various regions. The complexity of multi-tenant access means that a single exposed credential or a misconfigured vendor connection can provide attackers with unauthorized entry into core control layers.

Finally, the rapid digitalization of wind energy through AI-driven maintenance systems, digital twins, and cloud-based monitoring platforms expands the attack surface in ways that were previously unimaginable. Improperly configured cloud storage buckets, insecure APIs, or mismanaged digital twin environments can leak sensitive turbine performance data, maintenance logs, or proprietary algorithms. Since AI models rely on high-integrity input data, corrupted or tampered sensor values can degrade predictive algorithms and lead to incorrect turbine control decisions. Weak or inconsistent IAM policies further increase vulnerability by allowing unauthorized personnel, and potentially attackers, to access operational dashboards, download configuration files, or modify control logic. As a result, cybersecurity and privacy failures in wind infrastructures increasingly stem from poorly governed digital assets rather than mechanical defects alone. Table 4 shows the Vulnerabilities in Modern Wind Power Systems.

Table 4. Vulnerabilities in Modern Wind Power Systems

Vulnerability Category	Root Cause	Risk to Wind Operations
Legacy SCADA & Industrial Protocols	Outdated protocols (Modbus, DNP3) without encryption	Control hijacking, unsafe turbine commands
Distributed & Remote Deployment	Limited physical security across large areas	Hardware tampering, rogue device insertion
Third-Party Maintenance Access	Shared credentials, weak VPN policies	Unauthorized entry, supply chain compromise
Digitalization & Cloud Exposure	Misconfigured cloud services, insecure APIs	Data leakage, model tampering, operational disruption
Weak IAM & Credential Management	Default passwords, no MFA, poor logging	Privilege escalation, unauthorized control

Key Takeaways:

- Legacy SCADA systems remain one of the most critical vulnerabilities due to weak authentication and unencrypted protocols.
- Geographic dispersion and limited physical oversight increase the exposure of remote assets to tampering.
- Rapid digitalization requires strong IAM, secure configuration, and continuous monitoring to prevent cascading cyber failures.

4. DATA PRIVACY RISKS IN WIND POWER SYSTEMS

Wind power systems collect large volumes of sensitive operational data, including turbine performance logs, wind resource measurements, grid injection profiles, maintenance histories, and geolocation streams. As the frequency and granularity of these datasets increase through AIoT and digital twin technologies, the potential for misuse, unauthorized access, and cross-platform data leakage also grows. However, this introduces significant privacy challenges, particularly concerning data ownership, lawful processing, cross-border transfer compliance, and the risk that operational insights may reveal commercially sensitive or strategically valuable information about energy production patterns.

The increasing digitalization of wind farms has transformed operational data into a critical asset, powering analytics, optimization algorithms, and grid coordination. However, this data-rich environment introduces significant privacy challenges that extend beyond traditional cybersecurity considerations. Wind power systems continuously generate large volumes of information, including turbine performance metrics, geolocation data, environmental readings, grid frequency interactions, maintenance logs, and AI model outputs. Much of this information is commercially sensitive, strategically valuable, or subject to regulatory oversight.

When data privacy is not adequately protected, attackers or unauthorized entities can gain deep insight into wind farm operations, predictive maintenance schedules, power generation capabilities, or grid integration patterns. These risks elevate the need for comprehensive data governance frameworks that ensure confidentiality, control data sharing, and maintain compliance with jurisdiction-specific privacy laws such as the GDPR or national critical infrastructure regulations (Jurczuk & Suprynowicz, 2024).

4.1 Operational Data Privacy

Operational data produced by wind turbines, substations, and monitoring platforms provides detailed insight into real-time power generation, turbine availability, blade pitch cycles, rotor speed behavior, and grid injection patterns. This information reveals proprietary operational strategies, including how operators adjust turbine parameters to optimize performance under varying wind conditions. If accessed by competitors, such data could undermine competitive advantage by exposing turbine efficiency curves, maintenance timing, or advanced AI-driven control schemes. In more severe scenarios, adversaries could analyze operational data to identify vulnerabilities in grid synchronization or to map out "peak stress" periods in turbine operation. Because operational data flows between turbines, SCADA servers, and cloud-based dashboards, improper access control, unsecured APIs, and misconfigured storage buckets can turn this valuable information into a privacy liability. As wind power operators adopt AI-driven performance optimization, operational data becomes even more sensitive, as attackers who obtain it may infer proprietary algorithms or simulation parameters used in digital twins.

4.2 Geolocation and Environmental Data

Wind power systems rely heavily on precise geolocation and environmental data, including wind resource maps, terrain profiles, meteorological mast readings, offshore coordinates, and seasonal wind pattern analytics. Although often perceived as benign, these datasets carry significant commercial and strategic value. Detailed geospatial maps reveal ideal turbine siting locations, expected annual energy production (AEP), and microclimate conditions that influence turbine yield. If leaked, such information can influence land acquisition negotiations, bidding strategies, or competitor planning. In offshore wind farms, geolocation data can be even more sensitive, as it identifies the location of subsea cables, transformer platforms, and maintenance access routes. Unauthorized access to environmental datasets can also enable adversaries to manipulate AI forecasting models by injecting falsified inputs—for example, spoofed wind direction trends or temperature readings—which may degrade grid planning and operational scheduling. The privacy risks associated with geospatial data require strong encryption, anonymization, and clear boundaries around which data may be shared externally.

4.3 AI and Digital Twin Data Handling

AI models and digital twins are increasingly central to predictive maintenance, fault detection, energy forecasting, and optimization of turbine performance. These systems depend on large repositories of historical operational data, sensor readings, SCADA logs, and turbine-specific datasets.

The privacy risks arise from the fact that these AI models often embed sensitive operational knowledge within their trained parameters. If an attacker gains access to training datasets, they could reconstruct turbine performance signatures or learn maintenance vulnerabilities. Similarly, model extraction attacks can allow adversaries to recreate or steal AI models used for optimization, compromising intellectual property and strategic capabilities. Digital twin environments amplify this risk because they replicate the behavior of turbines, substations, and network interactions with high fidelity. A compromised twin reveals operational logic, equipment constraints, and failure thresholds. Furthermore, adversaries can poison the twin by inserting manipulated data inputs, causing incorrect maintenance forecasts or unsafe operational recommendations. Privacy protection for AI systems in wind energy therefore requires strict control over training data access, model versioning, and secure deployment pipelines.

As wind farms transition into digitally intensive infrastructures, the volume and sensitivity of operational, geospatial, and AI-derived data continue to increase. This growing data ecosystem introduces new privacy risks that must be managed alongside traditional cybersecurity concerns.

4.4 Cloud Storage and Cross-Border Transfers

As wind farms become more distributed and AI-driven, operators increasingly depend on cloud platforms for storage, analytics, and remote visualization. Cloud providers often store data across multiple geographic regions with varying data protection laws, making cross-border data transfers a major privacy concern. For example, European wind farms that store operational logs or SCADA snapshots on global cloud services may inadvertently violate GDPR restrictions on transferring personal or sensitive data outside the EU. Even in non-personal datasets, metadata—such as IP logs, maintenance access timestamps, or geolocation data from technician devices—can fall under privacy regulations. Misconfigured cloud storage buckets are one of the most common sources of privacy breaches in energy systems, enabling unauthorized access to control logs or performance data. Poorly secured APIs may expose wind farm dashboards or digital twin data to external parties. Additionally, cloud-based AI models are susceptible to privacy leakage through side-channel attacks, model inversion, or unauthorized model downloads. Ensuring compliance requires encryption, access control enforcement, and continuous auditing of cloud data flows. Hybrid cloud is a cost-effective way to address the problem of insufficient resources for satisfying its users' requirements in a private cloud by elastically scaling up or down its service capability by combining the private cloud and public clouds (Wang, Wang, Huang, Song, & Qin, 2021).

5. CYBERSECURITY ARCHITECTURE FOR WIND POWER SYSTEMS

A robust cybersecurity architecture is essential for safeguarding modern wind power systems, which increasingly operate as distributed cyber-physical infrastructures governed by AI-driven monitoring, digital twins, cloud integration, and remote operator access. Because wind farms interface with both industrial control networks and corporate IT environments, their architecture must address threats across operational technology (OT), IoT devices, edge gateways, and cloud environments. The goal of an effective cybersecurity architecture is not merely to block intrusions, but to ensure resilience, continuity of operations, and trustworthy data flows throughout the entire turbine lifecycle. As wind power becomes more intertwined with AI-powered forecasting and automated decision-making, architectural safeguards must evolve to detect anomalies, authenticate every system interaction, and enforce strict access boundaries across all layers of the infrastructure. Internet of Things (IoT) technologies allow building a digital representation of people, objects, or physical phenomena to be available on the Internet (Cañón-Clavijo, Montenegro-Marin, Gaona-Garcia, & Ortiz-Guzmán, 2023).

5.1 Network Segmentation

At the foundation of secure wind farm operations is the segmentation of networks into distinct security zones that isolate SCADA systems from corporate IT networks and public-facing services. Segmentation prevents attackers who breach one environment from moving laterally into turbine control systems. In a typical architecture, turbines communicate through a dedicated OT network protected by firewalls, intrusion detection systems, and strict routing rules that allow only essential communication paths. Remote access portals, maintenance interfaces, and cloud uplinks are placed in demilitarized zones (DMZs), ensuring additional scrutiny for traffic entering or leaving the SCADA domain. Without this layered structure, operators risk cascading breaches where a compromised laptop or cloud account grants attackers direct access to turbine controllers. As wind farms scale in size and integrate AI-driven diagnostics, segmentation becomes an even more critical strategy for ensuring operational integrity and minimizing attack propagation.

5.2 Zero Trust Architecture

Traditional security models assume that devices inside a wind farm network are trustworthy, but Zero Trust architectures reject this assumption entirely. In a Zero Trust model, every access request—whether from a technician, an OEM vendor, a digital twin application, or an AI inference engine—must be authenticated, authorized, and continually verified. Device health, behavioral context, and user identity are evaluated before granting access, and permissions are limited to the minimum required for specific tasks. In wind energy systems, Zero Trust prevents unauthorized remote entry via shared vendor credentials, mitigates insider threats, and restricts the movement of malicious actors if one subsystem becomes compromised. AI further strengthens Zero Trust by analyzing behavioral anomalies, such as unusual turbine command patterns or unexpected data download activity. Because wind farms operate across distributed geographies with many external access points, Zero Trust becomes an indispensable foundation for secure operations (Daah, Qureshi, Awan, & Konur, 2024).

5.3 Secure Communication Protocols

Given the reliance of wind power systems on remote monitoring, automated turbine commands, and cloud-based analytics, ensuring secure communication protocols is essential. Legacy industrial protocols like Modbus or DNP3 remain widely used but were never designed with encryption or authentication in mind. Modern cybersecurity architecture replaces or wraps these protocols in secure transport layers such as TLS, authenticated VPN tunnels, or encrypted IEC 62351-compliant channels. These protections prevent attackers from intercepting SCADA traffic or injecting malicious commands into turbine controllers. With the rise of AI-driven applications requiring large volumes of sensor and telemetry data, secure channels are even more critical—any intercepted or manipulated data can corrupt digital twins, degrade predictive models, or trigger unsafe turbine actions. Secure communication, therefore, acts as the backbone of trustworthy wind farm automation.

5.4 Intrusion Detection Systems (IDS)

Intrusion Detection Systems play a crucial role in identifying abnormal behaviors across turbine networks, sensor streams, and SCADA command flows. Network-based IDS inspect traffic patterns for anomalies, such as unexpected connections to turbine PLCs, unusual bandwidth consumption, or known attack signatures targeting industrial protocols. Host-based IDS monitor the integrity of turbine controllers, SCADA servers, and edge gateways by detecting unauthorized configuration changes or privilege escalations. As wind farms adopt AI-powered analytics and digital twins, IDS systems can integrate machine learning models capable of identifying subtle anomalies—such as minor deviations in pitch control or irregularities in condition-monitoring data—that conventional signature-based tools may overlook. Effective IDS deployment ensures early detection of attacks before they escalate into turbine shutdowns, grid instability, or widespread data leakage.

5.5 Patch Management and Firmware Validation

Wind turbines rely on a diverse collection of firmware-driven components, including PLCs, inverters, pitch systems, and communication modems. These devices often operate for years in remote locations, making the timely application of patches and firmware updates a persistent challenge. A strong cybersecurity architecture mandates secure update pipelines that validate digital signatures, verify firmware integrity, and prevent tampered or unauthorized updates from being installed. Attackers often exploit outdated firmware or known vulnerabilities in industrial components to gain access to critical control interfaces. Additionally, firmware validation protects against supply chain attacks in which compromised vendor updates introduce backdoors into turbine fleets. In an AI-enhanced wind environment, secure update processes ensure that models, firmware, and control logic remain aligned and protected from manipulation.

5.6 Access Control Hardening

Identity and access management (IAM) is a cornerstone of wind power cybersecurity. Many cyber incidents in renewable energy stem from weak access practices such as shared passwords, inactive accounts, generic administrator profiles, or insufficient monitoring of remote vendor sessions. Access control hardening enforces strict policies, including multi-factor authentication, role-based access control, session monitoring, and time-limited credentials. With turbines frequently maintained by third-party technicians, OEM engineers, and multiple internal teams, granular IAM becomes essential for preventing unauthorized entry and limiting the blast radius of compromised accounts. As operators increasingly rely on AI for troubleshooting and performance analytics, AI systems themselves require controlled access, ensuring that only authenticated models or services can interact with SCADA or digital twin environments. Hardened access control ensures that wind power systems maintain confidentiality, integrity, and governance across all operational layers.

6. AI-DRIVEN CYBER DEFENSE FOR WIND POWER

As wind power systems become increasingly digitized and interconnected, traditional cybersecurity tools alone are no longer sufficient to defend against sophisticated and adaptive cyber threats. The integration of artificial intelligence in wind energy operations creates opportunities not only for enhanced performance and predictive maintenance but also for advanced cybersecurity protection capable of detecting, responding to, and learning from cyber incidents in real time. AI-driven cyber defense transforms security from a reactive, manually supervised process into a proactive and autonomous layer of protection. Machine learning models can analyze vast volumes of SCADA logs, sensor streams, communication metadata, and user activity records to identify subtle deviations that indicate emerging threats. When combined with digital twins and edge analytics, AI creates a dynamic, self-adaptive defense system that strengthens resilience across the entire wind power ecosystem.

6.1 Anomaly Detection in SCADA and Sensor Data

AI-based anomaly detection is one of the most powerful tools for identifying cyber threats in wind power systems. Machine learning models are trained to recognize normal operational patterns—such as typical rotor speeds, pitch adjustments, yaw movements, temperature curves, generator vibrations, and SCADA communication intervals. When attackers tamper with sensor data, inject malicious SCADA commands, or attempt unauthorized reconfiguration of turbine controllers, these models detect deviations in real time. Unlike rule-based security tools, AI can identify attacks that do not match known signatures, including zero-day exploits or stealthy sensor manipulation. For example, models can detect abnormal power output correlations, uncharacteristic command timing, subtle shifts in communication frequency, or unexpected multi-sensor inconsistencies that indicate cyber intrusion. This capability is especially critical in remote or offshore farms where on-site monitoring is limited and real-time detection is essential to prevent turbine misbehavior.

6.2 Predictive Threat Intelligence

AI augments cyber defense by generating predictive threat intelligence, enabling operators to anticipate attacks before they escalate. By analyzing global cyberattack trends, malware signatures, dark-web activity, industrial threat reports, and the behavior of adversarial groups, AI models can forecast the likelihood of specific threats targeting wind energy assets. These predictive engines continuously collect telemetry from wind farm networks, correlating it with large external threat datasets to provide early warnings. This allows operators to strengthen controls when risk levels rise, such as tightening VPN monitoring during active exploitation of industrial protocol vulnerabilities or increasing IDS sensitivity when ransomware campaigns targeting energy infrastructures intensify. Predictive intelligence is particularly valuable for wind farms that rely on cloud-based monitoring and third-party remote access, as threat exposure can shift rapidly depending on global cyber events and the behavior of linked vendors.

6.3 Automated Incident Response

The speed of cyberattacks—especially those targeting high-availability environments like wind farms—makes manual incident response insufficient. AI-driven orchestration tools automate containment actions such as isolating compromised turbines, blocking malicious traffic flows, and disabling unauthorized accounts. When SCADA command sequences indicate potential hijacking or when sensor data reflects suspicious fluctuations, AI systems can immediately quarantine affected devices or segments of the network. Automated playbooks enable rapid execution of predefined response procedures, reducing damage and ensuring continuity of generation. For example, if an edge gateway is hijacked, AI can instantly re-route telemetry through redundant communication paths while locking down the compromised asset. In hybrid onshore-offshore wind farms, automated incident response is essential for minimizing downtime, as reaching physical assets for manual intervention can take hours or even days. Combined with continuous monitoring, AI-driven response significantly strengthens the resilience of wind energy systems.

6.4 Digital Twins for Cyber Simulation

Digital twins offer a transformative approach to cybersecurity by creating virtual replicas of turbines, substations, and entire wind farms that can be used to simulate attacks and test defensive strategies without risking physical assets. AI-powered cybersecurity digital twins mirror real-world conditions, including sensor dynamics, turbine control logic, network topologies, and operational behaviors. This allows operators to evaluate how different cyberattacks—such as sensor spoofing, SCADA hijacking, or communication interference—would affect turbine safety and grid stability. By experimenting in the digital environment, defenders can refine detection thresholds, tune anomaly detection models, and validate incident response workflows. Digital twins also help identify unforeseen cascading effects, such as how manipulating yaw control on one turbine might influence energy balancing across the farm or how delayed sensor data could compromise predictive maintenance accuracy. As wind farms adopt more AI-driven automation, digital twin cyber simulations become essential for ensuring that security measures remain effective against evolving threats.

7. REGULATORY AND STANDARDS FRAMEWORK

As wind power systems mature into critical national infrastructure, strong regulatory and standards frameworks have become essential for governing cybersecurity, data protection, and operational resilience. Wind farms now interface with national grids, cloud service providers, maintenance contractors, telecommunications networks, and AI-driven monitoring platforms. This multi-layered interdependence means that regulatory compliance is no longer limited to electrical performance standards; it now encompasses industrial cybersecurity, digital identity management, data privacy, cloud governance, and supply chain assurance. A comprehensive standards framework ensures wind operators adopt unified practices across disparate assets, reducing systemic vulnerability and supporting long-term grid stability. Figure 4 shows the Regulatory Mapping Across Wind Farm Cyber-Physical Architecture.

Figure 4. Regulatory Mapping Across Wind Farm Cyber-Physical Architecture

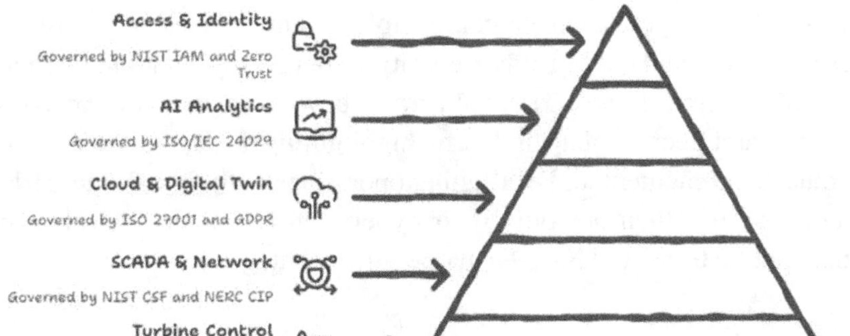

(Ouyang, Mo, Huang, & Chen, 2025)

The integration of AI, digital twins, and IoT sensors further increases the importance of regulatory clarity, since these technologies introduce new risk vectors related to model poisoning, data leakage, and algorithmic malfunction. Standards such as IEC 62443 explicitly address industrial automation and control system security, offering guidance on segmentation, access control, and risk assessment tailored for complex cyber-physical environments. Meanwhile, ISO 27001 focuses on information security management, helping operators protect sensitive operational and environmental data throughout its lifecycle. For wind farms operating within or supplying power to the European Union, GDPR applies whenever personal or pseudonymous data—such as technician access logs, IP addresses, or geolocation metadata—is collected or transmitted. In North America, NERC CIP standards impose strict requirements on entities connected to the Bulk Electric System (BES), including access control, incident reporting, and change management procedures.

Because wind energy operations rely heavily on third-party vendors, remote OEM access, and cloud-hosted dashboards, supply chain governance has also become a regulatory priority. Many frameworks now require operators to vet third-party cybersecurity practices, enforce role-based access control, and maintain detailed audit trails for remote access. This is particularly important for AI-driven wind farms, as external vendors may supply analytics engines, digital twin software, or predictive maintenance models whose data handling practices must be monitored. Regulations increasingly emphasize secure firmware update pipelines, model transparency, and trustworthy AI principles to ensure that wind systems remain resilient even as automation increases. Harmonizing these requirements across jurisdictions helps operators avoid compliance gaps and ensures consistent protection across all operational layers.

In practice, regulatory compliance should be integrated into the very architecture of wind power systems. Rather than treating standards as external requirements, operators must embed them into SCADA design, communication protocols, IAM policies, and AI lifecycle management. Table 5 summarizes the most relevant standards affecting wind energy cybersecurity, data privacy, and AI governance. A conceptual illustration (Figure 3) demonstrates how these standards map onto the cyber-physical architecture of a wind farm, highlighting the layers where regulatory controls must be implemented. By aligning operations with global standards, wind farm operators strengthen not only their cybersecurity posture but also investor confidence, public trust, and long-term operational integrity.

Table 5. Cybersecurity and Data Governance Standards for Wind Power Systems

Standard / Regulation	Primary Focus	Relevance to Wind Power
IEC 62443	Industrial control system (ICS/ OT) security	Protects SCADA, PLCs, network segmentation, device authentication
NIST Cybersecurity Framework	Risk management, protection, detection, response	Guides holistic cybersecurity programs for wind infrastructures
ISO/IEC 27001	Information security management	Secures operational logs, cloud data, technician credentials
GDPR	Data privacy for EU residents	Regulates geolocation metadata, remote access logs, cross-border transfers
NERC CIP	Critical infrastructure protection in North America	Enforces strict controls for grid-connected wind farms
IEC 62351	Secure communication protocols for power systems	Encrypts SCADA and telemetry between turbines and substations
ISO/IEC 24029	AI trustworthiness and robustness	Relevant for AI-driven monitoring, digital twins, and anomaly detection
ISO 55000	Asset management systems	Ensures lifecycle governance of turbine components and firmware

8. CASE STUDIES OF WIND POWER CYBERATTACKS

A well-known example occurred in 2022 when a ransomware attack disabled remote monitoring for more than 5,000 turbines across Europe, demonstrating how third-party communication hardware can become a single point of failure. Incidents like this illustrate the urgency of understanding the threat landscape described in the following sections. The ransomware attack on European wind farms in 2022 remains one of the most widely documented incidents in the renewable energy sector. In this event, over 5,000 wind turbines lost remote monitoring and control functionality when a third-party satellite modem provider was compromised. Although the turbines remained operational, operators were forced to switch to on-site manual control and halt predictive maintenance processes for several days. The attack demonstrated how vulnerabilities in external suppliers—rather than in the turbines themselves—can have widespread operational impact. The disruption highlighted the systemic risk introduced by cloud-connected OEM dashboards, remote firmware update channels, and external communication infrastructures, emphasizing the need for robust supply chain governance and continuous verification of vendor cybersecurity practices.

X-Force Threat Intelligence Index 2023 has observed that in 2022 energy sector encountered 10.7% of the recorded cyberattacks (Saraswat & Tiwari, 2025).

A second notable incident involved an attempted SCADA breach in Asia, where attackers exploited outdated VPN credentials belonging to a contractor to infiltrate a major wind farm's operational network. Once inside, the intruders attempted to modify turbine control logic, specifically targeting yaw angle settings and reactive power controls. Although the intrusion was detected early and no permanent damage occurred, the event exposed significant weaknesses in access control practices and remote maintenance pathways. It also underscored the importance of Zero Trust principles, multi-factor authentication, and real-time monitoring of SCADA command sequences. The incident illustrated how credential compromise can escalate rapidly into operational manipulation, especially when legacy industrial protocols and insufficient network segmentation are involved. The effects of external intentional attacks and internal cascade failures are anticipated to significantly impact smart power grids, particularly in the supervisory control and data acquisition (SCADA) system (Zhang, Fang, Dong, & Xiahou, 2026).

A third case involved GPS spoofing at an offshore wind facility, where interference with satellite-based geolocation caused several turbines and inspection drones to behave unpredictably. Turbine yaw systems began making abnormal directional adjustments, and autonomous drones used for blade inspection deviated from planned flight paths, leading to aborted missions and delays in scheduled maintenance. While the spoofing attack did not directly compromise turbine control logic, it demonstrated how wind farms dependent on external positioning systems are vulnerable to non-intrusive cyber threats that exploit environmental technologies rather than digital infrastructure. The incident raised awareness about securing timing and geolocation dependencies through redundant sensors, cross-validation algorithms, and AI-driven anomaly detection to prevent misalignment or false geospatial readings in offshore operations. Wind turbines are important for achieving renewable energy goals, but present a considerable threat to wildlife, especially birds and bats (Škrábal et al., 2025).

9. CYBERSECURITY BEST PRACTICES FOR WIND OPERATORS

Modern wind energy systems require a holistic and proactive cybersecurity strategy that extends across operational technology, information systems, cloud infrastructures, and third-party service environments. Because wind farms rely on continuous remote monitoring, AI-driven analytics, and distributed SCADA networks, security best practices must ensure the confidentiality, integrity, and availability of both data and control systems. Best practices combine technical safeguards, organizational governance, access control management, incident preparedness, and rigorous data handling procedures. By integrating these elements into daily operations, wind operators can strengthen resilience against emerging cyber threats while ensuring compliance with global standards such as IEC 62443, ISO 27001, and NERC CIP (Mishra, Alzoubi, Gill, & Anwar, 2022).

9.1 Robust IAM Policies

Identity and Access Management (IAM) represents the foundation of cyber governance in wind energy operations. Strong IAM programs enforce multi-factor authentication, strict least-privilege policies, and continuous session monitoring for both internal technicians and external vendors. Many wind farm incidents have originated from shared passwords, outdated contractor accounts, and weak VPN configurations. To mitigate these risks, operators must implement role-based access control (RBAC), limit the duration of remote access sessions, and maintain detailed logs of user activities across SCADA, turbine controllers, cloud dashboards, and digital twin environments. Effective IAM also requires periodic credential audits, automated revocation of unused accounts, and privileged access monitoring to prevent escalation or misuse. In an AI-driven environment, IAM also governs which AI tools or analytics engines may interact with operational data, preventing unauthorized model training or inference processes. PLC, SCADA, and fuzzy control ensure long-term water conservation in agriculture and reduction in the power consumption (Morsi & Ahmed, 2024).

9.2 Regular Penetration Testing

Penetration testing is essential for identifying vulnerabilities across SCADA networks, industrial controllers, edge devices, and communication infrastructures. Unlike traditional IT environments, wind farms require specialized OT penetration tests that examine PLC firmware, control logic pathways, industrial protocols, and remote access mechanisms used by turbine OEMs. Regular testing helps uncover weak segmentation policies, insecure API endpoints, unpatched controller vulnerabilities, and opportunities for lateral movement across OT–IT boundaries. Penetration tests should simulate realistic attack scenarios, including attempts to manipulate yaw commands, alter pitch sequences, or disrupt digital twin telemetry. Testing must also extend to cloud environments hosting analytics platforms, as insecure containers, unprotected buckets, and misconfigured orchestrators can expose sensitive operational data. By institutionalizing penetration testing, wind operators establish continuous visibility into evolving vulnerabilities and strengthen their defensive posture.

9.3 Data Governance Framework

A mature data governance framework ensures that wind farm data is classified, controlled, audited, and retained according to regulatory and operational requirements. Wind energy systems generate diverse datasets, including SCADA logs, turbine vibration metrics, geolocation streams, meteorological readings, and digital twin outputs. Each dataset carries different sensitivity levels and must be governed accordingly. Proper governance defines retention periods, encryption rules, access privileges, and cross-border transfer restrictions (for example, GDPR compliance for EU-based operators). Strong governance also prevents uncontrolled data proliferation, which can expose operational details to third parties or create inconsistencies across AI training pipelines. Logging and auditing mechanisms ensure that all data access, whether by maintenance staff, AI algorithms, remote OEMs, or cloud services, is tracked and validated. This level of oversight reduces privacy risks, prevents unauthorized data extraction, and supports forensic investigation in case of cyber incidents.

9.4 Secure AI Lifecycle

As wind farms adopt AI for predictive maintenance, forecasting, fault classification, and cybersecurity analytics, securing the AI lifecycle becomes mandatory. This requires robust protections across model training, validation, deployment, and continuous monitoring stages. Attackers may attempt to poison training data, extract proprietary models, or alter model parameters to disrupt turbine optimization algorithms. To counter these risks, operators must implement trustworthy AI practices, including dataset integrity validation, model explainability mechanisms, version control, and adversarial robustness testing. Continuous monitoring detects model drift or anomalous inference behavior, ensuring that AI-driven decisions remain safe and reliable. This is especially important where AI interacts directly with turbine controls or provides maintenance recommendations; compromised models can have immediate financial or physical consequences. Secure AI lifecycle governance therefore protects both operational performance and strategic intellectual property.

9.5 Disaster Recovery and Business Continuity Planning

Wind power systems must remain operational even under cyberattack conditions, requiring comprehensive disaster recovery (DR) and business continuity planning (BCP). Effective DR strategies include redundant SCADA control centers, backup communication channels for remote monitoring, and segmented failover networks capable of isolating compromised components while maintaining partial functionality. Operators must develop crisis response playbooks that define escalation protocols, communication procedures, and restoration priorities for critical assets such as substations, turbine clusters, or meteorological systems. Regular drills—both physical and simulated through cybersecurity digital twins—ensure that personnel are prepared to respond rapidly to ransomware incidents, network outages, or orchestrated attacks. Business continuity planning also involves safeguarding cloud-based data stores, enabling rapid reconstruction of historical logs, AI models, and turbine performance archives after an incident. By maintaining redundancy and operational readiness, wind farms can minimize downtime, prevent cascading grid disruptions, and maintain safety even during severe cyber events.

10. FUTURE CHALLENGES AND OUTLOOK

The future of wind power cybersecurity will be shaped by expanding offshore installations, increasing dependence on AI-driven automation, and the integration of hybrid wind–solar–storage ecosystems that significantly enlarge the cyber-physical attack surface. As turbines evolve toward autonomous operation and digital twins become central to maintenance planning and grid interaction, the complexity of defending these systems will grow, requiring more sophisticated anomaly detection, model validation, and secure data pipelines. Emerging technologies such as quantum computing threaten current cryptographic protocols, necessitating the adoption of quantum-resistant encryption to safeguard long-term confidential data. Meanwhile, global expansion of cross-border data flows, reliance on multinational cloud providers, and tightening regional privacy laws will demand stronger governance frameworks to ensure regulatory compliance across diverse jurisdictions.

In addition, the distributed nature of wind farms—often remote, unmanned, and dependent on third-party maintenance—creates persistent vulnerabilities that must be addressed through continuous monitoring, resilient communication architectures, and zero-trust operational models. Ultimately, the future of secure wind energy will hinge on integrating cybersecurity from initial system design through decommissioning, ensuring resilience, transparency, and public trust as renewable infrastructure becomes the backbone of global energy transitions. Table 6 shows the Emerging Future Challenges in Wind Power Cybersecurity.

Table 6. Emerging Future Challenges in Wind Power Cybersecurity

Challenge Area	Description	Implications for Wind Systems
Quantum-Resistant Security	Need for post-quantum cryptographic algorithms	Protect long-term SCADA data, firmware, and cloud communications
Autonomous & AI-Driven Turbines	Increased automation and AI control loops	Higher risk of model poisoning, drift, and unsafe decisions
Hybrid Renewable Integration	Wind combined with solar, storage, and microgrids	Larger attack surface and greater interdependence across assets
Offshore Expansion	Turbines in remote maritime locations	Limited physical security and higher reliance on satellite links
Cross-Border Data Governance	Cloud-based analytics spanning multiple regions	Complex compliance with GDPR, NERC CIP, and national privacy laws
Third-Party Access Risks	OEMs, contractors, and service vendors	Need for strict IAM, audit trails, and supply chain vetting
Digital Twin Proliferation	Increased use of simulation-based operations	Greater risk of twin tampering and inaccurate predictive modeling
Climate & Environmental Stress	Extreme weather and climate-driven disruptions	Higher likelihood of physical–cyber failure cascades

11. CONCLUSION

The accelerating digitalization of wind power systems has transformed the sector from a collection of standalone mechanical assets into a complex, interconnected cyber-physical ecosystem that depends on AI-driven analytics, cloud-based monitoring, IoT sensing, and remote-access operational technology. This transformation has significantly improved turbine efficiency, predictive maintenance capabilities, and situational awareness across geographically dispersed wind farms. However, it has also broadened the attack surface, exposing critical components such as SCADA networks, PLCs, edge sensors, and digital twin platforms to increasingly sophisticated cyber threats. The vulnerabilities described in this chapter—ranging from ransomware and DDoS attacks to GPS spoofing, supply chain intrusions, and data privacy breaches—demonstrate that cybersecurity in wind power is no longer a peripheral concern but a fundamental operational requirement.

As wind energy becomes a cornerstone of global decarbonization and grid modernization initiatives, ensuring the confidentiality, integrity, and availability of its digital infrastructure is essential for maintaining energy reliability, investor confidence, and public trust.

Looking ahead, securing wind power systems will require an integrated, lifecycle-oriented approach that embeds cybersecurity and data governance into every layer of turbine design, deployment, and operation. This includes adopting industry standards such as IEC 62443 and ISO 27001, deploying AI-driven anomaly detection and predictive threat intelligence, enforcing Zero Trust architectures, and strengthening identity and access management across all vendor and operator interfaces. Equally important is addressing emerging challenges posed by quantum computing, hybrid renewable integration, and the expanding reliance on cloud platforms and cross-border data flows.

As offshore wind capacity grows and autonomous turbine operation becomes more prevalent, the need for continuous monitoring, robust incident response strategies, and secure AI lifecycle management will only intensify. Ultimately, building resilience in wind power systems requires a proactive, adaptive cybersecurity posture that evolves in tandem with the technologies it protects, ensuring that renewable energy infrastructures remain secure, reliable, and capable of supporting the world's transition toward sustainable, intelligent energy grids.

REFERENCES

Ahatsi, E., Winkler, H., & Olanrewaju, O. (2025). Modified Teaching-Learning Based Optimisation for Resilience and Sustainability in the Renewable Energy Industry. In *2025 IEEE 13th International Conference on Smart Energy Grid Engineering, SEGE 2025* (pp. 225–230). DOI: 10.1109/SEGE65970.2025.11203381

Awad, H., & Bayoumi, E. H. E. (2025). Next-Generation Smart Inverters: Bridging AI, Cybersecurity, and Policy Gaps for Sustainable Energy Transition. *Technologies, 13*(4), 136. Advance online publication. DOI: 10.3390/technologies13040136

Biswas, P., Rashid, A., Al Masum, A., Al Nasim, M. A., Anas Ferdous, A. S. M., Gupta, K. D., & Biswas, A. (2025). *An Extensive and Methodical Review of Smart Grids for Sustainable Energy Management-Addressing Challenges with AI, Renewable Energy Integration and Leading-edge Technologies*. IEEE., DOI: 10.1109/ACCESS.2025.3537651

Cañón-Clavijo, R. E., Montenegro-Marin, C. E., Gaona-Garcia, P. A., & Ortiz-Guzmán, J. (2023). IoT Based System for Heart Monitoring and Arrhythmia Detection Using Machine Learning. *Journal of Healthcare Engineering, 6401673*(1), 6401673. Advance online publication. DOI: 10.1155/2023/6401673 PMID: 36818385

Daah, C., Qureshi, A., Awan, I., & Konur, S. (2024). Enhancing Zero Trust Models in the Financial Industry through Blockchain Integration: A Proposed Framework. *Electronics (Basel), 13*(5), 865. Advance online publication. DOI: 10.3390/electronics13050865

Demirtas, M., Sharkh, S. M., Gokyildirim, A., & Calgan, H. (2025). Secure operation of a stand-alone wind energy system based on an incommensurate fractional-order chaotic system. *Applied Energy, 384*, 125477. Advance online publication. DOI: 10.1016/j.apenergy.2025.125477

Faheem, M., & Al-Khasawneh, M. A. (2024). Multilayer cyberattacks identification and classification using machine learning in internet of blockchain (IoBC)-based energy networks. *Data in Brief, 54*, 110461. Advance online publication. DOI: 10.1016/j.dib.2024.110461 PMID: 38774244

Hachimi, M., El Gaabouri, I., El Bhiri, B., Motahhir, S., & Boussetta, M. (2025). Cybersecurity Attacks and Defense Techniques in Microgrid Networks: A Comprehensive Survey. In *Lecture Notes in Networks and Systems* (Vol. 1555 LNNS, pp. 182–199). DOI: 10.1007/978-3-031-99997-0_17

Jurczuk, M., & Suprynowicz, M. (2024). Consent in Data Privacy: A General Comparison of GDPR and HIPAA. *Adam Mickiewicz University Law Review*, *16*, 173–194. DOI: 10.14746/ppuam.2024.16.07

Mbasso, W. F., Harrison, A., Dagal, I., Jangir, P., Khishe, M., Kotb, H., Shaikh, M. S., Smerat, A., Fendzi Donfack, E., & Kumar, R. (2025). Digital twins in renewable energy systems: A comprehensive review of concepts, applications, and future directions. *Energy Strategy Reviews*, *61*, 101814. Advance online publication. DOI: 10.1016/j.esr.2025.101814

Mishra, A., Alzoubi, Y. I., Gill, A. Q., & Anwar, M. J. (2022). Cybersecurity Enterprises Policies: A Comparative Study. *Sensors (Basel)*, *22*(2), 1–35. DOI: 10.3390/s22020538 PMID: 35062504

Moayyed, H., Moradzadeh, A., Mohammadi-Ivatloo, B., Aguiar, A. P., & Ghorbani, R. (2022). A Cyber-Secure generalized supermodel for wind power forecasting based on deep federated learning and image processing. *Energy Conversion and Management*, *267*, 115852. Advance online publication. DOI: 10.1016/j.enconman.2022.115852

Morsi, I., & Ahmed, M. (2024). Smart irrigation system using PLC, SCADA, and fuzzy control. *Edelweiss Applied Science and Technology*, *8*(6), 9101–9120. DOI: 10.55214/25768484.v8i6.3947

Naderi, E., & Asrari, A. (2025). Cybersecurity Challenges in Power-to-X Integration. In *Power-to-X in Regional Energy Systems* (pp. 284–298). Planning, Operation, Control, and Market Perspectives., DOI: 10.1201/9781032719436-12

Ouyang, J., Mo, F., Huang, F., & Chen, Y. (2025). Risk Assessment Method for CPS-Based Distributed Generation Cluster Control in Active Distribution Networks Under Cyber Attacks. *Sensors (Basel)*, *25*(19), 6053. Advance online publication. DOI: 10.3390/s25196053 PMID: 41094876

Oyucu, S., Polat, O., Türkoğlu, M., Polat, H., Aksöz, A., & Ağdaş, M. T. (2024). Ensemble Learning Framework for DDoS Detection in SDN-Based SCADA Systems. *Sensors (Basel)*, *24*(1), 155. Advance online publication. DOI: 10.3390/s24010155 PMID: 38203015

Paul, S., Kumar, P., & Chatterjee, K. (2025). Detection and mitigation of cyber attack in DFIG based wind energy systems. *Sustainable Energy. Grids and Networks*, *44*, 102057. Advance online publication. DOI: 10.1016/j.segan.2025.102057

Ramchandani, M., Vahora, M. A., Wagle, S., & Dhiman, H. S. (2025). On Modeling False Data Injection Attacks for Wind Turbine Gearbox using Multivariate SCADA Data. In *2025 International Conference on Artificial Intelligence and Machine Vision, AIMV 2025*. DOI: 10.1109/AIMV66517.2025.11203394

Saraswat, A., & Tiwari, G. (2025). United Nations and Beyond: Legal Strategies for Defending Critical Energy Infrastructure Against Cyber Attacks. In *Studies in Computational Intelligence* (Vol. 1181, pp. 291–307). DOI: 10.1007/978-3-031-80557-8_13

Shafiee, M. (2025). Cyber Resilience of Wind Energy Infrastructure: Challenges and Mitigation Strategies. In *ICAC 2025 - 30th International Conference on Automation and Computing*. DOI: 10.1109/ICAC65379.2025.11196193

Škrábal, J., Raab, M., Raab, R., Grüebler, M. U., Kormann, U. G., Scherler, P., Sumasgutner, P., Åkesson, S., Bermejo, A., Chakarov, N., Fiedler, W., Godino, A., Haraszthy, L., Klein, K., Kolbe, M., Literák, I., Mammen, K., Mammen, U., Paquet, J.-Y., & Raab, R. (2025). Red kite (Milvus milvus) collision risk is higher at wind turbines with larger rotors and lower clearance, evidenced by GPS tracking. *Biological Conservation*, *312*, 111482. Advance online publication. DOI: 10.1016/j.biocon.2025.111482

Srivastava, A., Rawat, B. S., & Maurya, S. K. (2023). A Review on Protecting SCADA Systems from DDOS Attacks. In *2023 4th International Conference on Electronics and Sustainable Communication Systems, ICESC 2023 - Proceedings* (pp. 556–561). DOI: 10.1109/ICESC57686.2023.10193650

Wang, B., Wang, C., Huang, W., Song, Y., & Qin, X. (2021). Security-aware task scheduling with deadline constraints on heterogeneous hybrid clouds. *Journal of Parallel and Distributed Computing*, *153*, 15–28. https://doi.org/https://doi.org/10.1016/j.jpdc.2021.03.003. DOI: 10.1016/j.jpdc.2021.03.003

Yang, B., & Zhang, Y. (2022). Cybersecurity Analysis of Wind Farm Industrial Control System Based on Hierarchical Threat Analysis Model Framework. In *Proceedings - 2022 International Conference on Computing, Communication, Perception and Quantum Technology, CCPQT 2022* (pp. 6–13). DOI: 10.1109/CCPQT56151.2022.00008

Zhang, J., Fang, Z., Dong, W., & Xiahou, T. (2026). Deriving the optimal defense resource deployment: Modeling cascading effect and contest intensity in an attacker–defender game. *Reliability Engineering & System Safety*, *266*, 111779. Advance online publication. DOI: 10.1016/j.ress.2025.111779

Compilation of References

Abdullahi, I., Longo, S., & Samie, M. (2024). Towards a Distributed Digital Twin Framework for Predictive Maintenance in Industrial Internet of Things (IIoT). *Sensors (Basel)*, *24*(8), 1–30. DOI: 10.3390/s24082663 PMID: 38676279

Abdul-Ruhman, M. S., & Hawas, M. N. (2020). Optimal Pitch Angle Control for Wind Turbine Using Intelligent Controller. *IOP Conference Series. Materials Science and Engineering*, *745*(1), 012017. DOI: 10.1088/1757-899X/745/1/012017

Abimbola, O., Ojo, O. T., Fagbola, E., Idris, U. A., & Salman, M. B. (2025). IoT-driven predictive maintenance for wind turbines. *Path of Science*. DOI: 10.22178/pos.114-20

Adefarati, T., Bansal, R. C., Bettayeb, M., & Naidoo, R. (2021). Optimal energy management of a PV-WTG-BSS-DG microgrid system. *Energy*, *217*, 119358. DOI: 10.1016/j.energy.2020.119358

Afridi, Y. S., Ahmad, K., & Hassan, L. (2022). Artificial intelligence based prognostic maintenance of renewable energy systems: A review of techniques, challenges, and future research directions. *International Journal of Energy Research*, *46*(15), 21619–21642. DOI: 10.1002/er.7100

Aghaei, S. S., & Zamani, H. (2021). Review of IoT Security in Distributed Renewable Energy: Threats and Countermeasures. *Renewable Energy*, *180*, 133–150.

Ahatsi, E., Winkler, H., & Olanrewaju, O. (2025). Modified Teaching-Learning Based Optimisation for Resilience and Sustainability in the Renewable Energy Industry. In *2025 IEEE 13th International Conference on Smart Energy Grid Engineering, SEGE 2025* (pp. 225–230). DOI: 10.1109/SEGE65970.2025.11203381

Ahmed, I. A., & Asamoah, P. B. (2024). AI-driven predictive maintenance for energy infrastructure. *International Journal of Innovation and Scientific Research*, *XI*(IX), 507–528. DOI: 10.51244/IJRSI.2024.1109048

Ahmed, M., & Khan, M. R. (2025). Artificial intelligence-enabled digital twins for energy efficiency in smart grids. *Review of Applied Science and Technology*, *4*(02), 580–615. DOI: 10.63125/12kp9w74

Al Abdulwahid, A. (2024). Cyber-physical threat mitigation in wind energy systems: A novel secure architecture for industry 4.0 power grids. *Energy Informatics*, *7*(1), 141. DOI: 10.1186/s42162-024-00449-6

Alagha, N., Khairuddin, A., Haitaamar, Z., Al-Khatib, O., & Kanesan, J. (2025). Artificial Intelligence in Wind Turbine Fault Detection and Diagnosis: Advances and Perspectives. *Energies*, *18*(7), 1–23. DOI: 10.3390/en18071680

Alhamam, N., Hafizur Rahman, M., & Aljughaiman, A. (2025). A Comprehensive Review on Cybersecurity of Digital Twins Issues, Challenges, and Future Research Directions. *IEEE Access : Practical Innovations, Open Solutions*, *13*, 45106–45124. DOI: 10.1109/ACCESS.2025.3545004

Alshammari, A. (2025). Securing smart microgrids with a novel multi-layer cybersecurity framework for Industry 4.0 renewable energy systems. *Discover Computing*, *28*(1), 80. DOI: 10.1007/s10791-025-09600-7

Alves, D., Mendonça, F., Mostafa, S., & Morgado-Dias, F. (2024). A comprehensive IoT cloud-based wind station ready for real-time measurements and artificial intelligence integration. *e-Prime - Advances in Electrical Engineering. Electronics and Energy*, *10*, 100862. DOI: 10.1016/j.prime.2024.100862

Amini, A., Kanfoud, J., & Gan, T. H. (2022). An artificial intelligence neural network predictive model for anomaly detection and monitoring of wind turbines using SCADA data. *Applied Artificial Intelligence*, *36*(1), 2034718. DOI: 10.1080/08839514.2022.2034718

Amin, U., Kim, D., Ahmed, F. N., Ahmad, G., & Hossain, M. J. (2025). Digital twins for smart asset management in the energy industry: State-of-the-art. *Expert Systems with Applications*, *289*, 128358. DOI: 10.1016/j.eswa.2025.128358

An, J., Hu, X., Gong, L., Zou, Z., & Zheng, L. R. (2024). Fuzzy reliability evaluation and machine learning-based fault prediction of wind turbines. *Journal of Industrial Information Integration*, *40*, 100606. DOI: 10.1016/j.jii.2024.100606

Arroba, P., Buyya, R., Cárdenas, R., Risco-Martín, J., & Moya, J. (2024). Sustainable edge computing: Challenges and future directions. *Software, Practice & Experience*, *54*(11), 2272–2296. DOI: 10.1002/spe.3340

Astolfi, D., De Caro, F., & Vaccaro, A. (2023). Condition Monitoring of Wind Turbine Systems by Explainable Artificial Intelligence Techniques. *Sensors (Basel), 23*(12), 1–24. DOI: 10.3390/s23125376 PMID: 37420542

Avci, A. B. (2024). AI-driven approaches to enhance energy efficiency in heritage architecture: A review. *6th International Symposium on Innovation in Architecture, Planning and Design Proceedings*, 37–41. DOI: 10.36287/setsci.20.5.037

Awad, H., & Bayoumi, E. H. E. (2025). Next-Generation Smart Inverters: Bridging AI, Cybersecurity, and Policy Gaps for Sustainable Energy Transition. *Technologies, 13*(4), 136. Advance online publication. DOI: 10.3390/technologies13040136

Awal, M. A., & Husain, I. (2020). Unified Virtual Oscillator Control for Grid-Forming and Grid-Following Converters. *IEEE Journal of Emerging and Selected Topics in Power Electronics*. Advance online publication. 10.1109.

Aziz, U., Charbonnier, S., Bérenguer, C., Lebranchu, A., & Prevost, F. (2021). Critical comparison of power-based wind turbine fault-detection methods using a realistic framework for SCADA data simulation. *Renewable & Sustainable Energy Reviews, 144*, 110961. DOI: 10.1016/j.rser.2021.110961

Backhus, J., Rao, A., Venkatraman, C., Padmanabhan, A., Kumar, A., & Gupta, C. (2024). Equipment Health Assessment: Time Series Analysis for Wind Turbine Performance. *Applied Sciences (Basel, Switzerland), 14*(8), 1–20. DOI: 10.3390/app14083270

Badihi, H., Zhang, Y., Jiang, B., Pillay, P., & Rakheja, S. (2022). A comprehensive review on signal-based and model-based condition monitoring of wind turbines: Fault diagnosis and lifetime prognosis. *Proceedings of the IEEE, 110*(6), 754–806. DOI: 10.1109/JPROC.2022.3171691

Bakir, I., Yildirim, M., & Ursavas, E. (2021). An integrated optimization framework for multi-component predictive analytics in wind farm operations and maintenance. *Renewable & Sustainable Energy Reviews, 138*, 110639. DOI: 10.1016/j.rser.2020.110639

Bassey, K. E. (2022a). Enhanced design and development simulation and testing. Eng. Sci. Technol. j., 3(2), 18–31. DOI: 10.51594/estj.v3i2.1283

Bassey, K. E. (2022b). Optimizing wind farm performance using machine learning. Eng. Sci. Technol. j., 3(2), 32–44. DOI: 10.51594/estj.v3i2.1284

Bawane, S., Matange, G., Shrivastava, A., Qureshi, A., Sultan, S., & Shrivastava, D. (2025). Smart Wind Farm Management Using IoT and Predictive in Analytics. *International Journal of Environmental Sciences, 11*(3), 1079–1095.

Bebars, A. D., Eladl, A. A., Abdulsalam, G. M., & Badran, E. A. (2022). Internal electrical fault detection techniques in DFIG-based wind turbines: A review. *Protection and Control of Modern Power Systems*, *7*(2), 1–22. DOI: 10.1186/s41601-022-00236-z

Benbouzid, M., Berghout, T., Sarma, N., Djurović, S., Wu, Y., & Ma, X. (2021). Intelligent Condition Monitoring of Wind Power Systems: State of the Art Review. *Energies*, *14*(18), 5967. DOI: 10.3390/en14185967

Benítez, I. B., & Singh, J. G. (2025). A comprehensive review of machine learning applications in forecasting solar PV and wind turbine power output. *Journal of Electrical Systems and Information Technology*, *12*(1), 54. DOI: 10.1186/s43067-025-00239-4

Besigomwe, K. (2025). Closed-Loop Manufacturing with AI-Enabled Digital Twin Systems. *Cognizance Journal of Multidisciplinary Studies*, *5*(1), 18–38. DOI: 10.47760/cognizance.2025.v05i01.002

Bie, Y., Liu, X., Xu, T., Zhu, Z., & Li, Z. (2021). A review of the application of oil analysis in condition monitoring and life prediction of wind turbine gearboxes. *Insight (Northampton)*, *63*(5), 289–301. DOI: 10.1784/insi.2021.63.5.289

Bindingsbø, O. T., Singh, M., Øvsthus, K., & Keprate, A. (2023). Fault detection of a wind turbine generator bearing using interpretable machine learning. *Frontiers in Energy Research*, *11*, 1284676. Advance online publication. DOI: 10.3389/fenrg.2023.1284676

Bisu, C., Olaru, A., Olaru, S., Alexei, A., Mihai, N., & Ushaq, H. (2024). Monitoring the Wear Trends in Wind Turbines by Tracking Fourier Vibration Spectra and Density Based Support Vector Machines. *Mathematics*, *12*(9), 1307. DOI: 10.3390/math12091307

Biswas, P., Rashid, A., Al Masum, A., Al Nasim, M. A., Anas Ferdous, A. S. M., Gupta, K. D., & Biswas, A. (2025). *An Extensive and Methodical Review of Smart Grids for Sustainable Energy Management-Addressing Challenges with AI, Renewable Energy Integration and Leading-edge Technologies*. IEEE., DOI: 10.1109/ACCESS.2025.3537651

Boretti, A. (2019). Dependent performance of South Australian wind energy facilities with respect to resource and grid availability. *Energy Storage*, *1*(6), e97. DOI: 10.1002/est2.97

Bosch, C. (2022, February). Machine Learning for wind turbine fault prediction through the combination of datasets from same type turbines. Floating Offshore Energy Devices. DOI: 10.21741/9781644901731-7

Bose, B. K. (2017). Artificial Intelligence Techniques in Smart Grid and Renewable Energy Systems—Some Example Applications. *Proceedings of the IEEE*, *105*(11), 2262–2273. DOI: 10.1109/JPROC.2017.2756596

Bošnjaković, M., Martinović, M., & Đokić, K. (2025). Application of artificial intelligence in wind power systems. *Applied Sciences (Basel, Switzerland)*, *15*(5), 2443. DOI: 10.3390/app15052443

Bowen, T., Chernyakhovskiy, I., & Denholm, P. L. (2019). *Grid-scale battery storage: frequently asked questions*. National Renewable Energy Lab. (NREL), Golden, CO (United States). .DOI: 10.2172/1561843

Branlard, E., Jonkman, J., Brown, C., & Zhang, J. (2024). A digital twin solution for floating offshore wind turbines validated using a full-scale prototype. *Wind Energy Science*, *9*(1), 1–24. DOI: 10.5194/wes-9-1-2024

Buchberg, E. M., & Nielsen, K. V. (2025). Condition monitoring with machine learning: a data-driven framework for quantifying wind turbine energy loss. *arXiv preprint*. /arXiv.2506.13012DOI: 10.48550

Bullich-Massagué, E., Cifuentes-García, F.-J., Glenny-Crende, I., Cheah-Mañé, M., Aragüés-Peñalba, M., Díaz-González, F., & Gomis-Bellmunt, O. (2020). A review of energy storage technologies for large scale photovoltaic power plants. *Applied Energy*, *274*, 115213. DOI: 10.1016/j.apenergy.2020.115213

Bustamante, S., Pinzón, J. D., & Giraldo-Gómez, D. (2025). Evolution of real-time dynamics monitoring of Colombian power grid using wide-area monitoring system and high-speed big data analytics. *Sustainability (Basel)*, *17*(9), 3848. DOI: 10.3390/su17093848

Campoverde-Vilela, L., del Cisne Feijóo, M., Vidal, Y., Sampietro, J., & Tutivén, C. (2023). Anomaly-based fault detection in wind turbine main bearings. *Wind Energy Science*, *8*(4), 557–574. DOI: 10.5194/wes-8-557-2023

Cañón-Clavijo, R. E., Montenegro-Marin, C. E., Gaona-Garcia, P. A., & Ortiz-Guzmán, J. (2023). IoT Based System for Heart Monitoring and Arrhythmia Detection Using Machine Learning. *Journal of Healthcare Engineering*, *6401673*(1), 6401673. Advance online publication. DOI: 10.1155/2023/6401673 PMID: 36818385

Cao, Y., Tang, X., Gaidai, O., & Wang, F. (2022). Digital twin real-time monitoring method of turbine blade performance based on numerical simulation. *Ocean Engineering*, *263*, 112347. DOI: 10.1016/j.oceaneng.2022.112347

Casallas, J. S., Parra-Peñuela, H. G., & Gaona, E. E. (2023). Wireless real-time data acquisition system for aerodynamic characterization of a wind turbine. Ingenier\'\ ia, 28(Suppl), e18860. DOI: 10.14483/23448393.18860

Castellani, F., Natili, F., Astolfi, D., & Vidal, Y. (2024). Wind turbine gearbox condition monitoring through the sequential analysis of industrial SCADA and vibration data. *Energy Reports*, *12*, 750–761. DOI: 10.1016/j.egyr.2024.06.041

Castellani, F., Pandit, R., Natili, F., Belcastro, F., & Astolfi, D. (2023). Advanced Methods for Wind Turbine Performance Analysis Based on SCADA Data and CFD Simulations. *Energies*, *16*(3), 1081. Advance online publication. DOI: 10.3390/en16031081

Castillo, O., Andrade, V., Rivas, J., & González, R. (2023). Comparison of Power Coefficients in Wind Turbines Considering the Tip Speed Ratio and Blade Pitch Angle. *Energies*, *16*(6), 2774. Advance online publication. DOI: 10.3390/en16062774

Cavalcanti, D., Perez-Ramirez, J., Rashid, M., Fang, J., Galeev, M., & Stanton, K. (2019). Extending Accurate Time Distribution and Timeliness Capabilities Over the Air to Enable Future Wireless Industrial Automation Systems. *Proceedings of the IEEE*, *107*(6), 1132–1152. DOI: 10.1109/JPROC.2019.2903414

Chatterjee, J., & Dethlefs, N. (2022). Scientometric review of artificial intelligence for operations & maintenance of wind turbines: The past, present and future. *arXiv*. https://arxiv.org/abs/2204.02360

Chen, H., Liu, H., Chu, X., Liu, Q., & Xue, D. (2021). Anomaly detection and critical SCADA parameters identification for wind turbines based on LSTM-AE neural network. *Renewable Energy*, *172*, 829–840. DOI: 10.1016/j.renene.2021.03.078

Chen, X., Huang, L., Liu, J., Song, D., & Yang, S. (2022). Peak shaving benefit assessment considering the joint operation of nuclear and battery energy storage power stations: Hainan case study. *Energy*, *239*, 121897. DOI: 10.1016/j.energy.2021.121897

Chiang-Guizar, C. D., Hernandez-Martinez, J. I., Sevilla-Camacho, P. Y., & Solis-Cisneros, H. I. (2025). Artificial intelligence and integrated optimization in the energy sector: Advances in photovoltaic system. Energ\'\ias Renovables, 12(55). DOI: 10.59730/rer.v12n55a5

Chinnappan, C. C. (2025). Integrating data-driven and physics-based approaches for robust wind power prediction: A comprehensive ML-PINN-Simulink framework. *Scientific Reports*, *15*(1), 1–29. PMID: 39747330

Chu, Z., Markovic, U., Hug, G., & Teng, F. (2020). Towards optimal system scheduling with synthetic inertia provision from wind turbines. *IEEE Transactions on Power Systems*, *35*(5), 4056–4066. DOI: 10.1109/TPWRS.2020.2985843

Cid Montoya, M., Rubio-Medrano, C., & Kareem, A. (2024). On the cybersecurity of smart structures under wind. *Journal of Wind Engineering and Industrial Aerodynamics*, *251*, 105777. DOI: 10.1016/j.jweia.2024.105777

CIGRÉ Study Committee B5. (2020). Integration of Wind Power and Grid Services Using Digital Twins, CIGRÉ Brochure No. 806.

Civera, M., & Surace, C. (2022). Non-destructive techniques for the condition and structural health monitoring of wind turbines: A literature review of the last 20 years. *Sensors (Basel)*, *22*(4), 1627. DOI: 10.3390/s22041627 PMID: 35214529

Clifton, A., Barber, S., Bray, A., Enevoldsen, P., Fields, J., Sempreviva, A., Williams, L., Quick, J., Purdue, M., Totaro, P., & Ding, Y. (2023). Grand challenges in the digitalisation of wind energy. *Wind Energy Science*, *8*(6), 947–974. DOI: 10.5194/wes-8-947-2023

Coelho, P. (2023). The Betz limit and the corresponding thermodynamic limit. *Wind Engineering*, *47*(2), 491–496. DOI: 10.1177/0309524X221130109

Conradi Hoffmann, J. L., Horstmann, L. P., Martínez Lucena, M., Medeiros de Araujo, G., Fröhlich, A. A., & Nishioka, M. H. N. (2021). Anomaly detection on wind turbines based on a deep learning analysis of vibration signals. *Applied Artificial Intelligence*, *35*(12), 893–913. DOI: 10.1080/08839514.2021.1966879

Coombs, T. A., Wang, Q., Shah, A., Hu, J., Hao, L., Patel, I., Wei, H., Wu, Y., Coombs, T., & Wang, W. (2024). High-temperature superconductors and their large-scale applications. *Nature Reviews Electrical Engineering*, *1*(12), 788–801. DOI: 10.1038/s44287-024-00112-y

Cui, Y., Xie, J., & Zheng, K. (2021). Historical Inertia: An Ignored but Powerful Baseline for Long Sequence Time-series Forecasting. *ArXiv Preprint ArXiv:2103.16349*. DOI: 10.1145/3459637.3482120

Cullinane, M., Judge, F., O'Shea, M., Thandayutham, K., & Murphy, J. (2022). Subsea superconductors: The future of offshore renewable energy transmission? *Renewable & Sustainable Energy Reviews*, *156*, 111943. DOI: 10.1016/j.rser.2021.111943

Da Silva, T. C., Da Silva Antunes, F. A., Teixeira, J. C., & Contreras, R. C. (2025). Correlation Between Wind Turbine Failures and Environmental Conditions: A Machine Learning Approach. *IEEE Access : Practical Innovations, Open Solutions*, *13*, 50043–50058. DOI: 10.1109/ACCESS.2025.3551241

Daah, C., Qureshi, A., Awan, I., & Konur, S. (2024). Enhancing Zero Trust Models in the Financial Industry through Blockchain Integration: A Proposed Framework. *Electronics (Basel)*, *13*(5), 865. Advance online publication. DOI: 10.3390/electronics13050865

Dai, J., Yang, W., Cao, J., Liu, D., & Long, X. (2018). Ageing assessment of a wind turbine over time by interpreting wind farm SCADA data. *Renewable Energy*, *116*, 199–208. DOI: 10.1016/j.renene.2017.03.097

Dang, S., Amin, O., Shihada, B., & Alouini, M.-S. (2020). What should 6G be? *Nature Electronics*, *3*(1), 20–29. DOI: 10.1038/s41928-019-0355-6

Das, K., & Cutululis, N. A. (2017). *Offshore Wind Power Plant Technology Catalogue-Components of wind power plants, AC collection systems and HVDC systems.*

Datta, A., & Rahman, M. (2017). Cyber Threat Analysis Framework for the Wind Energy Based Power System. *Proceedings of the 2017 Workshop on Cyber-Physical Systems Security and PrivaCy*, pp. 81–92. DOI: 10.1145/3140241.3140247

De Kooning, J., Stockman, K., De Maeyer, J., Jarquin-Laguna, A., & Vandevelde, L. (2021). Digital Twins for Wind Energy Conversion Systems: A Literature Review of Potential Modelling Techniques Focused on Model Fidelity and Computational Load. *Processes (Basel, Switzerland)*, *9*(12), 1–26. DOI: 10.3390/pr9122224

de Souza Pereira Gomes, G., Moreira de Andrade Lopes, S., Carrijo Polonio Araujo, D., Andrade Flauzino, R., Marques Pinto, M., & Eduardo Guerra Alves, M. (2024). Wind turbine remaining useful life prediction using small dataset and machine learning techniques. Journal of Control. *Automation and Electrical Systems*, *35*(2), 337–345.

Demirtas, M., Sharkh, S. M., Gokyildirim, A., & Calgan, H. (2025). Secure operation of a stand-alone wind energy system based on an incommensurate fractional-order chaotic system. *Applied Energy*, *384*, 125477. Advance online publication. DOI: 10.1016/j.apenergy.2025.125477

Deutsche Windtechnik, T. (2022). *Cyber attack on Deutsche Windtechnik*. Retrieved January 29, 2026, from https://www.deutsche-windtechnik.com/en/news/press-releases/details/cyber-attack-on-deutsche-windtechnik/?

Díaz-González, F., Sumper, A., Gomis-Bellmunt, O., & Villafáfila-Robles, R. (2012). A review of energy storage technologies for wind power applications. *Renewable & Sustainable Energy Reviews, 16*(4), 2154–2171. DOI: 10.1016/j.rser.2012.01.029

Dibaj, A., Valavi, M., & Nejad, A. (2024). Unsupervised anomaly detection of permanent-magnet offshore wind generators through electrical and electromagnetic measurements. *Wind Energy Science, 9*(11), 2063–2086. DOI: 10.5194/wes-9-2063-2024

Divya, K. C., & Østergaard, J. (2009). Battery energy storage technology for power systems—An overview. *Electric Power Systems Research, 79*(4), 511–520. DOI: 10.1016/j.epsr.2008.09.017

Dolezilek, D., & Hussey, L. (2011). Requirements or recommendations? Sorting out NERC CIP, NIST, and DOE cybersecurity. In *2011 64th Annual Conference for Protective Relay Engineers* (pp. 328-333).

Dui, H., Cao, T., & Wang, F. (2025). Digital twin-based resilience evaluation and intelligent strategies of smart urban water distribution networks for emergency management. *Resilient Cities and Structures, 4*(1), 41–52. DOI: 10.1016/j.rcns.2025.02.001

Durgadevi, G., Nalawade, M. K., Sharma, K., Nishanthi, N., Shnain, A. H., Sutar, V., & Sujatha, M. S. (2024). Integration of Energy Storage with Wind Power Conversion Systems: Enhancing Grid Stability. *E3S Web of Conferences, 591*, 02004. .DOI: 10.1051/e3sconf/202459102004

Editorial. (2024). Online monitoring of wind power plants using digital twin models. *Energy Research.* DOI: 10.3389/fenrg.2024.1531689

Eidson, J. (2006). A Detailed Analysis of IEEE 1588. In *Measurement, Control, and Communication Using IEEE 1588* (pp. 61–131). Springer London.

Electronics360. (2025, May). Digital twins and AI for predictive maintenance in wind farms. *Electronics360.* https://electronics360.globalspec.com/article/22452/digital-twins-and-ai-for-predictive-maintenance-in-wind-farms

Elgebaly, H., Elhariry, B., Noureldin, A., & Stohy, D. (2025). Digital twin for maintenance and smart manufacturing: The mediating role of replacement maintenance in the Saudi industrial sector. *Jlsdgr, 5*(4), 6107. DOI: 10.47172/2965-730X.SDGsReview.v5.n04.pe06107

Elmousalami, H., Alnaser, A., & Hui, F. (2025). Sustainable AI-driven wind energy forecasting: Advancing zero-carbon cities and environmental computation. *Artificial Intelligence Review, 58*(6), 191. DOI: 10.1007/s10462-025-11191-0

Erturk, M., Al-Dubai, A., Gursu, K., & Canberk, B. (2025). A Digital Twin model for predicting wind turbine performance using federated learning. *Energy, 337*, 138644. DOI: 10.1016/j.energy.2025.138644

European Commission. (2020).Digitalisation of the Energy System – Report on AI, Big Data and IoT Technologies, EUR 29987 EN.

Faheem, M., & Al-Khasawneh, M. A. (2024). Multilayer cyberattacks identification and classification using machine learning in internet of blockchain (IoBC)-based energy networks. *Data in Brief, 54*, 110461. Advance online publication. DOI: 10.1016/j.dib.2024.110461 PMID: 38774244

Faheem, M., Raza, B., Bhutta, M., & Madni, S. (2024). A blockchain-based resilient and secure framework for events monitoring and control in distributed renewable energy systems. *IET Blockchain, 4*(S1), 644–658. DOI: 10.1049/blc2.12081

Fahim, M., Sharma, V., Cao, T. V., Canberk, B., & Duong, T. Q. (2022). Machine learning-based digital twin for predictive modeling in wind turbines. *IEEE Access : Practical Innovations, Open Solutions, 10*, 14184–14194. DOI: 10.1109/ACCESS.2022.3147602

Falekas, G., & Karlis, A. (2021). Digital twin in electrical machine control and predictive maintenance: State-of-the-art and future prospects. *Energies, 14*(18), 5933. DOI: 10.3390/en14185933

Farhan, M., Schneider, R., Thöns, S., & Gündel, M. (2025). Probabilistic cost modeling as a basis for optimizing inspection and maintenance of turbine support structures in offshore wind farms. *Wind Energy Science, 10*(2), 461–481. DOI: 10.5194/wes-10-461-2025

Fei, X., Jiang, H., Zhou, J., & Zhou, Y. (2025). Digital twin modeling method for wind turbine generator system based on operational mechanism and scenario fusion. The International Journal of Advanced Manufacturing Technology 2025 139:3, 139(3), 1979–2001. DOI: 10.1007/s00170-025-15892-2

Fernandez-Navamuel, A.. (2025). Gaussian mixture autoencoder for FOWT damage identification. *Wind Energy Science, 10*, 857–885. DOI: 10.5194/wes-10-857-2025

Ferreira da Silva, M., Nunes Masson, J. E., Santos, M. F., Rodrigues Silva, W., Wladimir Molina, I., & Martins, G. M. C. (2025). Audible Noise Evaluation in Wind Turbines Through Artificial Intelligence Techniques. *Sensors (Basel), 25*(5), 1492. DOI: 10.3390/s25051492 PMID: 40096305

Finn, N. (2018). Introduction to Time-Sensitive Networking. *IEEE Communications Standards Magazine, 2*(2), 22–28. DOI: 10.1109/MCOMSTD.2018.1700076

Finn, N., Thubert, P., Varga, B., & Farkas, J. (2019). Deterministic networking architecture. In *RFC 8655*. IETF. DOI: 10.17487/RFC8655

Foukas, X., Patounas, G., Elmokashfi, A., & Marina, M. (2017). Network Slicing in 5G: Survey and Challenges. *IEEE Communications Magazine*, *55*(5), 94–100. DOI: 10.1109/MCOM.2017.1600951

Frederiksen, R., Bocewicz, G., Radzki, G., Banaszak, Z., & Nielsen, P. (2024). Cost-Effectiveness of Predictive Maintenance for Offshore Wind Farms: A Case Study. *Energies*, *17*(13), 1–24. DOI: 10.3390/en17133147

Freudenreich, K., & Argyriadis, K. (2007). The Load Level of Modern Wind Turbines according to IEC 61400-1. *Journal of Physics: Conference Series*, *75*(1), 012075. DOI: 10.1088/1742-6596/75/1/012075

Fuller, A., Fan, Z., Day, C., & Barlow, C. (2020). Digital Twin: Enabling Technologies, Challenges and Open Research. *IEEE Access : Practical Innovations, Open Solutions*, *8*, 108952–108971. DOI: 10.1109/ACCESS.2020.2998358

Gan, X., Zuo, Y., Zhang, A., Li, S., & Tao, F. (2023). Digital twin-enabled adaptive scheduling strategy based on deep reinforcement learning. *Science China. Technological Sciences*, *66*(7), 1937–1951. DOI: 10.1007/s11431-022-2413-5

Gao, Z., & Odgaard, P. (2023). Real-time monitoring, fault prediction and health management for offshore wind turbine systems. *Renewable Energy*, *218*, 119258. DOI: 10.1016/j.renene.2023.119258

Garan, M., Tidriri, K., & Kovalenko, I. (2022). A Data-Centric Machine Learning Methodology: Application on Predictive Maintenance of Wind Turbines. *Energies*, *15*(3), 826. DOI: 10.3390/en15030826

Garcia-Espinosa, J., Di Capua, D., & Pastor Sanchez, A. (2024). A digital twin for assessing the remaining useful life of offshore wind turbine structures. *Journal of Marine Science and Engineering*, *12*(4), 573. DOI: 10.3390/jmse12040573

Gaurav, R. S. (2021). Rathi, Arvind Ramprakash: Predictive Maintenance for Wind Turbines via Deep Learning and SCADA Data. *IEEE Access : Practical Innovations, Open Solutions*, *9*, 60244–60255.

Gbashi, S., Olatunji, O., Adedeji, P., & Madushele, N. (2024). An Explainable AI Approach Using SHapley Additive exPlanations for Feature Selection in Vibration-Based Fault Diagnostics of Wind Turbine Gearbox. *2024 IEEE 5th International Conference on Electro-Computing Technologies for Humanity (NIGERCON)*, pp. 1-5.

Gielen, D., Boshell, F., Saygin, D., Bazilian, M. D., Wagner, N., & Gorini, R. (2019). The role of renewable energy in the global energy transformation. *Energy Strategy Reviews*, *24*, 38–50. DOI: 10.1016/j.esr.2019.01.006

Gigoni, L., Betti, A., Tucci, M., & Crisostomi, E. (2019). *A Scalable Predictive Maintenance Model for Detecting Wind Turbine Component Failures Based on SCADA Data. 2019 IEEE Power & Energy Society General Meeting*. PESGM., DOI: 10.1109/PESGM40551.2019.8973898

Gijón, A., Eiraudo, S., Manjavacas, A., Salvatore, D., Molina Solana, M., & Gómez Romero, J. (2025). Integrating physics and data-driven approaches: An explainable and uncertainty-aware hybrid model for wind turbine power prediction. *arXiv*. https://arxiv.org/abs/2502.07344

González, I., Tutivén, C., & Vidal, Y. (2023). Advancing wind turbine reliability: Machine learning-based early detection of rotor imbalance for proactive maintenance strategies. Latin American Workshop on Structural Health Monitoring. December, 5-7 2023. https://www.ndt.net/?id=31544

Gonzalez, E., Stephen, B., Infield, D., & Melero, J. (2019). Using high-frequency SCADA data for wind turbine performance monitoring: A sensitivity study. *Renewable Energy*, *131*, 841–853. DOI: 10.1016/j.renene.2018.07.068

Gonzalez-Longatt, F. (2012). Impact of synthetic inertia from wind power on the protection/control schemes of future power systems: Simulation study. *11th IET International Conference on Developments in Power Systems Protection (DPSP 2012)*, 74. DOI: 10.1049/cp.2012.0030

Granados, D., Ruiz, M., Acosta, J., Lara, S., Domínguez, R., & Kañetas, P. (2023). A Wind Turbine Vibration Monitoring System for Predictive Maintenance Based on Machine Learning Methods Developed under Safely Controlled Laboratory Conditions. *Energies*, *16*(5), 1–17. DOI: 10.3390/en16052290

Grataloup, A., Jonas, S., & Meyer, A. (2024). A review of federated learning in renewable energy applications: Potential, challenges, and future directions. *Energy and AI*, *17*, 100375. DOI: 10.1016/j.egyai.2024.100375

Grieves, M. (2014). Digital Twin: Manufacturing Excellence through Virtual Factory Replication. *White Paper*.

Grieves, M., & Vickers, J. (2017). Digital Twin: Mitigating Unpredictable, Undesirable Emergent Behavior in Complex Systems. Transdisciplinary Perspectives on Complex Systems: New Findings and Approaches, 85–113. DOI: 10.1007/978-3-319-38756-7_4

Gungor, V., Sahin, D., Kocak, T., Ergut, S., Buccella, C., Cecati, C., & Hancke, G. (2011). Smart Grid Technologies: Communication Technologies and Standards. *IEEE Transactions on Industrial Informatics*, *7*(4), 529–539. DOI: 10.1109/TII.2011.2166794

Guo, T., & Wang, Y. (2021). IoT-based Monitoring and Analytics for Wind Turbine Operations. *Journal of Cleaner Production*, *299*, 126828.

Gursoy, M., Zhuo, G., Lozowski, A. G., & Wang, X. (2021). Photovoltaic energy conversion systems with sliding mode control. *Energies*, *14*(19), 6071. DOI: 10.3390/en14196071

Gwabavu, M., Bansal, R. C., & Naidoo, R. M. (2024). A review of onshore wind farm battery energy storage systems for techno-economic optimization. *International Journal of Modelling and Simulation*, 1–20. .DOI: 10.1080/02286203.2024.2320583

GWEC. (2024). *Global wind report 2024*. Global Wind Energy Council. Retrieved from https://img.saurenergy.com/2024/05/gwr-2024_digital-version_final-1-compressed.pdf

Habbouche, H., Amirat, Y., & Benbouzid, M. (2025). Leveraging Digital Twins and AI for Enhanced Gearbox Condition Monitoring in Wind Turbines: A Review. *Applied Sciences (Basel, Switzerland)*, *15*(10), 5725. Advance online publication. DOI: 10.3390/app15105725

Hachimi, M., El Gaabouri, I., El Bhiri, B., Motahhir, S., & Boussetta, M. (2025). Cybersecurity Attacks and Defense Techniques in Microgrid Networks: A Comprehensive Survey. In *Lecture Notes in Networks and Systems* (Vol. 1555 LNNS, pp. 182–199). DOI: 10.1007/978-3-031-99997-0_17

Haghshenas, A., Hasan, A., Osen, O., & Mikalsen, E. (2023). Predictive digital twin for offshore wind farms. *Energy Informatics*, *6*(1), 1–26. DOI: 10.1186/s42162-023-00257-4

Hagmann, F., Fuchs, T., & Behrens, R. (2022). Remaining Useful Life Prediction Using Deep Learning for Wind Turbines. *Renewable Energy*, *188*, 147–156.

Hamdan, A., Ibekwe, K. I., Ilojianya, V. I., Sonko, S., & Etukudoh, E. A. (2024). AI in renewable energy: A review of predictive maintenance and energy optimization. *International Journal of Science and Research Archive*, *11*(1), 718–729. DOI: 10.30574/ijsra.2024.11.1.0112

Han, T., Xie, W., & Pei, Z. (2023). Semi-supervised adversarial discriminative learning approach for intelligent fault diagnosis of wind turbine. *Information Sciences*, *648*, 119496. DOI: 10.1016/j.ins.2023.119496

Haroon, R., Erfan, K., Jawad, R., & Canras, B. (2020). Fault Prediction of Wind Turbine Gearbox Based on SCADA Data and Machine Learning. *2020 10th International Conference on Advanced Computer Information Technologies (ACIT)*, pp. 391-395. DOI: 10.1109/ACIT49673.2020.9208884

Hasan, A. K. M. K., Haque, M. H., & Aziz, S. M. (2019). Application of battery energy storage systems to enhance power system inertia. *2019 29th Australasian Universities Power Engineering Conference (AUPEC)*, 1–6. DOI: 10.1109/AUPEC48547.2019.211941

Hasan, S. K. K., & Crawford, C. W. (2025). *A new era for digital twins: progress and industry adoption*. Digital Twin., DOI: 10.1080/27525783.2025.2555877

Hassani, S., & Dackermann, U. (2023). A Systematic Review of Advanced Sensor Technologies for Non-Destructive Testing and Structural Health Monitoring. Sensors 2023, Vol. 23, Page 2204, 23(4), 2204. DOI: 10.3390/s23042204

Hassan, Q., Viktor, P., Al-Musawi, T. J., Ali, B. M., Algburi, S., Alzoubi, H. M., Al-Jiboory, A. K., Sameen, A. Z., Salman, H. M., & Jaszczur, M. (2024). The renewable energy role in the global energy Transformations. *Renewable Energy Focus*, *48*, 100545. DOI: 10.1016/j.ref.2024.100545

Heydari, A., Garcia, D. A., Fekih, A., Keynia, F., Tjernberg, L. B., & De Santoli, L. (2021). A hybrid intelligent model for the condition monitoring and diagnostics of wind turbines gearbox. *IEEE Access : Practical Innovations, Open Solutions*, *9*, 89878–89890. DOI: 10.1109/ACCESS.2021.3090434

Huang, H., Lin, Y., Zhou, Y., Zhao, Y., Zhang, P., & Fan, L. (2023). Data-driven modeling of power system dynamics: Challenges, state of the art, and future work. iEnergy, 2(3), 200–221.

Huang, J., Zhou, S., Li, G., & Shen, Q. (2025). Real-time monitoring and optimization methods for user-side energy management based on edge computing. *Scientific Reports*, *15*(1), 24890. DOI: 10.1038/s41598-025-07592-4 PMID: 40640293

Huang, Y., Liang, S., Cui, T., Mu, X., Luo, T., Wang, S., & Wu, G. (2024). Edge Computing and Fault Diagnosis of Rotating Machinery Based on MobileNet in Wireless Sensor Networks for Mechanical Vibration. *Sensors (Basel)*, *24*(16), 1–14. DOI: 10.3390/s24165156 PMID: 39204852

Hussain, M., Hussain Mirjat, N., Shaikh, F., Luxmi Dhirani, L., Kumar, L., & Sleiti, A. (2024). Condition Monitoring and Fault Diagnosis of Wind Turbine: A Systematic Literature Review. *IEEE Access : Practical Innovations, Open Solutions*, *12*, 190220–190239. DOI: 10.1109/ACCESS.2024.3514747

Hu, W., Jiao, Q., Liu, H., Wang, K., Jiang, Z., Wu, J., Cong, F., & Hao, G. (2024). A transferable diagnosis method with incipient fault detection for a digital twin of wind turbine. *Digital Engineering*, *1*, 100001. DOI: 10.1016/j.dte.2024.100001

Hwang, S., & Yoo, C. (2021). Health Monitoring and Diagnosis System for a Small H-Type Darrieus Vertical-Axis Wind Turbine. *Energies*, *14*(21), 7246. DOI: 10.3390/en14217246

Ibrahim, M., Rassõlkin, A., Vaimann, T., Kallaste, A., Zakis, J., Hyunh, V. K., & Pomarnacki, R. (2023). Digital twin as a virtual sensor for wind turbine applications. *Energies*, *16*(17), 6246. DOI: 10.3390/en16176246

IEA. (2024). *Renewables 2024 – Market update.* International Energy Agency. Retrieved from https://iea.blob.core.windows.net/assets/17033b62-07a5-4144-8dd0-651cdb6caa24/Renewables2024.pdf

Intergovernmental Panel on Climate Change (IPCC). (2022). *Special Report on Renewable Energy Sources and Climate Change Mitigation*. IPCC.

International Energy Agency (IEA). (2023). Wind Power Data and Projections, IEA Technology Report.

Inturi, V., Shreyas, N., Chetti, K., & Sabareesh, G. R. (2021). Comprehensive fault diagnostics of wind turbine gearbox through adaptive condition monitoring scheme. *Applied Acoustics*, *174*, 107738. DOI: 10.1016/j.apacoust.2020.107738

Ismail, L., Abdelmoti, A., Basu, A., Berini, D. E., & Naouss, M. (n.d.). A Systematic Review of Digital Twin-Driven Predictive Maintenance in Industrial Engineering: Taxonomy, Architectural Elements, and Future Research Directions. Retrieved December 2, 2025, from https://arxiv.org/pdf/2509.24443v1

Jankauskas, M., Serackis, A., Šapurov, M., Pomarnacki, R., Baskys, A., Hyunh, V. K., Vaimann, T., & Zakis, J. (2023). Exploring the limits of early predictive maintenance in wind turbines applying an anomaly detection technique. *Sensors (Basel)*, *23*(12), 5695. DOI: 10.3390/s23125695 PMID: 37420861

Järvinen, M., Paulomäki, H., Floors, R., Pansch, C., Thompson, P., Apajalahti, E.-L., & Kangasoja, J. (2025). Wind Energy. In Järvinen, M., & Paulomäki, H. (Eds.), *Designing Renewable Energy Systems within Planetary Boundaries: A Textbook for Energy Engineers* (pp. 345–460). Springer Nature Switzerland., DOI: 10.1007/978-3-031-69856-9_5

Jenkel, L., Jonas, S., & Meyer, A. (2023). Privacy-Preserving Fleet-Wide Learning of Wind Turbine Conditions with Federated Learning. *Energies*, *16*(17), 6377. Advance online publication. DOI: 10.3390/en16176377

Jeong, D.-Y., Baek, M.-S., Lim, T.-B., & Lee, I.-B. (2022). Digital twin: Evolution stages and implementation layers with technology elements. *Computers in Industry*, *136*, 103594. DOI: 10.1016/j.compind.2022.103594

Jiang, H., Qin, S., Fu, J., Zhang, J., & Ding, G. (2021). How to model and implement connections between physical and virtual models for digital twin application. *Journal of Manufacturing Systems*, *58*, 36–51. DOI: 10.1016/j.jmsy.2020.05.012

Jin, X., Liu, Z., & Zhang, J. (2020). Wind Turbine Fault Detection Using Neural Networks and Physical Modelling. *Energies*, *13*(5), 1146.

Jørgensen, B., & Ma, Z. (2025). Digital Twin of the European Electricity Grid: A Review of Regulatory Barriers, Technological Challenges, and Economic Opportunities. *Applied Sciences (Basel, Switzerland)*, *15*(12), 1–32. DOI: 10.3390/app15126475

Jorgensen, J., Hodkiewicz, M., Cripps, E., & Hassan, G. M. (2023). Requirements for the application of the Digital Twin Paradigm to offshore wind turbine structures for uncertain fatigue analysis. *Computers in Industry*, *145*, 103806. DOI: 10.1016/j.compind.2022.103806

Juarez, M. G. J., Giret, A., & Botti, V. (2025). Semantic and modular orchestration of AI-driven digital twins for industrial interoperability and optimization. *Journal of Industrial Information Integration*, *48*, 100959. DOI: 10.1016/j.jii.2025.100959

Jurczuk, M., & Suprynowicz, M. (2024). Consent in Data Privacy: A General Comparison of GDPR and HIPAA. *Adam Mickiewicz University Law Review*, *16*, 173–194. DOI: 10.14746/ppuam.2024.16.07

Justo, J. J., Demidova, G., Lukichev, D., Poliakov, N., Makolo, P., & Bansal, R. C. (2025). Synthesis of Dynamic Virtual-Inertia for Microgrid System Applications. *2025 International Conference on Intelligent Control, Computing and Communications (IC3)*, 177–182. DOI: 10.1109/IC363308.2025.10957416

Justo, J. J., Mwasilu, F., & Jung, J.-W. (2015). Doubly-fed induction generator-based wind turbines: A comprehensive review of fault ride-through strategies. *Renewable & Sustainable Energy Reviews*, *45*, 447–467. DOI: 10.1016/j.rser.2015.01.064

Justo, J. J., Mwasilu, F., Lee, J., & Jung, J.-W. (2013). AC-microgrids versus DC-microgrids with distributed energy resources: A review. *Renewable & Sustainable Energy Reviews*, *24*, 387–405. DOI: 10.1016/j.rser.2013.03.067

Kabir, M. R., Halder, D., & Ray, S. (2024). Digital twins for IoT-driven energy systems: A survey. *IEEE Access : Practical Innovations, Open Solutions*, *12*, 177123–177145. DOI: 10.1109/ACCESS.2024.3506660

Kamarzarrin, M., Refan, M. H., Amiri, P., & Dameshghi, A. (2022). A new intelligent fault diagnosis and prognosis method for wind turbine doubly-fed induction generator. *Wind Engineering*, *46*(1), 308–340. DOI: 10.1177/0309524X211027808

Kandemir, E., Hasan, A., Kvamsdal, T., & Abdel-Afou Alaliyat, S. (2024). Predictive digital twin for wind energy systems: A literature review. *Energy Informatics*, *7*(1), 68. DOI: 10.1186/s42162-024-00373-9

Karad, S., & Thakur, R. (2021). Efficient monitoring and control of wind energy conversion systems using Internet of Things (IoT): A comprehensive review. *Environment, Development and Sustainability*, *23*(10), 14197–14214. DOI: 10.1007/s10668-021-01267-6

Karthik Selva Kumar, K., Brintha, R., Aruna Devi, K., Balaji, P. S., & Jaiswal, K. K., Sangmesh, ..., & Krishna, B. V. V. K. (2023, December). AI-enabled condition monitoring for a wind energy conversion system—Vertical Axis Wind Turbine (VAWT). In Conference on Fluid Mechanics and Fluid Power (pp. 705–714). Springer Nature Singapore.

Kathirvel, A., Gobalakrishnan, N., & Naren, A. K. (2025a). Cyber security in industrial automation using AI. In *AI-enhanced cybersecurity for industrial automation*. IGI Global.

Kathirvel, A., Gobinath, V. M., & Naren, A. K. (2025e). AI-based healthcare systems for pets and birds. In *Cases on AI-driven solutions to environmental challenges* (pp. 419–447). IGI Global.

Kathirvel, A., Gobinath, V. M., Naren, A. K., & Maheswaran, C. P. (2025d). AI-based automated disease detection based on symptoms healthcare for pets and birds. In *AI and the revival of big data* (pp. 311–345). IGI Global.

Kathirvel, A., Gobinath, V. M., Sathya, V., Naren, A. K., & Raj, A. A. (2025). The role of robotics in smart manufacturing: Increasing efficiency and reducing costs. In *The quantum AI era of neuromarketing*. IGI Global.

Kestel, K., & Chesterman, X. Zappal\'a, D., Watson, S., Li, M., Hart, E., . . . Helsen, J. (2025). Condition monitoring of wind turbine drivetrains: State-of-the-art technologies, recent trends, and future outlook. *Wind Energy Science Discussions*, 1--85. DOI: 10.5194/wes-2025-168

Khan, M. R. (2024). Digital twin-driven optimization of bioenergy production from waste materials. *Innovatech Engineering Journal*, *1*(01), 187–204. DOI: 10.70937/itej.v1i01.19

Khan, P. W., & Byun, Y. C. (2024). A review of machine learning techniques for wind turbine's fault detection, diagnosis, and prognosis. *International Journal of Green Energy*, *21*(4), 771–786. DOI: 10.1080/15435075.2023.2217901

Khan, P. W., & Byun, Y. C. (2024). Fault prognosis of wind turbines using multimodal machine learning. *Energy Reports*, *12*, 4938–4952. DOI: 10.1016/j.egyr.2024.10.043

Kim, J., Chen, R., & Gao, L. (2025). Research and prospects of digital twin-based fault diagnosis of electric motors, including wind-turbine generators. *Applied Sciences (Basel, Switzerland)*, *15*, 4089. DOI: 10.3390/app1504089

Kini, K., Harrou, F., Madakyaru, M., & Sun, Y. (2023). Enhancing Wind Turbine Performance: Statistical Detection of Sensor Faults Based on Improved Dynamic Independent Component Analysis. *Energies*, *16*(15), 1–25. DOI: 10.3390/en16155793

Koblischka, M. R., Půst, L., Chang, C.-S., Hauet, T., & Koblischka-Veneva, A. (2023). The paramagnetic Meissner effect (PME) in metallic superconductors. *Metals*, *13*(6), 1140. DOI: 10.3390/met13061140

Kong, L., Liang, H., Liu, G., & Liu, S. (2023). Research on Wind Turbine Fault Detection Based on the Fusion of ASL-CatBoost and TtRSA. *Sensors (Basel)*, *23*(15), 6741. DOI: 10.3390/s23156741 PMID: 37571525

Kostecka, E., Miller, T., Durlik, I., & Nerć, A. (2025). Artificial Intelligence in Floating Offshore Wind Turbines: A Critical Review of Applications in Design, Monitoring, Control, and Digital Twins. *Energies*, *18*(22), 5937. Advance online publication. DOI: 10.3390/en18225937

Kühn, M., & Muskulus, M. (2022). State of the Art in Wind Farm Digital Twin Technology. *Wind Energy Science*, *7*, 185–200.

Kull, K., Asad, B., Khan, M. A., Naseer, M. U., Kallaste, A., & Vaimann, T. (2025). Faults, Failures, Reliability, and Predictive Maintenance of Grid-Connected Solar Systems: A Comprehensive Review. Applied Sciences 2025, Vol. 15, Page 11461, 15(21), 11461. DOI: 10.3390/app152111461

Lebranchu, A., Charbonnier, S., Bérenguer, C., & Prévost, F. (2019). A combined mono- and multi-turbine approach for fault indicator synthesis and wind turbine monitoring using SCADA data. *ISA Transactions*, *87*, 272–281. DOI: 10.1016/j.isatra.2018.11.041 PMID: 30545768

Lee, C.-Y., & Maceren, E. (2024). Wind energy system fault classification and detection using deep convolutional neural network and particle swarm optimization-extreme gradient boosting. *IET Energy Systems Integration*, *6*(4), 479–497. DOI: 10.1049/esi2.12144

Lei, Y., He, Z., & Hu, B. Q. (2020). A review of deep learning-based fault diagnosis methods for rotating machinery. *Mechanical Systems and Signal Processing, 138*, 106583. DOI: 10.1016/j.ymssp.2019.106587

Lei, Y., Li, N., Guo, L., Li, N., Yan, T., & Lin, J. (2018). Machinery health prognostics: A systematic review from data acquisition to RUL prediction. *Mechanical Systems and Signal Processing, 104*, 799–834. DOI: 10.1016/j.ymssp.2017.11.016

Leon-Medina, J. X., ... (2025). Digital twin technology in wind turbine components: A review. *Elsevier.* https://www.sciencedirect.com/science/article/pii/S2667305325000614

Leon-Medina, J., Tibaduiza, D., Parés, N., & Pozo, F. (2025). Digital twin technology in wind turbine components: A review. *Intelligent Systems with Applications, 26*, 200535. DOI: 10.1016/j.iswa.2025.200535

Le, T. D., Kim, J. H., Kim, D. J., Boo, C. J., & Kim, H. M. (2015). Status of the technology development of large scale HTS generators for wind turbine. *Progress in Superconductivity and Cryogenics, 17*(2), 18–24. DOI: 10.9714/psac.2015.17.2.018

Letzgus, S., & Müller, K.-R. (2024). An explainable AI framework for robust and transparent data-driven wind turbine power curve models. *Energy and AI, 100328*, 100328. Advance online publication. DOI: 10.1016/j.egyai.2023.100328

Li, B., Wang, R., & Hu, X. (2021). Edge AI-Driven Wind Turbine Anomaly Forecasting. *IEEE Industrial Electronics Magazine, 15*(4), 89–102.

Lim, S., Seo, K., Park, J.-W., & Lee, K. Y. (2025). New frequency stability assessment based on contribution rates of wind power plants. *International Journal of Electrical Power & Energy Systems, 164*, 110388. DOI: 10.1016/j.ijepes.2024.110388

Li, S., Li, X., Jiang, Y., Yang, Q., Lin, M., Peng, L., & Yu, J. (2025). A novel frequency-domain physics-informed neural network for accurate prediction of 3D spatio-temporal wind fields in wind turbine applications. *Applied Energy, 386*, 125526. Advance online publication. DOI: 10.1016/j.apenergy.2025.125526

Liu, X., Wu, X., He, J., & Gupta, R. (2025). Intelligent Operation and Maintenance and Prediction Model Optimization for Improving Wind Power Generation Efficiency. *2025 7th International Congress on Human-Computer Interaction, Optimization and Robotic Applications (ICHORA)*, pp. 1-7. DOI: 10.1109/ICHORA65333.2025.11017307

Liu, A., & Zhou, Z. (2021). A Survey on Communication Protocols for Large-Scale Wind Farm Monitoring. *IEEE Internet of Things Journal, 8*(3), 2053–2066.

Liu, S., Ren, S., & Jiang, H. (2023). Predictive maintenance of wind turbines based on digital twin technology. *Energy Reports*, *9*, 1344–1352. DOI: 10.1016/j.egyr.2023.05.052

Liu, Y., & Yu, D. (2021). LoRaWAN-Based Sensor Networks for Remote Wind Turbine Monitoring. *Sensors (Basel)*, *21*(18), 6074. PMID: 34577288

Li, X., Zou, Y., & Xia, X. (2024). *Unlocking the potential of digital twins for a greener grid*. IEEE PES Electrification.

Li, Y., & Shen, X. (2021). A novel wind speed-sensing methodology for wind turbines based on digital twin technology. *IEEE Transactions on Instrumentation and Measurement*, *71*, 1–13.

Lloberas, J., Sumper, A., Sanmarti, M., & Granados, X. (2014). A review of high temperature superconductors for offshore wind power synchronous generators. *Renewable & Sustainable Energy Reviews*, *38*, 404–414. DOI: 10.1016/j.rser.2014.05.003

Lu, B., Li, Y., & Wu, X. A Review of Recent Advances in Wind Turbine Condition Monitoring and Fault Diagnosis, *Proceedings of the IEEE Power Engineering Society General Meeting*, 2019, Pages 1-8.

Lui, S. (2024). Wind power short-term prediction based on digital twin technology. *Frontiers in Energy Research*, *12*, 1–9. DOI: 10.3389/fenrg.2024.1365237

Madani, S. S., Shabeer, Y., Fowler, M., Panchal, S., Chaoui, H., Mekhilef, S., Dou, S. X., & See, K. (2025). Artificial intelligence and digital twin technologies for intelligent lithium-ion battery management systems: A comprehensive review. *Batteries*, *11*(8), 298. DOI: 10.3390/batteries11080298

Madeswaran, A., Bisht, D., Yuvaraj, S., Reedy, M. U., Al-Attabi, K., & Dhablia, A. (2024). AI-controlled wind turbine systems: Integrating IoT and machine learning for smart grids. *E3S Web of Conferences*, *540*, 03008. .DOI: 10.1051/e3sconf/202454003008

Mahaveera, A., & Pasuk, S. (2021). Development of smart monitoring system for wind energy system. International Journal of Systems Applications. *Engineering*, *15*, 132–134. DOI: 10.46300/91015.2021.15.18

Mahmoud, M., Semeraro, C., Abdelkareem, M. A., & Olabi, A. G. (2024). Designing and prototyping the architecture of a digital twin for wind turbine. *International Journal of Thermofluids*, *22*, 100622. DOI: 10.1016/j.ijft.2024.100622

Ma, J., Yuan, Y., & Chen, P. (2022). A fault prediction framework for doubly-fed induction generator under time-varying operating conditions driven by digital twin. *IET Electric Power Applications*, *16*(11), 1421–1430. DOI: 10.1049/elp2.12345

Ma, J., Yuan, Y., & Chen, P. (2023). A fault prediction framework for Doubly-fed induction generator under time-varying operating conditions driven by digital twin. *IET Electric Power Applications*, *17*(4), 499–521. DOI: 10.1049/elp2.12280

Makolo, P., Cheah-Mañe, M., & Gomis-Bellmunt, O. (2025). Quantification of virtual inertia from derivative control using a recursive least square with a weighted forgetting method. *Electric Power Systems Research*, *241*, 111363. DOI: 10.1016/j.epsr.2024.111363

Makolo, P., Justo, J. J., Mwasilu, F., & Zamora, R. (2018). Fault Ride Through Technique for DFIG-based Wind Turbines Under Grid Three-phase Faults. *2018 Australasian Universities Power Engineering Conference (AUPEC)*, 1–5. DOI: 10.1109/AUPEC.2018.8757926

Makolo, P., Zamora, R., & Lie, T. (2021a). Online inertia estimation for power systems with high penetration of RES using recursive parameters estimation. *IET Renewable Power Generation*, *15*(12), 2571–2585. Advance online publication. DOI: 10.1049/rpg2.12181

Makolo, P., Zamora, R., & Lie, T.-T. (2021b). The role of inertia for grid flexibility under high penetration of variable renewables-A review of challenges and solutions. *Renewable & Sustainable Energy Reviews*, *147*, 111223. DOI: 10.1016/j.rser.2021.111223

Maksimović, M., Jokić, S., & Bošković, M. (2025). Innovative Horizons for Sustainable Smart Energy: Exploring the Synergy of 5G and Digital Twin Technologies. Process Integration and Optimization for Sustainability 2025 9:2, 9(2), 431–470. DOI: 10.1007/s41660-024-00478-4

Maldonado-Correa, J., Martín-Martínez, S., Artigao, E., & Gómez-Lázaro, E. (2020). Using SCADA Data for Wind Turbine Condition Monitoring: A Systematic Literature Review. *Energies*, *13*(12), 3132. DOI: 10.3390/en13123132

Maldonado-Correa, J., Torres-Cabrera, J., Martín-Martínez, S., Artigao, E., & Gómez-Lázaro, E. (2023). Wind turbine fault detection based on the transformer model using SCADA data. *Engineering Failure Analysis*, *162*, 108354. DOI: 10.1016/j.engfailanal.2024.108354

Maldonado-Correa, J., Valdiviezo-Condolo, M., Artigao, E., Martín-Martínez, S., & Gómez-Lázaro, E. (2024). Classification of Highly Imbalanced Supervisory Control and Data Acquisition Data for Fault Detection of Wind Turbine Generators. *Energies*, *17*(1), 1–20. DOI: 10.3390/en17071590

Maron, J., Anagnostos, D., Brodbeck, B., & Meyer, A. (2022). Artificial intelligence-based condition monitoring and predictive maintenance framework for wind turbines. *Journal of Physics: Conference Series*, *2151*(1), 012007. DOI: 10.1088/1742-6596/2151/1/012007

Marykovskiy, Y., Clark, T., Deparday, J., Chatzi, E., & Barber, S. (2024). Architecting a digital twin for wind turbine rotor blade aerodynamic monitoring. *Frontiers in Energy Research*, *12*, 1428387. Advance online publication. DOI: 10.3389/fenrg.2024.1428387

Masenge, I. H., & Mwasilu, F. (2020). Modeling and control of solar PV with battery energy storage for rural electrification. *Tanzania Journal of Engineering and Technology*, *39*(1), 47–58. DOI: 10.52339/tjet.v39i1.518

Masoumi, M. (2023). Machine learning solutions for offshore wind farms: A review of applications and impacts. *Journal of Marine Science and Engineering*, *11*(10), 1855. DOI: 10.3390/jmse11101855

Matsui, T., Yamamoto, K., & Ogata, J. (2022). Anomaly detection using a SCADA feature extractor and machine learning to detect lightning damage on wind turbine blades. *IEEJ Transactions on Electrical and Electronic Engineering*, *17*(6), 945–951. DOI: 10.1002/tee.23599

May, G. J., Davidson, A., & Monahov, B. (2018). Lead batteries for utility energy storage: A review. *Journal of Energy Storage*, *15*, 145–157. DOI: 10.1016/j.est.2017.11.008

May, G., Taylor, G., & Lee, A. (2022). Analysis of Remaining Useful Life (RUL) for Wind Turbine Gearboxes using AI-Oriented Approaches. *Mechanical Systems and Signal Processing*, *168*, 108669.

Ma, Z., Xu, J., Yang, G., & Zhu, W. (2021). Digital Twin-Driven Smart Manufacturing: Connotation, Reference Model, Applications and Research Issues. *Robotics and Computer-integrated Manufacturing*, *68*, 102102.

Mbasso, W. F.. (2025). *Digital twins in renewable energy systems*. Renewable Energy Journal.

Mbasso, W. F., Harrison, A., Dagal, I., Jangir, P., Khishe, M., Kotb, H., Shaikh, M. S., Smerat, A., Fendzi Donfack, E., & Kumar, R. (2025). Digital twins in renewable energy systems: A comprehensive review of concepts, applications, and future directions. *Energy Strategy Reviews*, *61*, 101814. DOI: 10.1016/j.esr.2025.101814

Mchirgui, N., Quadar, N., Kraiem, H., & Lakhssassi, A. (2024). The Applications and Challenges of Digital Twin Technology in Smart Grids: A Comprehensive Review. *Applied Sciences (Basel, Switzerland)*, *14*(23), 1–32. DOI: 10.3390/app142310933

McKenna, R., Ostman, P., & Fichtner, W. (2016). Key challenges and prospects for large wind turbines. *Renewable & Sustainable Energy Reviews*, *53*, 1212–1221. DOI: 10.1016/j.rser.2015.09.080

Mehlan, F. C., Nejad, A. R., & Gao, Z. (2022). Digital twin-based virtual sensor for online fatigue damage monitoring in offshore wind turbine drivetrains. *Journal of Offshore Mechanics and Arctic Engineering*, *144*(6), 060901. DOI: 10.1115/1.4055551

Mehlan, F., Keller, J., & Nejad, A. (2023). Virtual sensing of wind turbine hub loads and drivetrain fatigue damage. *Forschung im Ingenieurwesen*, *87*(1), 207–218. DOI: 10.1007/s10010-023-00627-0

Messenger, J. (2018). Time-Sensitive Networking: An Introduction. *IEEE Communications Standards Magazine*, *2*(2), 29–33. DOI: 10.1109/MCOMSTD.2018.1700047

Mikołajewska, E., Mikołajewski, D., Mikołajczyk, T., & Paczkowski, T. (2025). Generative AI in AI-based digital twins for fault diagnosis for predictive maintenance in Industry 4.0/5.0. *Applied Sciences (Basel, Switzerland)*, *15*(6), 3166. DOI: 10.3390/app15063166

Milani, S., Leoni, J., Cacciola, S., Croce, A., & Tanelli, M. (2025). A machine-learning-based approach for active monitoring of blade pitch misalignment in wind turbines. *Wind Energy Science*, *10*(3), 497–510. DOI: 10.5194/wes-10-497-2025

Mishra, A., Alzoubi, Y. I., Gill, A. Q., & Anwar, M. J. (2022). Cybersecurity Enterprises Policies: A Comparative Study. *Sensors (Basel)*, *22*(2), 1–35. DOI: 10.3390/s22020538 PMID: 35062504

Mkoi, P., Makolo, P. M., & Mwasilu, F. A. (2025). Synthetic Inertia Provision for Load Frequency Control in Networks with High Penetration of Renewable Energy Sources. *Tanzania Journal of Engineering and Technology*, *44*(1), 245–256. DOI: 10.52339/tjet.v44i1.1201

Moayyed, H., Moradzadeh, A., Mohammadi-Ivatloo, B., Aguiar, A. P., & Ghorbani, R. (2022). A Cyber-Secure generalized supermodel for wind power forecasting based on deep federated learning and image processing. *Energy Conversion and Management, 267*, 115852. Advance online publication. DOI: 10.1016/j.enconman.2022.115852

Mohamed, A., & Chul-Hwan, K. (2017). Communication Architecture for Grid Integration of Cyber Physical Wind Energy Systems. *Applied Sciences (Basel, Switzerland), 7*(10), 1–20. DOI: 10.3390/app7101034

Morsi, I., & Ahmed, M. (2024). Smart irrigation system using PLC, SCADA, and fuzzy control. *Edelweiss Applied Science and Technology, 8*(6), 9101–9120. DOI: 10.55214/25768484.v8i6.3947

Mousavi, Z., ... (2024). A digital twin-based framework for damage detection of a floating wind turbine. *Elsevier*. https://www.sciencedirect.com/science/article/abs/pii/S0029801823029475

Mulders, S., Gallo, A., & Rotea, M. (2024). Analysis of Extremum Seeking Control for Wind Turbine Torque Controller Optimization by Aerodynamic and Generator Power Objectives. In *2024 American Control Conference (ACC)* (pp. 1030-1037). Toronto: IEEE. DOI: 10.23919/ACC60939.2024.10644899

Muñoz-Palomeque, E., Sierra-García, J., & Santos, M. (2022). Wind turbine maximum power point tracking control based on unsupervised neural networks. *Journal of Computational Design and Engineering, 10*(1), 108–121. DOI: 10.1093/jcde/qwac132

Muñoz-Palomeque, E., Sierra-García, J., & Santos, M. (2024). Enhancing offshore wind turbines performance with hybrid control strategies using neural networks and conventional controllers. *Journal of Computational Design and Engineering, 12*(3), 80–97. DOI: 10.1093/jcde/qwae103

Murray, J., Huang, C., Parkinson, S., Bolton, R., Gledhill, D., Parkinson, H., Basher, D., Vulovic, V., Yang, W., & Ball, A. (2025). Unified Digital Twin Approach for Energy Grid Condition and Cybersecurity Monitoring. *IEEE Access : Practical Innovations, Open Solutions, 13*, 180710–180724. DOI: 10.1109/ACCESS.2025.3622468

Naderi, E., & Asrari, A. (2025). Cybersecurity Challenges in Power-to-X Integration. In *Power-to-X in Regional Energy Systems* (pp. 284–298). Planning, Operation, Control, and Market Perspectives., DOI: 10.1201/9781032719436-12

Nasrallah, A., Thyagaturu, A., Alharbi, Z., Wang, C., Shao, X., Reisslein, M., & ElBakoury, H. (2019). Ultra-Low Latency (ULL) Networks: The IEEE TSN and IETF DetNet Standards and Related 5G ULL Research. *IEEE Communications Surveys and Tutorials, 21*(1), 88–145. DOI: 10.1109/COMST.2018.2869350

Ng, E. Y. K., & Lim, J. T. (2022). Machine learning on fault diagnosis in wind turbines. *Fluids (Basel, Switzerland), 7*(12), 371. DOI: 10.3390/fluids7120371

Nordex. (2022). *Update on cyber security incident.* Retrieved January 29, 2026, from https://www.nordex-online.com/en/2022/04/update-on-cyber-security-incident/?

Nycander, E., & Söder, L. (2018). Review of european grid codes for wind farms and their implications for wind power curtailments. *17th International Wind Integration Workshop Stockholm,* Swedenl 17–19 October 2018.

Odgaard, P. F., & Stoustrup, J. (2013, August). Fault tolerant wind farm control — A benchmark model. *2013 IEEE International Conference on Control (CCA).* DOI: 10.1109/CCA.2013.6662784

Olabi, A. G., Wilberforce, T., Sayed, E. T., Abo-Khalil, A. G., Maghrabie, H. M., Elsaid, K., & Abdelkareem, M. A. (2022). Battery energy storage systems and SWOT (strengths, weakness, opportunities, and threats) analysis of batteries in power transmission. *Energy, 254,* 123987. DOI: 10.1016/j.energy.2022.123987

Olatunji, O. O., Adedeji, P. A., Madushele, N., & Jen, T. C. (2021, May). Overview of digital twin technology in wind turbine fault diagnosis and condition monitoring. In 2021 IEEE 12th International Conference on Mechanical and Intelligent Manufacturing Technologies (ICMIMT) (pp. 201–207). IEEE. DOI: 10.1109/ICMIMT52186.2021.9476186

Oliveira-Filho, A., Comeau, M., Cave, J., Nasr, C., Côté, P., & Tahan, A. (2024). Wind Turbine SCADA Data Imbalance: A Review of Its Impact on Health Condition Analyses and Mitigation Strategies. *Energies, 18*(1), 1–23. DOI: 10.3390/en18010059

Ouyang, J., Mo, F., Huang, F., & Chen, Y. (2025). Risk Assessment Method for CPS-Based Distributed Generation Cluster Control in Active Distribution Networks Under Cyber Attacks. *Sensors (Basel), 25*(19), 6053. Advance online publication. DOI: 10.3390/s25196053 PMID: 41094876

Overbye, T. J., Wan, Q., & Cheema, K. (2022). Edge Computing for Grid-Integrated Wind Farms: Challenges and Solutions. *Electric Power Systems Research, 209,* 108077.

Oyucu, S., Polat, O., Türkoğlu, M., Polat, H., Aksöz, A., & Ağdaş, M. T. (2024). Ensemble Learning Framework for DDoS Detection in SDN-Based SCADA Systems. *Sensors (Basel)*, *24*(1), 155. Advance online publication. DOI: 10.3390/s24010155 PMID: 38203015

Ozturk, S., Fthenakis, V., & Faulstich, S. (2018). Assessing the Factors Impacting on the Reliability of Wind Turbines via Survival Analysis—A Case Study. *Energies*, *11*(11), 1–20. DOI: 10.3390/en11113034

Pandit, R. K., Astolfi, D., & Durazo Cardenas, I. (2023). A Review of Predictive Techniques Used to Support Decision Making for Maintenance Operations of Wind Turbines. Energies 2023, Vol. 16, Page 1654, 16(4), 1654. DOI: 10.3390/en16041654

Pandit, R., Astolfi, D., Hong, J., Infield, D., & Santos, M. (2023). SCADA data for wind turbine data-driven condition/performance monitoring: A review on state-of-art, challenges and future trends. *Wind Engineering*, *47*(2), 422–441. DOI: 10.1177/0309524X221124031

Pandit, R., Santos, M., & Sierra-García, J. E. (2024). Comparative analysis of novel data-driven techniques for remaining useful life estimation of wind turbine high-speed shaft bearings. *Energy Science & Engineering*, *12*(10), 4613–4623. DOI: 10.1002/ese3.1911

Pao, L., Pusch, M., & Zalkind, D. (2024). Control Co-Design of Wind Turbines. *Annual Review of Control, Robotics, and Autonomous Systems*, *7*(1), 201–226. DOI: 10.1146/annurev-control-061423-101708

Park, J. W., Youn, T., & Shin, K. (2022). Practical Implementation of a Wind Turbine Digital Twin Using Physics-Based and AI-Based Data Fusion. *Renewable Energy*, *184*, 825–840.

Pascoe, C. (2023). *Public draft: The NIST cybersecurity framework 2.0.* National Institute of Standards and Technology.

Paul, S., Kumar, P., & Chatterjee, K. (2025). Detection and mitigation of cyber attack in DFIG based wind energy systems. *Sustainable Energy*. *Grids and Networks*, *44*, 102057. Advance online publication. DOI: 10.1016/j.segan.2025.102057

Pereira, A., Hovstø, A., Rodrigues, E., Silvia, F., & Suciu, G., AL-Naday, M., ... & Atmojo, U. D. (2024). Edge driven digital twins in distributed energy systems: Role and opportunities for hybrid data driven solutions. *Applied Sciences (Basel, Switzerland)*.

Pérez, E. (2022). A simulation-driven online scheduling algorithm for the maintenance and operation of wind farm systems. *Simulation*, *98*(1), 47–61. DOI: 10.1177/00375497211028605

Pliuhin, V., Tsegelnyk, Y., & Slovikovskyi, O. (2024). Digital twins of different types electrical machines. *Lighting Engineering & Power Engineering*, *63*(2), 35–45. DOI: 10.33042/2079-424X.2024.63.2.01

Popovski, P., Trillingsgaard, K., Simeone, O., & Durisi, G. (2018). 5G Wireless Network Slicing for eMBB, URLLC, and mMTC: A Communication-Theoretic View. *IEEE Access : Practical Innovations, Open Solutions*, *6*, 55765–55779. DOI: 10.1109/ACCESS.2018.2872781

Prabhavathi, K., Prabhakar, P. E., Arunadevi, M., & Shanthi, D. (2023). Design and Implementation of a Machine Learning-Based Wind Turbine Control System. In *E3S Web of Conferences* (Vol. 387, p. 01008). EDP Sciences.

Prasath, C. A. (2025). AI-enabled digital twin framework for predictive maintenance in smart urban infrastructure. *Journal of Smart Infrastructure and Environmental Sustainability*, *2*(1), 1–10.

Preuster, P., Alekseev, A., & Wasserscheid, P. (2017). Hydrogen Storage Technologies for Future Energy Systems. *Annual Review of Chemical and Biomolecular Engineering*, *8*(1), 445–471. DOI: 10.1146/annurev-chembioeng-060816-101334 PMID: 28592172

Rahman, M. A., Shahrior, M. F., Iqbal, K., & Abushaiba, A. A. (2025). Enabling Intelligent Industrial Automation: A Review of Machine Learning Applications with Digital Twin and Edge AI Integration. Automation 2025, Vol. 6, Page 37, 6(3), 37. DOI: 10.3390/automation6030037

Ramchandani, M., Vahora, M. A., Wagle, S., & Dhiman, H. S. (2025). On Modeling False Data Injection Attacks for Wind Turbine Gearbox using Multivariate SCADA Data. In *2025 International Conference on Artificial Intelligence and Machine Vision, AIMV 2025*. DOI: 10.1109/AIMV66517.2025.11203394

Rasheed, A., Stadtmann, F., Fonn, E., et al. (2024). Digital twins in wind energy: Emerging technologies and industry-informed future directions. *arXiv*.

Rasheed, A., San, O., & Kvamsdal, T. (2020). Digital Twin: Values, Challenges and Enablers From a Modeling Perspective. *IEEE Access : Practical Innovations, Open Solutions*, *8*, 21980–22012. DOI: 10.1109/ACCESS.2020.2970143

Rengasamy, P., & Ramachandran, R. (2025). Explainable artificial intelligence framework for wind turbine fault detection using random forest – Extreme gradient boosting hybrid model. *Results in Engineering*, *28*, 107446. DOI: 10.1016/j.rineng.2025.107446

Rithik, G., Kathirvel, A., & Naren, A. K. (2024). The IoT of robotics: The frontier of automation. In *Artificial intelligence-enabled digital twin for smart manufacturing*. Wiley–Scrivener Publishing LLC. DOI: 10.1002/9781394303601.ch13

Robertson, D., & McCarthy, B. (2018). Portland General Electric Driving Collaboration for Sustainable Legislation. In *Sustainable Electricity II: A Conversation on Tradeoffs* (pp. 143–159). Springer.

Rojas, L., Peña, Á., & Garcia, J. (2025). AI-driven predictive maintenance in mining: A systematic literature review on fault detection, digital twins, and intelligent asset management. *Applied Sciences (Basel, Switzerland)*, *15*(6), 3337. DOI: 10.3390/app15063337

Rositani, M. (2023). Physics-informed learning and data-driven control of an Airborne Wind Energy system.

Ru, L., & Yuan, C. (2022). Economic Impact Assessment of Predictive Maintenance in Wind Power. *Renewable & Sustainable Energy Reviews*, *154*, 111787.

Saad, W., Bennis, M., & Chen, M. (2020). A Vision of 6G Wireless Systems: Applications, Trends, Technologies, and Open Research Problems. *IEEE Network*, *34*(3), 134–142. DOI: 10.1109/MNET.001.1900287

Sadeghi, A., Seyyedbarzegar, S., & Yazdani-Asrami, M. (2022). Investigation on the electrothermal performance of a high-temperature superconducting cable in an offshore wind farm integrated power system: Fault and islanding conditions. *IEEE Transactions on Applied Superconductivity*, *32*(8), 1–11. DOI: 10.1109/TASC.2022.3196770

Saleh, A. M., István, V., Khan, M. A., Waseem, M., & Ali Ahmed, A. N. (2024). Power system stability in the Era of energy Transition: Importance, Opportunities, Challenges, and future directions. Energy Conversion and Management: X, 24, 100820. DOI: 10.1016/j.ecmx.2024.100820

Sanchez, H. E., Escobet, T., Puig, V., & Odgaard, P. F. (2015). Fault diagnosis of advanced wind turbine benchmark using interval-based ARRs and observers. *IEEE Transactions on Industrial Electronics*, *1*, 1. Advance online publication. DOI: 10.1109/TIE.2015.2399401

Santiago, R., Barbosa, N., Mergulhão, H., Carvalho, T., Santos, A., Medrado, R., Filho, J. B. M., Pinheiro, O. R., & Nascimento, E. (2024). Data-Driven Models Applied to Predictive and Prescriptive Maintenance of Wind Turbine: A Systematic Review of Approaches Based on Failure Detection, Diagnosis, and Prognosis. *Energies*, *17*(5), 1–24. DOI: 10.3390/en17051010

Saraswat, A., & Tiwari, G. (2025). United Nations and Beyond: Legal Strategies for Defending Critical Energy Infrastructure Against Cyber Attacks. In *Studies in Computational Intelligence* (Vol. 1181, pp. 291–307). DOI: 10.1007/978-3-031-80557-8_13

Sarić, A. A., Transtrum, M. K., Sarić, A. T., & Stanković, A. M. (2022). Integration of physics- and data-driven power system models in transient analysis after major disturbances. *IEEE Systems Journal*, *17*(1), 479–490. DOI: 10.1109/JSYST.2022.3150237

Sasinthiran, A., Gnanasekaran, S., & Ragala, R. (2024). A review of artificial intelligence applications in wind turbine health monitoring. *International Journal of Sustainable Energy*, *43*(1), 2326296. DOI: 10.1080/14786451.2024.2326296

Sethi, M. R., Sahoo, S., Dhanraj, J. A., & Sugumaran, V. (2023). Vibration signal-based diagnosis of wind turbine blade conditions for improving energy extraction using machine learning approach. *Smart and Sustainable Manufacturing Systems*, *7*(1), 14–40. DOI: 10.1520/SSMS20220023

Shafiee, M. (2025). Cyber Resilience of Wind Energy Infrastructure: Challenges and Mitigation Strategies. In *ICAC 2025 - 30th International Conference on Automation and Computing*. DOI: 10.1109/ICAC65379.2025.11196193

Shah, S., Daoliang, T., & Kumar, S. (2024). RUL forecasting for wind turbine predictive maintenance based on deep learning. *Heliyon*, *10*(20), e39268. DOI: 10.1016/j.heliyon.2024.e39268 PMID: 39678167

Shao, K., He, Y., Xing, Z., & Du, B. (2023). Detecting wind turbine anomalies using nonlinear dynamic parameters-assisted machine learning with normal samples. *Reliability Engineering & System Safety*, *233*, 109092. DOI: 10.1016/j.ress.2023.109092

Sheibani, M., & Akbari, A. A. (2015). Finite element modeling of a wind turbine blade. *Journal of Vibroengineering*, *17*(7), 3774–3791. https://www.extrica.com/article/16033

Sheng, S. (2012). *Wind Turbine Gearbox Condition Monitoring Round Robin Study – Vibration Analysis.* United States: National Renewable Energy Laboratory (NREL), Golden, CO (United States). DOI: 10.2172/1048981

Sheng, S., Link, H., LaCava, W., van Dam, J., McNiff, B., Veers, P., . . . Oyague, F. (2011). Wind Turbine Drivetrain Condition Monitoring During GRC Phase 1 and Phase 2 Testing. United States: National Renewable Energy Laboratory (NREL), Golden, CO (United States). DOI: 10.2172/1027157

Sheng, S., & Veers, P. (2020). Accelerated Life Testing and Performance Modelling for Large Wind Turbine Systems. *Wind Energy Science*, *3*, 1001–1018.

Shen, L. (2014). The NIST cybersecurity framework: Overview and potential impacts. *Scitech Lawyer*, *10*(4), 16.

Shepherd, A., Roberts, S., Sünnenberg, G., Lovett, A., & Hastings, A. F. S. (2021). Scotland's onshore wind energy generation, impact on natural capital & satisfying no-nuclear energy policy. *Energy Reports*, *7*, 7106–7117. DOI: 10.1016/j.egyr.2021.10.063

Shi, P., Jia, L., Yi, S., & Han, D. (2024). Wind turbines fault diagnosis method under variable working conditions based on AMVMD and deep discrimination transfer learning network. *Measurement Science & Technology*, *35*(4), 46120. DOI: 10.1088/1361-6501/ad1fcd

Shittu, H. A., & Shittu, M. A. (2024). Leveraging digital twins and AI for enhanced gearbox condition monitoring in wind turbines. *Applied Sciences (Basel, Switzerland)*, *15*(10), 5725.

Shu, Y., Yuan, Q., Ke, W., & Kou, L. (2022). Security Access Control Method for Wind-Power-Monitoring System Based on Agile Authentication Mechanism. *Electronics (Basel)*, *11*(23), 1–12. DOI: 10.3390/electronics11233938

Siddiqi, M., Yu, H., & Joung, J. (2019). 5G Ultra-Reliable Low-Latency Communication Implementation Challenges and Operational Issues with IoT Devices. *Electronics (Basel)*, *8*(9), 981. DOI: 10.3390/electronics8090981

Sifat, M. M. H., & Das, S. K. (2024). Proactive and reactive maintenance strategies for self-healing digital twin islanded microgrids using fuzzy logic controllers and machine learning techniques. *IEEE Transactions on Power Systems*, *40*(1), 178–187. DOI: 10.1109/TPWRS.2024.3408096

Sikdar, S., Kalgutkar, A. P., & Banerjee, S. (2024). Guided wave based structural health monitoring of composite wind turbine blades. *Proceedings of the Annual British Conference on Non-Destructive*, 2024(1), 1–11. DOI: 10.1784/ndt2024.4b4

Singh, V., Sharma, R., & Govindarasu, M. (2020). *Testbed-based Performance Evaluation of Attack Resilient Control for Wind Farm SCADA System. 2020 IEEE Power & Energy Society General Meeting.* PESGM., DOI: 10.1109/PES-GM41954.2020.9281431

Škrábal, J., Raab, M., Raab, R., Grüebler, M. U., Kormann, U. G., Scherler, P., Sumasgutner, P., Åkesson, S., Bermejo, A., Chakarov, N., Fiedler, W., Godino, A., Haraszthy, L., Klein, K., Kolbe, M., Literák, I., Mammen, K., Mammen, U., Paquet, J.-Y., & Raab, R. (2025). Red kite (Milvus milvus) collision risk is higher at wind turbines with larger rotors and lower clearance, evidenced by GPS tracking. *Biological Conservation*, *312*, 111482. Advance online publication. DOI: 10.1016/j.biocon.2025.111482

Sliz, H., & Gudmestad, O. T. (2022). Machine Learning for Wind Turbine Condition Monitoring: A Comparative Study. *Renewable & Sustainable Energy Reviews*, *161*, 112385.

Smith, A., & Lee, B. (2025). AI-enhanced digital twins in maintenance: Systematic review of Industry 4.0 applications. *Computers & Industrial Engineering*, *177*, 109333. DOI: 10.1016/j.cie.2025.109333

Sousa, J., Lagarto, J., & Fonseca, M. (2024). The role of storage and flexibility in the energy transition: Substitution effect of resources with application to the Portuguese electricity system. *Renewable Energy*, *228*, 120694. DOI: 10.1016/j.renene.2024.120694

Srivastava, A., Rawat, B. S., & Maurya, S. K. (2023). A Review on Protecting SCADA Systems from DDOS Attacks. In *2023 4th International Conference on Electronics and Sustainable Communication Systems, ICESC 2023 - Proceedings* (pp. 556–561). DOI: 10.1109/ICESC57686.2023.10193650

Stadtmann, F., Rasheed, A., Kvamsdal, T., Johannessen, K., San, O., Kölle, K., Tande, J. O., Barstad, I., Benhamou, A., Brathaug, T., Christiansen, T., Firle, A.-L., Fjeldly, A., Frøyd, L., Gleim, A., Høiberget, A., Meissner, C., Nygård, G., Olsen, J., & Skogås, J. (2023). Digital Twins in Wind Energy: Emerging Technologies and Industry-Informed Future Directions. *IEEE Access : Practical Innovations, Open Solutions*, *11*, 110762–110795. DOI: 10.1109/ACCESS.2023.3321320

Sun, S., Wang, T., Yang, H., & Chu, F. (2022). Condition monitoring of wind turbine blades based on self-supervised health representation learning. *Applied Energy*, *313*, 118882. DOI: 10.1016/j.apenergy.2022.118882

Sun, T., Yu, G., Gao, M., Zhao, L., Bai, C., & Yang, W. (2021). Fault diagnosis methods based on machine learning and its applications for wind turbines: A review. *IEEE Access : Practical Innovations, Open Solutions*, *9*, 147481–147511. DOI: 10.1109/ACCESS.2021.3124025

Suvire, G. O., & Mercado, P. E. (2010). DSTATCOM with flywheel energy storage system for wind energy applications: Control design and simulation. *Electric Power Systems Research*, *80*(3), 345–353. DOI: 10.1016/j.epsr.2009.09.020

Takanashi, M., Sato, S.-I., Indo, K., Nishihara, N., Ichikawa, H., & Watanabe, H. (2021). Anomaly prediction for wind turbines using an autoencoder based on power-curve filtering. IEICE Trans. Inf. Syst., E104.D(9), 1506–1509. DOI: 10.1587/transinf.2020EDL8127

Tambunan, H. B., Hakam, D. F., Prahastono, I., Pharmatrisanti, A., Purnomoadi, A. P., Aisyah, S., Wicaksono, Y., & Sandy, I. G. R. (2020). The challenges and opportunities of renewable energy source (RES) penetration in Indonesia: Case study of Java-Bali power system. *Energies*, *13*(22), 5903. DOI: 10.3390/en13225903

Tang, M., Zhao, Q., Wu, H., Wang, Z., Meng, C., & Wang, Y. (2021). Review and perspectives of machine learning methods for wind turbine fault diagnosis. *Frontiers in Energy Research*, *9*, 751066. DOI: 10.3389/fenrg.2021.751066

Tao, F., Zhang, H., Liu, A., & Nee, A. (2019). Digital Twin in Industry: State-of-the-Art. *IEEE Transactions on Industrial Informatics*, *15*(4), 2405–2415. DOI: 10.1109/TII.2018.2873186

Tataria, H., Shafi, M., Molisch, A., Dohler, M., Sjöland, H., & Tufvesson, F. (2021). 6G Wireless Systems: Vision, Requirements, Challenges, Insights, and Opportunities. *Proceedings of the IEEE*, *109*(7), 1166–1199. DOI: 10.1109/JPROC.2021.3061701

Tautz-Weinert, J., & Watson, S. J. (2017). Using SCADA data for wind turbine condition monitoring - A review. *IET Renewable Power Generation*, *11*(4), 382–394. DOI: 10.1049/iet-rpg.2016.0248

Thangamuthu, S., Concer, N., Cuijpers, P., & Lukkien, J. (2015). *Analysis of Ethernet-switch traffic shapers for in-vehicle networking applications. 2015 Design, Automation & Test in Europe Conference & Exhibition*. DATE.

Udo, W. S., Kwakye, J. M., Ekechukwu, D. E., & Ogundipe, O. B. (2023). Optimizing wind Energy Systems using Machine Learning for Predictive Maintenance and Efficiency Enhancement. Comput. Sci. IT Res. j., 4(3), 386–397. DOI: 10.51594/csitrj.v4i3.1398

Uhanto, U., Yandri, E., Hilmi, E., Saiful, R., & Hamja, N. (2024). Predictive maintenance with machine learning: A comparative analysis of wind turbines and PV power plants. *Heca J. Appl. Sci.*, *2*(2), 87–98. DOI: 10.60084/hjas.v2i2.219

Ukoba, K., Olatunji, K. O., Adeoye, E., Jen, T.-C., & Madyira, D. M. (2024). Optimizing renewable energy systems through artificial intelligence: Review and future prospects. *Energy & Environment*, *35*(7), 3833–3879. DOI: 10.1177/0958305X241256293

Ulmer, M., Jarlskog, E., Pizza, G., & Goren Huber, L. (2020). Cross-turbine training of convolutional neural networks for SCADA-based fault detection in wind turbines. *Proceedings of the Annual Conference of the Prognostics and Health Management Society. Prognostics and Health Management Society. Conference*, *12*(1), 10. DOI: 10.36001/phmconf.2020.v12i1.1205

V. R., & K. A. (2025). Improving EEG signal classification in brain–computer interfaces via Grey Wolf Optimizer-based deep learning. In *Proceedings of the 3rd International Conference on Sustainable Computing and Data Communication Systems (ICSCDS)* (pp. 1312–1318). IEEE.

Vattenfall Press. (2024, January). Digital twins – A road to more profitable offshore wind. *Vattenfall Group*.

Vestas. (2021). *Third update on cyber incident*. Retrieved January 29, 2026, from https://www.vestas.com/en/media/company-news/2021/third-update-on-cyber-incident-c3466518?

Vieira, J., Farias, F., Ochoa, A., de Menezes, F., Costa, A., da Costa, J., de Novaes Pires Leite, G., Vilela, O. C., de Souza, M. G. G., & Michima, P. (2024). Remaining Useful Life Estimation Framework for the Main Bearing of Wind Turbines Operating in Real Time. *Energies*, *17*(6), 1–17. DOI: 10.3390/en17061430

Vives, J. (2022). Incorporating machine learning into vibration detection for wind turbines. *Modelling and Simulation in Engineering*, *2022*, 1–8. DOI: 10.1155/2022/6572298

Wanasinghe, T. R., Wroblewski, L., Petersen, B. K., Gosine, R. G., James, L. A., De Silva, O., Mann, G. K. I., & Warrian, P. J. (2020). Digital Twin for the Oil and Gas Industry: Overview, Research Trends, Opportunities, and Challenges. *IEEE Access : Practical Innovations, Open Solutions*, *8*, 104175–104197. DOI: 10.1109/ACCESS.2020.2998723

Wang, B., Wang, C., Huang, W., Song, Y., & Qin, X. (2021). Security-aware task scheduling with deadline constraints on heterogeneous hybrid clouds. *Journal of Parallel and Distributed Computing, 153*, 15–28. https://doi.org/https://doi.org/10.1016/j.jpdc.2021.03.003. DOI: 10.1016/j.jpdc.2021.03.003

Wang, F., Zhang, L., Feng, X., & Guo, H. (2018). An adaptive control strategy for virtual synchronous generator. *IEEE Transactions on Industry Applications, 54*(5), 5124–5133. DOI: 10.1109/TIA.2018.2859384

Wang, L., Fan, W., Jiang, G., & Xie, P. (2023). An efficient federated transfer learning framework for collaborative monitoring of wind turbines in IoE-enabled wind farms. *Energy, 284*, 128518. DOI: 10.1016/j.energy.2023.128518

Wang, N., Chen, Q., Zhu, L., & Sun, H. (2022). Integration of data-driven and physics-based modeling of wind waves in a shallow estuary. *Ocean Modelling (Oxford), 172*, 101978. DOI: 10.1016/j.ocemod.2022.101978

Wang, S., Vidal, Y., & Pozo, F. (2026). Recent advances in wind turbine condition monitoring using SCADA data: A state-of-the-art review. *Reliability Engineering & System Safety, 267*, 111838. DOI: 10.1016/j.ress.2025.111838

Wang, X., Zhou, D., Zhuang, X., Ge, J., & Xiang, J. (2024). *SGG-DGCN: Wind turbine anomaly identification by using deep graph convolutional networks with similarity graph generation strategy. Journal of Dynamics*. Monitoring and Diagnostics., DOI: 10.37965/jdmd.2024.558

Wang, Y., Sun, W., Liu, L., Wang, B., Bao, S., & Jiang, R. (2023). Fault diagnosis of wind turbine planetary gear based on a digital twin. *Applied Sciences (Basel, Switzerland), 13*(8), 4776. DOI: 10.3390/app13084776

Wang, Y., Wang, X., Wang, Y., & Fang, S. (2025). Edge–Cloud Intelligence for Sustainable Wind Turbine Blade Transportation: Machine-Vision-Driven Safety Monitoring in Renewable Energy Systems. *Energies, 18*, 1–22. DOI: 10.3390/en19010001

Wang, Z., & Bu, S. (2024). Probabilistic frequency stability analysis considering dynamics of wind power generation with different control strategies. *IEEE Transactions on Power Systems, 39*(5), 6412–6425. DOI: 10.1109/TPWRS.2024.3362056

Williams, L., Phillips, C., Sheng, S., Dobos, A., & Wei, X. (2020). Scalable Wind Turbine Generator Bearing Fault Prediction Using Machine Learning: A Case Study. In *2020 IEEE International Conference on Prognostics and Health Management (ICPHM)* (pp. 1-9). Detroit: IEEE. DOI: 10.1109/ICPHM49022.2020.9187050

WindEurope. (2023). End-of-Life and Circular Economy Solutions for Wind Turbine Blades, WindEurope Report.

Worden, K., Manson, G., & Dervilis, N. (2021). Modern Machine Learning Methods for Wind Turbine Structural Health Monitoring. *Structural Health Monitoring*, *21*(4), 1765–1782.

Wu, H., Liu, J., Liu, J., Cui, M., Liu, X., & Gao, H. (2019). Power Grid Reliability Evaluation Considering Wind Farm Cyber Security and Ramping Events. *Applied Sciences (Basel, Switzerland)*, *9*(15), 1–18. DOI: 10.3390/app9153003

Wu, Y., Tang, B., Deng, L., & Li, Q. (2022). Distillation-enhanced fast neural architecture search method for edge-side fault diagnosis of wind turbine gearboxes. *Expert Systems with Applications*, *118049*, 118049. Advance online publication. DOI: 10.1016/j.eswa.2022.118049

Xiang, L., Wang, P., Yang, X., Hu, A., & Su, H. (2021). Fault detection of wind turbine based on SCADA data analysis using CNN and LSTM with attention mechanism. *Measurement*, *175*, 109094. DOI: 10.1016/j.measurement.2021.109094

Xiao, C., Liu, Z., Zhang, T., & Zhang, X. (2021). Deep learning method for fault detection of wind turbine converter. *Applied Sciences (Basel, Switzerland)*, *11*(3), 1280. DOI: 10.3390/app11031280

Xie, Z., Zou, Z., & Song, G. (2022). Digital Twin Modeling and Cybersecurity for Wind Farm Operations. *IEEE Transactions on Smart Grid*, *13*(3), 1626–1634.

Xu, Q., Lu, S., Zhai, Z., & Jiang, C. (2020). Adaptive fault detection in wind turbine via RF and CUSUM. *IET Renewable Power Generation*, *14*(10), 1789–1796. DOI: 10.1049/iet-rpg.2019.0913

Xu, T., Jang, W., & Overbye, T. (2018). Commitment of Fast-Responding Storage Devices to Mimic Inertia for the Enhancement of Primary Frequency Response. *IEEE Transactions on Power Systems*, *33*(2), 1219–1230. DOI: 10.1109/TPWRS.2017.2735990

Yan, C., Xu, S., Sun, Z., Lutz, T., Guo, D., & Yang, G. (2024). A framework of data assimilation for wind flow fields by physics-informed neural networks. *arXiv*. https://arxiv.org/abs/2401.17001

Yan, D., Sadeghi, A., Yazdani-Asrami, M., & Song, W. (2024). Artificial-Intelligence-Driven Model for Resistive Superconducting Fault Current Limiter in Future Electric Aircraft. *IEEE Transactions on Applied Superconductivity*, *34*(7), 1–16. DOI: 10.1109/TASC.2024.3421903

Yang, B., & Zhang, Y. (2022). Cybersecurity Analysis of Wind Farm Industrial Control System Based on Hierarchical Threat Analysis Model Framework. In *Proceedings - 2022 International Conference on Computing, Communication, Perception and Quantum Technology, CCPQT 2022* (pp. 6–13). DOI: 10.1109/CCPQT56151.2022.00008

Yang, B., Wang, J., Chen, Y., Li, D., Zeng, C., Chen, Y., Guo, Z., Shu, H., Zhang, X., Yu, T., & Sun, L. (2020). Optimal sizing and placement of energy storage system in power grids: A state-of-the-art one-stop handbook. *Journal of Energy Storage*, *32*, 101814. DOI: 10.1016/j.est.2020.101814

Yang, L., Huang, Z., Mo, X., & Luo, T. (2025). Enhanced GAIN-Based Missing Data Imputation for a Wind Energy Farm SCADA System. *Electronics (Basel)*, *14*(8), 1–26. DOI: 10.3390/electronics14081590

Yang, L., Liang, G., Yang, Y., Ruan, J., Yu, P., & Yang, C. (2024). Adversarial false data injection attacks on deep learning-based short-term wind speed forecasting. *IET Renewable Power Generation*, *18*(7), 1370–1379. DOI: 10.1049/rpg2.12853

Yang, L., & Zhang, Z. (2021). Wind Turbine Gearbox Failure Detection Based on SCADA Data: A Deep Learning-Based Approach. *IEEE Transactions on Instrumentation and Measurement*, *70*, 1–11. DOI: 10.1109/TIM.2020.3045800

Yang, R., Meng, F., & Li, X. (2022). Blockchain-Enabled Peer-to-Peer Wind Energy Trading. *Applied Energy*, *313*, 118871.

Yan, J., Liu, Y., & Jiang, Q. (2021). Hybrid CNN-LSTM Approach for Accurate Fault Prognosis in Wind Turbines. *IEEE Transactions on Power Electronics*, *33*(5), 3936–3947.

Yenra. (2025, March). AI predictive maintenance for wind turbines: 20 advances. *Yenra*. https://yenra.com/ai20/predictive-maintenance-for-wind-turbines/

Yin, S., Luo, H., & Ding, S.-X. (2018). Real-time Fault Detection for Wind Turbines Using Support Vector Machines and Big Data Architecture. *IEEE Transactions on Industrial Informatics*, *14*(8), 3644–3653.

Yousefzadeh, M., & Ahmadi Kamarposhti, M. (2024). Wind Energy Conversion Systems: A Review on Aerodynamic, Electrical and Control Aspects, Recent Trends, Comparisons and Insights. *Green Energy and Technology*, (Part F2899), 25–60. DOI: 10.1007/978-3-031-18389-8_2

Yu, Q., Patriksson, M., & Sagitov, S. (2021). Optimal scheduling of the next preventive maintenance activity for a wind farm. *Wind Energy Science*, *6*(3), 949–959. DOI: 10.5194/wes-6-949-2021

Zhang, R., Zhu, H., Chang, Q., & Mao, Q. (2025). A Comprehensive Review of Digital Twins Technology in Agriculture. Agriculture 2025, Vol. 15, Page 903, 15(9), 903. DOI: 10.3390/agriculture15090903

Zhang, J., Fang, Z., Dong, W., & Xiahou, T. (2026). Deriving the optimal defense resource deployment: Modeling cascading effect and contest intensity in an attacker–defender game. *Reliability Engineering & System Safety*, *266*, 111779. Advance online publication. DOI: 10.1016/j.ress.2025.111779

Zhang, J., & Zhao, X. (2023). Digital twin of wind farms via physics-informed deep learning. *Energy Conversion and Management*, *293*, 117507. DOI: 10.1016/j.enconman.2023.117507

Zhang, K., Cao, J., & Zhang, Y. (2021). Adaptive digital twin and multiagent deep reinforcement learning for vehicular edge computing and networks. *IEEE Transactions on Industrial Informatics*, *18*(2), 1405–1413. DOI: 10.1109/TII.2021.3088407

Zhang, L., Zhang, H., & Cai, G. (2022). The multiclass fault diagnosis of wind turbine bearing based on multisource signal fusion and deep learning generative model. *IEEE Transactions on Instrumentation and Measurement*, *71*, 1–12. DOI: 10.1109/TIM.2022.3178483

Zhang, Y., Lin, F., & Wang, K. (2020). Robustness of Short-Term Wind Power Forecasting against False Data Injection Attacks. *Energies*, *13*(15), 1–22. DOI: 10.3390/en13153780

Zhang, Z., Liu, X., Zhao, D., Post, S., & Chen, J. (2023). Overview of the development and application of wind energy in New Zealand. *Energy and Built Environment*, *4*(6), 725–742. DOI: 10.1016/j.enbenv.2022.06.009

Zhang, Z., Xiao, Y., Ma, Z., Xiao, M., Ding, Z., Lei, X., Karagiannidis, G. K., & Fan, P. (2019). 6G Wireless Networks: Vision, Requirements, Architecture, and Key Technologies. *IEEE Vehicular Technology Magazine*, *14*(3), 28–41. DOI: 10.1109/MVT.2019.2921208

Zheng, M., Man, J., Wang, D., Chen, Y., Li, Q., & Liu, Y. (2023). Semi-supervised multivariate time series anomaly detection for wind turbines using generator SCADA data. *Reliability Engineering & System Safety*, *235*, 109235. DOI: 10.1016/j.ress.2023.109235

Zhou, J., Yang, M., Zhan, Y., & Xu, L. (2023). Digital twin application for reinforcement learning-based optimal scheduling and reliability management enhancement of systems. *Solar Energy*, *252*, 29–38. DOI: 10.1016/j.solener.2023.01.042

Zhou, R., Li, Y., & Lin, X. (2025). A clustered federated learning framework for collaborative fault diagnosis of wind turbines. *Applied Energy*, *377*, 124532. DOI: 10.1016/j.apenergy.2024.124532

Zhou, Y., Zhou, J., Cui, Q., Wen, J., & Fei, X. (2024). Digital twin-driven online intelligent assessment of wind turbine gearbox. *Wind Energy (Chichester, England)*, *27*(8), 797–815. DOI: 10.1002/we.2912

Zhu, J., Hu, J., Hung, W., Wang, C., Zhang, X., Bu, S., Li, Q., Urdal, H., & Booth, C. D. (2017). Synthetic inertia control strategy for doubly fed induction generator wind turbine generators using lithium-ion supercapacitors. *IEEE Transactions on Energy Conversion*, *33*(2), 773–783. DOI: 10.1109/TEC.2017.2764089

Zhu, X., Wu, Y., Zhao, X., Yang, Y., Liu, S., Shi, L., & Wu, Y. (2024). Overview of Wind and Photovoltaic Data Stream Classification and Data Drift Issues. *Energies*, *17*(17), 1–24. DOI: 10.3390/en17174371

Zhu, Z., Zhang, J., Zhu, S., & Yang, J. (2023). Digital twin technology for wind turbine towers based on joint load–response estimation: A laboratory experimental study. *Applied Energy*, *352*, 121953. DOI: 10.1016/j.apenergy.2023.121953

About the Contributors

Jackson J. Justo (PhD): Currently working as a Senior Researcher at the Faculty of Control Systems and Robotics, ITMO University in St. Petersburg, Russia and a Senior Lecturer with the Department of Electrical Engineering, University of Dar es Salaam, Tanzania. Received his PhD Degree from Dongguk University, Seoul, South Korea (2015). He was with the University of Pretoria, Pretoria, South Africa (2016 – 2018) as Postdoctoral Research Fellow. His current research interests on: High power converter topology, control systems, condition monitoring and fault detection, grid Integration of renewable power generation, AC and DC microgrid, virtual power plants (VPPs) configurations, Intelligent Control Systems, Electrical Machine Drives, Wireless Power Transfer for Electric Vehicle (EVs) Autonomous Charging Systems. Virtual Energy Storage Systems, and Power System Control, Operation and Dynamics.

Galina Demidova is currently with Hangzhou Dianzi University, China. Previously worked as an Associate Professor with the Faculty of Control Systems and Robotics, ITMO University, and was a Post-Doctoral with the Tallinn University of Technology, Tallinn, Estonia. Her current research interests include control systems, wind turbines, adaptive control, fuzzy logic control, motion control, machine learning, hybrid intelligent systems, genetic algorithms, and electromechanical systems. Her experience in the field of control systems, condition monitoring and modeling of electromechanical systems will be an advantage during the editorial of the proposed book. Prof. Galina will lead the review process of the book chapters by analyzing the review comments, making decisions and proving timely response to the contributors.

Francis Mwasilu is currently working with Department of Electrical Engineering, University of Dar es Salaam, Tanzania. From 2008 to 2009 he was an Instrumentation Engineer with Illovo Sugar Ltd, Tanzania. From 2009 to 2011, he was a Utilities Engineer with the Japan Tobacco International, Tanzania. Since

July 2016, he has been a Lecturer with the Department of Electrical. His current research interests include digital signal processor-based electric machine drives, electric vehicles, smart grid enabling technologies, and renewable energy sources integration into modern power systems.

Dmitry Lukichev is leading lector and scientific researcher, who works with industrial projects in various areas of industry who works at ITMO University in St. Petersburg, Russia. He is also visiting Professor of Changchun University of Technology and Hangzhou Dianzi University, China. His current research interests encompass control systems for both DC and AC drives, adaptive control modelling and application and the implementation of neuro-fuzzy algorithms in the control systems of complex dynamic objects. He is author of about 70 scientific articles. Additionally, Professor Dmitry is actively involved in projects focused on the control of wind turbines and generators, further expanding his contributions to innovative control systems.

Ramesh Bansal has more than 25 years of diversified experience of scholarship of teaching and learning, accreditation, research, industrial, and academic leadership in several countries. Currently he is a Professor in the Department of Electrical Engineering at University of Sharjah and extraordinary Professor at the University of Pretoria. Previously he was Professor and Group Head (Power) in the ECE Department at University of Pretoria (UP), South Africa. Prior to his appointment at UP, he was employed by the University of Queensland, Australia; University of the South Pacific, Fiji; and BITS Pilani, India. Prof. Bansal has significant experience of collaborating with industry and Government organizations. He has made significant contribution to the development and delivery of BS and ME programmes for Utilities. He has extensive experience in the design and delivery of CPD programmes for professional engineers. He has carried out research and consultancy and attracted significant funding from Industry and Government Organisations. Prof. Bansal has published over 450 journal articles, presented papers at conferences, books, and chapters in books. He has Google citations of over 20,000 with h-index of 67. He has supervised 25 PhD, 5 Post Docs and currently supervising several PhD students. His diversified research interests are in the areas of Renewable Energy (Wind, PV, DG, Micro Grid) and Smart Grid. Professor Bansal is an editor of several highly regarded journals including IEEE Systems Journal, ECPS, SGSE. He is a Fellow and Chartered Engineer IET-UK, Fellow Institution of Engineers (India), Fellow SAIEE, and Senior Member of IEEE-USA.

***

Suresh Kumar Annam received the B.Tech degree in 2005, the M.Tech degree from JNTU College of Engineering, Anantapur, India, in 2007. The Ph.D. from Jawaharlal Nehru Technological University, Hyderabad (JNTUH), India, in 2024. He is currently an Assistant Professor of Electrical and Electronics Engineering, Malla Reddy University, Hyderabad. He is a Reviewer for IEEE ACCESS and Scopus Journal. His research interests include Neural Networks, Fuzzy logic System, Estimation Techniques, AI techniques to Power System and Drives Applications, Fault detection Techniques, System Identification, Adaptive Control.

Bali Ram Gupta is currently working as an Assistant Professor of Mathematics at Jaypee University of Engineering and Technology, Guna. He completed his M. Sc. From the University of Allahabad and was awarded Ph.D. from University of Allahabad in 2010. He qualified NET (JRF) in 2005, conducted by CSIR-UGC, and also cleared GATE in 2004. With a strong research background, he has published over 21 research papers in National and International journals, and holds 2 patents to his credit. He has also presented his research papers in many national and international conferences. Dr. Gupta has successfully guided three Ph.D. scholars. With 15 years of teaching and research experience, he has served in reputed colleges and universities, holding various academic and administrative positions throughout his career.

Deepak Gupta is a distinguished academician with over 20 years of teaching experience and expertise in Computer Science and Engineering. He holds a Ph.D. in Computer Engineering and has an extensive research portfolio, including many international journal papers, multiple conference proceedings, and several books. His research interests span machine learning, artificial intelligence, and data mining. Currently, Dr. Gupta serves as an Associate Professor and Head of Department of CSE (Emerging Branches) at ITM Gwalior, MP, India. He has also contributed significantly to the field through patents and copyrights and actively engages as a reviewer, conference chair and editor.

Robinson Joel received the Doctorate Ph.D degree in Computer and Information Technology from the Manonmaniam Sundaranar University, Tirunelveli in March 2019. He completed his M.Tech and MCA from the same university year 2008 and 2002 respectively. He is currently working as an Associate Professor in the Department of Computer Science and Engineering at KCG College of Technology, Chennai. He has 16 years of teaching experience.

Bekzod Madaminov is a faculty member in the Department of General Professional Sciences at Mamun University, Urgench, Uzbekistan. His academic interests include general professional education, applied social and management sciences, and interdisciplinary approaches to skill development in higher education. He is actively involved in teaching, academic research, and curriculum development, with a focus on strengthening professional competencies and aligning educational programs with contemporary societal and institutional needs.

Peter Makolo is a Lecturer and Researcher in the Department of Electrical Engineering at the University of Dar es Salaam, Tanzania. He holds a Ph.D. in Electrical Engineering from Auckland University of Technology (2022), an MSc in Electric Power Engineering from Chalmers University of Technology (2013), and a BSc (Hons.) in Electrical Engineering from the University of Dar es Salaam (2008). With over 16 years of experience in research, teaching, and consultancy, his expertise spans grid integration of renewable energy, energy storage systems, synthetic inertia, and power system dynamics. Dr. Makolo has published extensively in high-impact journals and conferences, focusing on inertia estimation and grid flexibility for renewable energy systems. His research interests include intelligent grid control, power quality, and AI-driven optimization for modern power systems.

Rampelli Manojkumar received his Ph.D. in Power Engineering from the Indian Institute of Technology Guwahati in 2023. He obtained his Master of Technology in Power and Energy Systems from the National Institute of Technology Karnataka, Surathkal, India, in 2015. He is currently an Assistant Professor in the Department of Electrical and Electronics Engineering at BVRIT Hyderabad College of Engineering for Women. He is currently the Editorial Board Member of Scientific Reports, Early Career Editor for the International Journal of Ambient Energy and the Guest Editor for PLOS ONE. He has been a speaker at the IEEE Smart Grid webinar series and has contributed to multiple funded research projects on smart transformer-based microgrid systems and distributed generation management. He has published 9 SCI journal papers, including articles in IEEE Transactions on Power Systems, IEEE Transactions on Industrial Informatics, and Sustainable Cities and Society. His research expertise spans energy management, hybrid AC/DC distribution systems, energy storage, and power electronics applications in modern distribution systems. Additionally, he is an active reviewer for around 20 SCI journals, including IEEE Transactions on Industrial Informatics, IEEE Systems Journal, Journal of Energy Storage, Applied Energy, etc. He has also been actively involved in faculty training programs focused on teaching and learning enhancement, including instructional design, formative assessment, and teaching analytics, thereby strengthening his interdisciplinary expertise in education and technology integration.

Shakhboz Meylikulov is affiliated with the Department of Information Technology and Exact Sciences at Termez University of Economics and Service,

Termez, Uzbekistan. His academic interests include information technology, applied computing, and quantitative methods, with a focus on integrating modern digital tools into teaching and research. He is actively engaged in academic instruction and scholarly activities that support the development of technical skills and analytical thinking in higher education.

Navamani C is an Assistant Professor in the Department of Computer Science and Engineering at Nandha Engineering College, Erode, India. Her academic interests include programming languages, data structures, artificial intelligence, and emerging computational technologies. She is dedicated to providing quality engineering education and actively engages in mentoring students through academic projects and technical activities. She contributes to departmental development initiatives and promotes a research-oriented learning environment that prepares students for industry and higher studies.

Shabana Pathan is currently serving as an Assistant Professor in the Department of Information Technology at St. Vincent Pallotti College of Engineering & Technology, Nagpur, where she has been contributing for the past 19 years. She earned her Ph.D. in Computer Science and Engineering from Oriental University, Indore (M.P.). With a strong academic and research background in Information Technology, Dr. Pathan has authored and published more than 34 research papers in reputed SCI, Scopus, and UGC-indexed international and national journals. Her scholarly contributions also extend to book chapters and copyrights, reflecting her active engagement in advancing knowledge. Her expertise lies in the domains of Artificial Intelligence, Machine Learning, Augmented Reality (AR), and Virtual Reality (VR). She is particularly interested in exploring the intersection of technology and society, ensuring her teaching and research address contemporary challenges in meaningful ways. Through her holistic approach to research and pedagogy, Dr. Pathan continues to make a significant impact on students, peers, and the wider academic community.

P. Ramesh Babu is a seasoned academic professional with over 20 years of experience in teaching, research, and academic administration. He specializes in Data Mining and Information Retrieval, holding a Ph.D. in Geo-Spatial Data Mining from Rayalaseema University and an M.Tech in Software Engineering from JNTUK, where he secured a university gold medal. As an Associate Professor at Anurag University, he has previously held key positions at KL Deemed to be University and KLR College of Engineering and Technology, serving as Vice Principal and Head of the Department. His research focuses on artificial intelligence, machine learning, cloud computing, and geospatial data mining, with numerous publications in

SCOPUS and SCI-indexed journals. He is also a reviewer for prestigious journals like the *Journal of Big Data*. Dr. Babu possesses expertise in AWS, Linux, DevOps tools, and machine learning, with certifications including IBM DB2 and Enterprise Design Thinking Practitioner. His contributions extend to patents in natural language processing and intelligent systems, demonstrating his commitment to technological innovation. Recognized for academic excellence, he has received accolades such as a gold medal for his M.Tech performance and a Letter of Appreciation for his long-standing contributions to academia. His administrative experience includes NBA accreditation and curriculum design, further reinforcing his leadership in higher education.

Anji Reddy Polu has completed his PhD from Dr. H S Gour Central University (2013) and postdoctoral studies from Sogang University, South Korea. ¬ Dr. Reddy currently working as Associate Dean R & D and Associate Professor in the Department of Physics, BVRIT Hyderabad College of Engineering for Women (UGC Autonomous), a premier engineering college in Hyderabad. ¬ Before joining BVRIT HYDERABAD, Dr. Reddy worked in VIT University and K L University as Assistant Professor. ¬ He has more than 16 years of experience with Battery materials and their application in EVs. ¬ Dr. Anji Reddy Polu featured among the World's TOP 2% SCIENTISTS LIST published by Stanford University, USA, and Elsevier in 2025, 2024, 2023 and 2022. ¬ He has published more than 90 research papers in reputed journals and has been serving as a reviewer for many reputed journals (ACS, RSC, Elsevier, Wiley, Springer etc.). ¬ He has given more than 35 invited talks and keynote lectures in various countries at reputed international conferences. ¬ Dr. Reddy has published 20 patents for his Credit. Dr. Reddy has an h-index of 30 and has more than 3400 citations.

Ravikumar R. Natarajan is an experienced academician with over 12+ years of teaching and research expertise across institutions in India, the UAE, and the Maldives. He currently serves as an Assistant Professor at Marwadi University, Rajkot, specializing in Machine Learning, Recommender Systems, and AI-driven applications in healthcare and security. He holds a Ph.D. in Machine Learning and Recommender Systems from Amity University, Jaipur, a Master of Engineering in CSE from VMRF University, Salem, and a Bachelor of Engineering in CSE from Anna University. He also has a MiniMBA from the Swiss eLearning Institute. Dr. Ravikumar has authored more than 20 IEEE conference papers, multiple SCIE/ Scopus-indexed journal articles, and co-authored a book on Artificial Intelligence. He has also published 9 patents and contributed several book chapters in emerging domains such as Federated Learning, Blockchain, and Digital Twins. He actively mentors students through initiatives like the Research Activities Club (RAC) and

Student Research Internship Training (SRIT), promoting innovation and research culture.

K. Balakrishna Reddy is an accomplished academic and researcher in the field of Mechanical Engineering, currently serving as an Assistant Professor. His research endeavors are primarily focused on the integration of cutting-edge technologies into advanced manufacturing systems and intelligent robotics. At present, he is actively engaged in two significant research projects. The first involves the implementation of digital twinning in Wire Arc Additive Manufacturing (WAAM), presenting a novel and efficient alternative to conventional metal 3D printing by leveraging the capabilities of industrial robotic systems. This project aims to enhance process monitoring, control, and predictive maintenance in additive manufacturing through the use of virtual replicas and real-time data synchronization. His key research contributions also extend to the application of Internet of Things (IoT) technologies in advanced manufacturing processes, facilitating smart, connected production environments. Additionally, he has developed machine learning-based models for the prediction and optimization of machining parameters, leading to improved productivity, precision, and energy efficiency in manufacturing operations. Another major thrust area is focused on robot manipulator and mobile robot path planning techniques, including the hybridization of classical and intelligent algorithms. His work aims to achieve optimal path planning solutions, improve navigation efficiency, and enhance robot process automation in dynamic and complex environments.

Aarthi S is a dedicated Assistant Professor in Computer Science with over six years of teaching experience. She is currently pursuing her Ph.D. at Marwadi University, Gujarat, and holds an M.Phil. and MCA from Bharathiar University, Tamil Nadu. Throughout her academic career, Aarthi has specialized in programming languages such as C++, Java, and C, and contributed to research in recommender systems and AI-related data privacy. She has published journal articles, contributed to book chapters, and actively participated in technical workshops. Known for her student mentorship and engagement in academic events, Aarthi remains committed to fostering innovation and excellence in computer science education.

Binastya Anggara Sekti has been a lecturer at the Faculty of Computer Science, Esa Unggul University, Jakarta, since 2016. He is actively involved in the Artificial Intelligence and Information Security Study Group. He holds a Bachelor's degree in Informatics Engineering from the Bandung Institute of Technology and a Master's degree in Information Systems Management from Trisakti University, Jakarta. Prior to entering academia, he accumulated over 25 years of professional experience as an IT practitioner in various companies across Jakarta. His work focused on developing data centers for internet service providers and delivering IT solutions for the manufacturing and oil & gas sectors. His research interests include artificial intelligence, cybersecurity, and the development of IT infrastructure.

Eva I. Shayo is a Lecturer at the University of Dar es Salaam in the Department of Electronics and Telecommunications Engineering. Having earned her PhD in Telecommunications Engineering, she has dedicated her research to advancing the capabilities of the Internet of Things (IoT), particularly in the areas of LoRaWAN scalability and energy-efficient systems. Her work is characterized by the development of sophisticated synchronization and scheduling algorithms, contributing significantly to the field of industrial wireless communication networks.

Manikandan T., born in Erode, India, is currently working as an Associate Professor in the Department of Electronics and Communication Engineering (ECE) at Vivekanandha College of Engineering for Women (Autonomous), Namakkal. He has over 12 years of teaching experience and has published several papers in reputed national and international journals and conferences. His research interests include Image Processing, Embedded Systems, Communication Systems, and Wireless Networks. He has guided more than 50 Projects and has successfully obtained grant funding through YUKTI–Innovation Repository, MSME, and other government-supported schemes. He is dedicated to fostering innovation, entrepreneurship, and applied research among students to bridge the gap between academia and industry.

T. M. Saravanan is an Associate Professor in the Department of Computer Technology at Kongu Engineering College. He brings extensive academic and research experience in computer science, with expertise spanning emerging technologies, system design, and advanced computing methodologies. He is actively involved in teaching, mentoring students, and contributing to curriculum enhancement. His research interests include innovative computing solutions, data-driven applications, and modern IT advancements. Through his academic contributions and commitment to excellence, he supports the department's mission of fostering quality technical education and research-driven learning.

Nguyen Duc Thuan received the M.Sc. degree in Control Engineering and Automation from the Hanoi University of Science and Technology (HUST), Hanoi, Vietnam, and the Ph.D. degree in Artificial Intelligence and Computer Engineering from the University of Ulsan, Ulsan, Republic of Korea. He is currently a Lecturer with the School of Electrical and Electronic Engineering, Hanoi University of Science and Technology. His current research interests include AI architecture for industrial systems, digital-twin-based predictive maintenance, and AIoT-driven smart manufacturing and energy systems. His broader research goal is to bridge physics-based modeling, machine learning, and cyber-physical system design toward trustworthy and autonomous industrial intelligence.

Gundu Venu received his B.Tech. Degree in Electrical and Electronics Engineering from Jawaharlal Nehru Technological University, Hyderabad in 2006. M.Tech. Degree in Power Electronics and Industrial Drives from Jawaharlal Nehru Technological University, Hyderabad in 2010. Obtained Ph.D. from Jawaharlal Nehru Technological University 2024, Hyderabad. Currently he is working as a Associate Professor, Dept. of EEE, Malla Reddy (MR) Deemed to Be University,Dhullapally,via Kompally, Medchal Dist, Telangana. His area of interest is Advanced Control strategies of Electric Drives, Fuzzy logic & ANN applications, and Network Analysis. Have professional society memberships in ISTE (LM) and IAENG (M).

T. Yuvaraj (IEEE Member) is currently working as an Associate Professor in department of EEE and Research member of Centre for Centre for Smart Energy Systems at Chennai Institute of Technology, Chennai, India. He received the B.E. degree in Electrical and Electronics Engineering and the M.E. degree in Power Electronics and Drives from Anna University, Chennai, India, in 2011 and 2013, respectively, and the Ph.D. degree from VIT University, Vellore, India, in 2017. He has published more than 100 papers in Web of Science/Scopus indexed journals. His research interests include the power system optimization, allocation of distributed energy resources,electric vehicle technology, power system resilience, and renewable energy. He is a member of IEEE and IAENG. He has also served as a reviewer for various reputed journals such as IEEE, Elsevier, Springer, Taylor and Francis and Inderscience journals.

Index

E

F

S

U

W

Z

www.ingramcontent.com/pod-product-compliance
Lightning Source LLC
LaVergne TN
LVHW062324300926
849161LV00007B/1557